Dynamics in Logistics

Hans-Jörg Kreowski · Bernd Scholz-Reiter ·
Klaus-Dieter Thoben
Editors

Dynamics in Logistics

Second International Conference,
LDIC 2009, Bremen, Germany, August 2009
Proceedings

 Springer

Editors
Prof. Dr. Hans-Jörg Kreowski
FB Mathematik und Informatik
AG Theoretische Informatik
Universität Bremen
Linzer Str. 9a
28359 Bremen
Germany
e-mail: kreo@informatik.uni-bremen.de

Prof. Dr. Bernd Scholz-Reiter
BIBA GmbH
Bremer Institut für Produktion
  und Logistik GmbH
Universität Bremen
Hochschulring 20
28359 Bremen
Germany
e-mail: bsr@biba.uni-bremen.de

Dr. Klaus-Dieter Thoben
BIBA GmbH
Bremer Institut für Produktion
  und Logistik GmbH
Universität Bremen
Hochschulring 20
28359 Bremen
Germany
e-mail: tho@biba.uni-bremen.de

ISBN 978-3-642-11995-8   e-ISBN 978-3-642-11996-5

DOI 10.1007/978-3-642-11996-5

Springer Heidelberg Dordrecht London New York

Library of Congress Control Number: 2010936697

© Springer-Verlag Berlin Heidelberg 2011

Cover design: eStudio Calamar S.L., Girona

Printed on acid-free paper

Springer is part of Springer Science+Business Media (www.springer.com)

Preface of the Editors

LDIC 2009 was the second International Conference on Dynamics in Logistics held in Bremen in August 2009. As the first conference in 2007, it was organized by the Research Cluster for Dynamics in Logistics (Log*Dynamics*) at the BIBA (Bremer Institut für Produktion und Logistik GmbH), which is a scientific engineering research institute affiliated to the University of Bremen.

The scope of the conference was concerned with the identification, analysis, and description of the dynamics of logistic processes and networks. The spectrum reached from the planning and modelling of processes over innovative methods like autonomous control and knowledge management to the new technologies provided by radio frequency identification, mobile communication, and networking. The growing dynamic confronts the area of logistics with completely new challenges: It must become possible to rapidly and flexibly adapt logistic processes and networks to continuously changing conditions. LDIC 2009 provided a forum for the discussion of advances in that matter. The conference addressed scientists in logistics, operations research, and computer science and aimed at bringing together both researchers and practitioners interested in dynamics in logistics.

The proceedings of LDIC 2009 consist of 48 contributions. The first paper was invited, the others selected by a strong reviewing process. The volume is organized into the following six subject areas: *Mathematical Modelling in Transport and Production Logistics* with ten papers, *Routing, Collaboration, and Control* with nine papers, *Information, Communication, Autonomy, Adaption and Cognition* with seven papers, *Radio Frequency Identification* with eight papers, *Production Logistics* with six papers, and *Ports, Container Terminals, Regional Logistics and Services* with eight papers.

We would like to thank the members of the program and organization committee and the secondary reviewers Christoph Angerer, Eleonora Bottani, Zeng-Yu Cai, Nick Chung, Gianluigi Ferrari, Christian Gorldt, Nicolau Gualda, Harry Halfar, Tilo Hamann, Carl Hans, Bernd Hellingrath, Karl Hribernik, Michael Huhns, Jens Kamenik, Hamid Reza Karimi, Thorsten Klaas-Wissing, Renate Klempien-Hinrichs, Sabine Kuske, Björn Kvarnström, Xue Li, Gerben Meyer, Dennis Ommen, Jürgen Sauer, Jörn Schönberger, Steffen Sowade, Thomas

Suding, Hauke Tönnies, and Shu-Jen Wang for their help in the selection process. We are also grateful to Martina Cornels, Renate Klempien-Hinrichs, Jakub Piotrowski, Ingrid Rügge, Jörn Schönberger, Aleksandra Slaby, Tobias Warden and several other colleagues for their support in the local organization and for the technical assistance in running the conference system. Special thanks go to Marcus Seifert for organizing the doctoral workshop of LDIC 2009. We are particularly indebted to Caroline von Totth for her support in editing the volume and in careful unification of the print files of all the contributions. Moreover, we would like to acknowledge the financial support by the BIBA, the Research Cluster for Dynamics in Logistics (Log*Dynamics*), the Nolting-Hauff Stiftung, the Center for Computing and Communication Technologies (TZI), and the University of Bremen. Finally, we appreciate once again the excellent cooperation with the Springer-Verlag.

Bremen, March 2010   Hans-Jörg Kreowski
                                                          Bernd Scholz-Reiter
                                                          Klaus-Dieter Thoben

Committees

Program

Neil A. Duffie, Madison (Wisconsin, USA)
Kai Furmans, Karlsruhe (Germany)
Axel Hahn, Oldenburg (Germany)
Bill C. Hardgrave, Arkansas (USA)
Bonghee Hong, Pusan (Korea)
Kap Hwan Kim, Pusan (Korea)
Paul George Maropoulos, Bath (United Kingdom)
Dirk Mattfeld, Braunschweig (Germany)
Antônio G.N. Novaes, Florianópolis (Brazil)
Kulwant Pawar, Nottingham (United Kingdom)
Mykhailo Postan, Odessa (Ukraine)
Antonio Rizzi, Parma (Italy)
Janat Shah, Bangalore (India)
Wilfried Sihn, Wien (Austria)
Alexander Smirnov, St. Petersburg (Russian Federation)
Wolfgang Stölzle, St. Gallen (Switzerland)

Organisation

Martina Cornels
Sergey Dashkovskiy
Carmelita Görg
Hans-Dietrich Haasis
Michael Hülsmann
Herbert Kopfer
Jakub Piotrowski

Ingrid Rügge
Marcus Seifert
Aleksandra Slaby
Hauke Tönnies
Dieter Uckelmann
Katja Windt

Contents

Part IV Radio Frequency Identification

Part V Production Logistics

Part I
Mathematical Modeling in Transport and Production Logistics

Structural Properties of Third-Party Logistics Networks

Dieter Armbruster, M. P. M. Hendriks, Erjen Lefeber
and Jan T. Udding

1 Introduction

Third party logistic providers are the middle men between producers and their customers. They typically have contracts with a producer to take over whatever is produced in a certain factory during a certain time window and to provide all customers with the goods that they require. In that way, third party logistic providers relieve both the customer and the producer from keeping large inventories to balance fluctuations in production and demand. Simultaneously, the third party logistic provider assumes the distribution risk, both in terms of aggregating shipments from and to different sources and for delays in the actual deliveries.

The logistic network and flows resulting from such a business model are significantly different from a usual supply network. In particular, there is only a general framework specifying the location and number of producers as well as their average production rate and the location, number and demand rate of their customers. In contrast to a supply or a distribution network, there are no demands funneled up the supply chain from the customer to the producer. Here, producers request pickup of finished goods and customers request delivery at a certain time

D. Armbruster (✉)
School of Mathematical and Statistical Sciences, Arizona State University, Tempe, USA
e-mail: armbruster@asu.edu

D. Armbruster, M. P. M. Hendriks, E. Lefeber and J. T. Udding
Department of Mechanical Engineering, Eindhoven University of Technology,
Eindhoven, The Netherlands
e-mail: m.p.m.hendriks@tue.nl

E. Lefeber
e-mail: a.a.j.lefeber@tue.nl

J. T. Udding
e-mail: j.t.udding@tue.nl

H.-J. Kreowski et al. (eds.), *Dynamics in Logistics*,
DOI: 10.1007/978-3-642-11996-5_1, © Springer-Verlag Berlin Heidelberg 2011

and it is the job of the third party logistic provider to store excess production and deliver requests on time. While there are long term contractual obligations to service specific producers and their customers, the actual quantities fluctuate stochastically and are only known a few days ahead of actual delivery. A typical scenario is the just in time (JIT) production common in the automobile industry. Parts are produced by a big plastics company at different locations throughout Europe and needed at different assembly plants in Europe. Neither the plastics producer nor the assembly factory keep any sizeable inventory—however the customers expect delivery JIT for assembly. Keeping the right size inventory and shipping on time is the business of the third party provider.

There are three management levels to operate a third party logistic provider:

1. At the *strategic level* the number of warehouses, their geographic location and the geographic range of the logistic network have to be determined.
2. At the *tactical level* transportation links between the established set of customers, producers and warehouses are set up.
3. At the *operational level* size and direction of shipments are established.

These three levels correspond to different timescales—warehouses and geographic operations correspond to long term plans and investments, the operational level covers day to day operations and the tactical level lives on a monthly to yearly timescale.

The current paper is specifically concerned with the tactical level. We assume that the strategic decisions have been made and we assume a fixed set of warehouses, producers and customers. We are studying the network of links. In particular, we focus on the influence of the operational level on the network structure, i.e. we are interested to determine the topology of the network as a function of the operational parameters.

We will study the influence of the following operational parameters on the tactical logistics networks:

• Cost of delivery at the wrong time. Typically there are contractual penalties for early or late delivery.
• The actual transportation costs—we assume costs that are linear in the amount that is shipped.
• Variation in demand and production.

To determine the network structure we develop an optimization scheme that optimizes the operational level. Given a network structure, transportation cost coefficients, missed delivery penalties and given stochastic production and demand rates over a time interval, we will determine the optimal allocation of shipments to direct shipping and to warehousing. We then repeat the optimization for a network with a smaller number of links and continue this process until the costs for the logistic provider go up significantly. Establishing transportation links is costly: local shippers have to be contracted, customs and other licensing issues may arise, etc. A reduction in the number of links reduces the complexity of the network and with that the complexity of the organizational task of the third party logistic provider.

In addition reducing the number of links pushes the flow of goods into the remaining links leading to thicker flows which lead to economy of scales (more fully loaded trucks or railroad wagons). In principle, the economy of scales and the costs of establishing links could be put into a cost function and an optimal link structure could be derived. However, the resulting optimization scheme will become extremely complicated and very hard to solve for any reasonable size network. In addition, the structural costs of establishing a link are very hard to quantify.

Our study shows:

- The cost of operating a third party logistic network can be quantified and optimized.
- A heuristics is developed that reduces the number of links based on their usage.
- As the number of links in a network is reduced the cost of running the operation on the network stays almost constant. Most links are redundant and can be deleted without any influence on the operation of the network until a critical network size and structure is reached. At that point the operational cost explode by several orders of magnitude reminiscent of a physical phase transition. Hence there exists a network with a very limited number of links that has close to optimal operational cost called the reduced network.
- The reduced network is largely insensitive to variations in the production and the demand rate.
- The parameters of the operational cost function determine the topological structure of the reduced network. In particular, JIT shipping requires direct links between some producers and some customers whereas time insensitive deliveries lead to a network where all links go through warehouses.

This paper gives a short summary of our work: details on the connection between the heuristics and exact optimization via a branch and bound algorithm, about the relationship between continuous production flows and transportation via trucks leading to a stepwise cost function, on extensions to more than one product flow and other issues can be found in Hendriks et al. (2008).

2 Related Approaches

Some related work focuses on optimizing the strategic level: for instance Meepetchdee and Shah (2007) determines the optimal number and location of warehouses, given the location of a plant producing one type of product and given the number and location of the customers. Results show that there is a tradeoff between the robustness of the network and its efficiency and its complexity (Cordeau et al. 2008). Typical studies investigating supply chain policies place orders to manufacturing facilities to keep inventory positions in warehouses at a desired level and to satisfy customer demand (Kwon et al. 2006; Hax and Candea 1984; Wasner and Zapfel 2004). While the decision space on an operational level in a supply chain network is crucially different from the decision space for a third party logistic provider, some of the approaches are similar to ours. For instance Tsiakis et al. (2001) studies the strategic design of a multi-echelon, multi-product supply chain network under demand uncertainty. The objective is to minimize total costs taking infrastructure as well as operational costs into consideration. A finite set of demand scenarios is generated and the objective function is expanded by adding weighted costs for each of the possible scenarios. A process very similar to our heuristic process is employed in Kwon et al. (2006) to determine the number and locations of transshipment hubs in a supply chain network as a function of the product flows. The costs of operational activities in a network in which all potential hubs are present are minimized. Deleting one hub at a time the decrease in cost is determined and the hub which causes the largest decrease is deleted from the network. This process is continued until deleting a hub does not significantly affect the costs.

The dependency of the operational costs in a multi-echelon supply chain on the number of warehouses connected to each customer is studied in Cheung and Powell (1996). In a standard distribution network, each customer desires to be connected to one warehouse for practical reasons. The study suggests that large improvements are made when a portion of the customers is served by two warehouses (and the rest by one). An interesting recommendation for future research which we pick up in this paper, is to investigate the sensitivity of the optimal network to the ratio of different operating costs.

The most interesting work on optimizing the operations of a third party logistic provider is Topaloglu (2005). Here a company owns several production plants and has to distribute its goods to different regional markets. In each time period, a random (uncontrollable) amount of goods becomes available at each of these plants. Before the random amount of the customers becomes available, the company has to decide which proportion of the goods should be shipped directly and which proportion should be held at the production plants. Linear costs are assigned to transportation, storage and backlog. A look-ahead mechanism is introduced by using approximations of the value function and improving these approximations using samples of the random quantities. It is numerically shown that the method yields high quality solutions. Again it is found that improvements for the operational costs are made when a part of the customers is served by two plants, while the other part is served by one.

3 Methodology

To fix ideas we consider the graph in Fig. 1: production facilities supply goods and push them into the network while customers have a demand for them. Goods can be either sent directly from manufacturer(s) to customer(s) or stored temporarily in warehouses. We do not allow shipments from one warehouse to another.

We consider the following problem: given m production facilities for the same product, n warehouses, c customers and their spatial location, and given a time series

Fig. 1 A typical small
logistic network

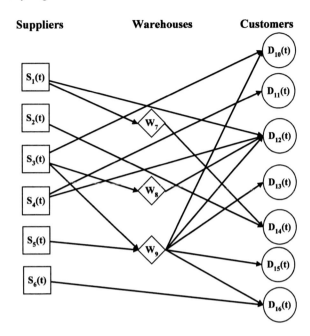

of daily production and daily demands at each location over a time period T, determine a network topology such that both, the costs of operating the network, and the number of links are close to minimal. We assume the total supply to be equal to the total demand over the time period T. The time series for production and demand are fixed and randomly generated. However, they are only known to the third party logistic provider (the optimization scheme) over a time horizon S where $S \ll T$.

We will set up a linear programming (LP) scheme, which describes the flows through such a logistics network for a fixed topology. The decision variables of this model are the flows each day through each of the links. The modeling is in discrete time (days) and is based on the following conservation laws:

- Every day the production at each factory has to be distributed among the links, connected to that particular factory.
- The storage level in a warehouse is updated daily, dependent on incoming and outgoing flows. Flows going through a warehouse incur a delay: products arriving in a warehouse at a certain time instance are shipped out 1 day later.

An objective function is constructed by assigning costs per unit of goods per day for transportation, storage and early and late delivery. The amount of goods, which are not delivered to a certain customer at a certain time instance, is added to the original demand of that particular customer for the next time instance. The conservation laws are equality constraints for this optimization problem. Since forecasts of supplies and demands are available over the time horizon S, routing decisions are based on this time horizon. This leads to a model predictive control (MPC) scheme with receding time horizon which is executed each day during the time period T to determine the most cost-effective routing schedule for the given period T and the fixed network topology.

Since the number of possible topologies grows very large when realistic values for m, n and c are considered, we propose a heuristic to determine the network that has the lowest number of links while still generating a close to optimal schedule.

4 Optimization Scheme for the Operational Level

Let $k = 0, 1, 2, 3...$ represent time in days, $u_{i,j}(k)$ is the amount of goods transported from node i to node j on day k, and $w_p(k)$ is the inventory position in warehouse p on day k. Although the total supply on day k can differ from the total demand on day k, we require that the following conservation law holds for period T:

$$\sum_{k=0}^{T}\sum_{i=1}^{m} M_i(k) = \sum_{k=0}^{T}\sum_{j=1}^{c} C_j(k). \tag{1}$$

Here $M_i(k)$ is the production of supplier i on day k and $C_j(k)$ is the demand of customer j on day k. This assumption implies that the total amount of goods produced and consumed over the time period T is balanced. In addition

$$M_i(k) = \sum_{j\in S_i} u_{ij}(k), \tag{2}$$

where the set S_i contains the indices of all the nodes to which manufacturing facility i is connected. A similar constraint exists for the warehouse:

$$w_p(k) = w_{p-1}(k) + \sum_{i\in I_p} u_{ij}(k-1) - \sum_{j\in O_p} u_{pj}(k), \tag{3}$$

where the set I_p contains the indices of all the suppliers of warehouse p and O_p contains the indices of all the nodes that are its customers. The total cost $A(u)(k)$ of operating the logistic enterprise at day k for the flows $u_{ij}(k)$ represented by u consists of the sum of the transportation costs, the warehouse costs and the backlog costs

$$A(u)(k) = \sum_{j=1}^{c} b_j B_j^2(k) + \sum_{i,j\in U} a_{ij} u_{ij}(k) + \sum_{p\in W} s_p w_p(k); \tag{4}$$

where b_j is the backlog cost per item per day for customer j, a_{ij} represents the unit cost of transportation from location i to location j, and s_p is the unit cost per day of storing an item in warehouse p. $B_j(k)$ is the backlog at customer j on day k given by

$$B_j(k) = D_j(k) - \sum_{i\in Q_j} u_{ij}(k) \tag{5}$$

where Q_j is the set of all nodes connecting to customer j and

$$D_j(k) = C_j(k) + B_j(k-1). \tag{6}$$

Hence old backlog is added to the original customer demand. Note that we are penalizing early and late delivery equally which is most certainly not the case in reality but makes our calculations much easier.

Equation (4) reflects the cost on 1 day. However, forecasts of supplies and demands are available over a time horizon of length S and hence routing decisions should be made based on the time interval $[k, k+S]$. To do this we use MPC with rolling horizon (MPC) (Garcia et al. 1989) and have the following cost function:

$$\min_{u(k),\dots,u(k+S-1)} \sum_{q=0}^{S-1} A(u)(k+q) \tag{7}$$

subject to all the previous constraints over the time interval $[k, k+S]$.

The above MPC determines the optimal routing schedule for day k. Subsequently the MPC for day $k+1$ is executed, rolling the time horizon forward. To find the most cost-effective routing schedule for the time period T thus entails solving the optimization problem (Eq. 7) T times leading to a total minimal cost over that time period.

5 Optimization Scheme for the Network Topology

The MPC presented in the previous section uses a fixed topology to make the best operational decisions given a set of time series for supply and demand. For small enough networks we can now change the network topology and determine the best network for the optimal operational decisions. A branch and bound method is described in Hendriks et al. (2008). However, since the problem has integer state variables (i.e. whether a link exists or not) we can only solve rather small problems. We therefore introduce the following heuristics:

- Start with a fully connected network and run the MPC.
- Until the network is minimally connected (i.e. each supplier has at least one customer, each customer has at least one supplier), do:

  - delete all links whose total flow over the time period T is below a threshold.
  - Run the MPC for the truncated network.

Plot the costs of the optimal operation against the number of deleted links.

Figure 2 shows a typical result for a network of 8 suppliers, 4 warehouses and 75 customers over a time interval of $T = 100$ and a horizon of $S = 3$. The operating costs for the optimal operations are minimal for the fully connected network, since we do not figure in the economics of scale. Typically about 95% of the links can be deleted without changing the operational cost. At some critical

Fig. 2 Cost of optimal
scheduling a network as a
function of the number of
deleted links

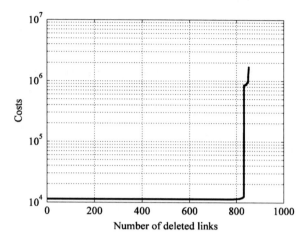

link number the costs explode. Choosing a network that stays just below this critical link number allows us to find a network whose operational cost are very close to optimal and at the same time have a very small number of links. We call such a network the *reduced network*.

We show in Hendriks et al. (2008) that the reduced network has costs that are very close to the costs of the optimal network and that the topology of the reduced network is equivalent to the topology of the optimal network determined by a branch and bound method.

6 The Influence of the Cost Parameters on the Topology of the Reduced Network

We study the dependence of the reduced network topology on the ratio of backlog costs to transportation cost. This reflects different business scenarios: high backlog costs are associated with JIT production methods or with the delivery of very time sensitive material, e.g. pharmaceutical products. Low backlog costs are associated with the delivery of commodity items.

Figures 3 and 4 show the resulting close to optimal network topologies for a test case with 20 suppliers, 4 warehouses, and 25 customers for low backlog costs and high backlog costs, respectively. Two differences between the structures are very obvious: (1) low backlog costs contains far fewer links than the topology for the JIT-product, (2) low backlog costs generate a near optimal topology that has only indirect links, while a lot of direct links are present for the JIT scenario. A posteriori it is not hard to interpret these topologies: JIT production needs both, the speed of the direct links as well as the buffering capacity of the warehouses. That also leads to a significant increase of links. In a scenario with negligible backlog costs, the network is determined by the structure of the transportation

Fig. 3 The reduced network for a backlog/transportation cost ratio of 10,000/1

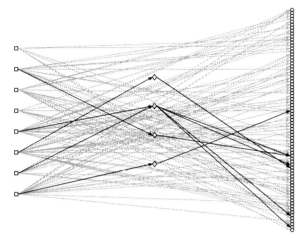

Fig. 4 The reduced network for a backlog cost/transportation cost ratio of 0.0001/1

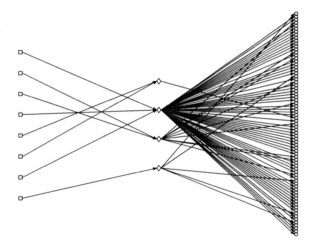

costs which here favor the links through the warehouses since they are the ones with the thick flows. We have to stress here that for relatively small backlog costs goods are still delivered to the customers. Although it might be cheaper to not satisfy customer demands rather than shipping goods from warehouses to customers, high future storage costs, forecasted over the time window S induce the system to ship goods out of the warehouses.

It is instructive to study how the topology changes as the ratio of backlog/ transportation cost changes. We generate 60 different supply and demand samples and run the MPC and subsequent heuristics for each of these samples. The number and type of links (either through a warehouse or through direct links) are determined for each of these samples as a function of the cost ratio. Figure 5 shows the mean number of direct and indirect links and a 95% confidence interval for the 60 samples. Figure 6 shows the average number of links (with confidence interval) as

Fig. 5 Mean number of
direct and indirect links as a
function of the cost ratios

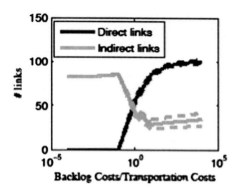

Fig. 6 Mean number of links
connecting to a customer

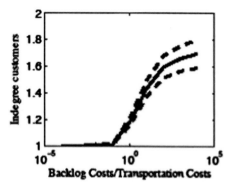

a function of the cost ratios. In all cases we used 4 suppliers, 3 warehouses and 30 customers.

We find that

- There is a clear crossover between systems that are dominated by direct links and systems that are dominated by indirect links. The crossover occurs approximated when backlog costs and transportation costs are equal.
- The number of direct (indirect) links increases (decrease) monotonically with the backlog to transportation cost ratio.
- For extremely low backlog costs, the number of direct links is zero and the number of indirect links is minimal, i.e. everybody has one connection with a warehouse which provides all the redundancy.
- For JIT networks, the redundancy is provided by more than one link on average to every customer and most of these links are direct links.

7 Conclusions

We have studied the structure of the distribution networks for a third party logistic provider. Using MPC and a linear program we find optimal operational decisions that allow us to heuristically trim underused links in networks until the operational cost explode. A network just before the transition is called the reduced network and is the target of our study. It combines low cost operations with a small number of links. The typical number of links of the reduced network does not exceed 5% of a minimally connected network. Experimental results reported in Hendriks et al. (2008) suggest that the reduced network is insensitive to the second and higher moments of the supply and demand distributions. This suggests that a robust reduced network can be found by a random choice of supply and demand distributions with the correct means. Very plausible topological structures for the reduced networks have been created for the extreme cases of JIT delivery and commodity delivery. Open problems include the study of the evolution of the topology of these networks as the cost parameters change continuously, the inclusion of transshipments between warehouses, the inclusion of multiple type products and the possibility of high cost and low cost links (airfreight and trucks).

Acknowledgments This study was supported by the Koninklijke Frans Maas Groep. D. A. was supported by NSF grant DMS 0604986 and a grant by the Stiftung Volkswagenwerk under the program on Complex Networks.

References

Cheung, R., & Powell, W. (1996). Models and algorithms for distributions problems with uncertain demands. Transport Sci, 30 (1), 43–59.

Cordeau, J., Laporte, G., & Pasin, F. (2008). An iterated local search heuristic for the logistics network design problem with single assignment. Int J Prod Econ, 113, 624–640.

Garcia, C., Prett, D., & Morari, M. (1989). Model predictive control: theory and practice. Automatica, 25 (3), 335–348.

Hax, A., & Candea, D. (1984). Production and inventory management (L. Mason, Ed.). Prentice Hall, Inc.

Hendriks, M.P.M., Armbruster, D., Laumanns, M., Lefeber, E., & Udding, J.T. (2008). Design of robust distribution networks run by fourth party logistic service providers, preprint TU Eindhoven, submitted.

Kwon, S., Park, K., Lee, C., Kim, S., Kim, H., & Liang, Z. (2006). Supply chain network design and transshipment hub location for third party logistics providers. In Lecture notes in computer science (Vol. 3982, p. 928–933). Springer Berlin/Heidelberg.

Meepetchdee, Y., & Shah, N. (2007). Logistical network design with robustness and complexity considerations. Int J Phys Distrib, 37 (3), 201–222.

Topaloglu, H. (2005). An approximate dynamic programming approach for a product distribution problem. IIE Trans, 37, 697–710.

Tsiakis, P., Shah, N., & Pantelides, C. (2001). Design of multi-echelon supply chain networks under demand uncertainty. Ind Eng Chem Res, 40, 3585–3604.

Wasner, M., & Zapfel, G. (2004). An integrated multi-depot hub-location vehicle routing model for network planning of parcel service. Int J Prod Econ, 90, 403–419.

Development of a Computational System to Determine the Optimal Bus-stop Spacing in order to Minimize the Travel Time of All Passengers

Homero F. Oliveira, Mirian B. Gonçalves, Eduardo S. Cursi
and Antonio G. Novaes

1 Introduction

One of the main concerns regarding urban planning nowadays is public transportation. The great number of vehicles in the main cities has been causing many problems, from infrastructure (number of vehicles over street capability), trough safety (high accident rates) and environmental issues (high pollution rates), among others. In infrastructure, one of the problems caused by the large number of vehicles on the streets is the travel time between two locations.

These problems aggravate specially in big cities, where traffic jams have already become part of the urban landscape.

However, one of the main aspects to be considered in a public transportation system is the travel time of the passenger using a bus line. The number of stops affects the total travel time deeply. In this manner, the number stops must be chosen very carefully, in a way that the bus lines become more appealing to the

H. F. Oliveira (✉)
Universidade Estadual do Oeste do Paraná–Campus de Toledo, Caixa Postal 520,
Toledo, Paraná, CEP 85903-000, Brazil
e-mail: homero2@uol.com.br

M. B. Gonçalves and A. G. Novaes
Departamento de Engenharia de Produção e Sistemas, Universidade Federal de Santa
Catarina–UFSC, Florianópolis, SC, Brazil

M. B. Gonçalves
e-mail: mirianbuss@deps.ufsc.br

A. G. Novaes
e-mail: novaes@deps.ufsc.br

E. S. Cursi
Laboratoire de Mécanique de Rouen (LMR), Institut National des Sciences Appliquées
de Rouen–INSA Rouen, 76801 Saint-Étienne-du-Rouvray Cedex, France
e-mail: eduardo.souza@insa-rouen.fr

H.-J. Kreowski et al. (eds.), *Dynamics in Logistics*,
DOI: 10.1007/978-3-642-11996-5_2, © Springer-Verlag Berlin Heidelberg 2011

users. With a large number of stops, the user walks very little, but it makes the trip too long and unpleasant for those who travel a long distance in that line. On the other hand, too few stops make the trip faster, but the passengers also have to walk more to get to the bus-stop, as well as to the final destination.

Ammons (2001) studied various spacing patterns between bus-stops around the world and concluded that the average spacing is from 200 to 600 m in urban areas. Reilly (1997) noticed that the European traffic departments have different standards to determine the spacing between bus-stops. In Europe, there are 2–3 stops per kilometer, which means that the spacing is from 330 to 500 m, in opposition to United States standards, where the stops are spaced from 160 to 250 m.

These studies show that the distance between stops does not follow a scientific procedure, or even based on predefined methodological studies. According to Kehoe (2004), in many routes along the USA, the bus-stops were defined trough time, as a result of user's requests to authorities and/or bus companies. Because the stops were based on citizen's needs, altering the distance between stops becomes a complicated process, for the population has already grown accustomed to the original spacing.

These remarks lead to the following question:

How will the ideal number of stops be determined in order to optimize the line for the users?

To answer this question we combined the concepts of non-linear programming (Frielander 1994) and Voronoi diagram. Voronoi diagrams has been around for at least four centuries, and many relevant material can be found in many areas, such as anthropology, archeology, astronomy, biology, cartography, chemistry, computational geometry, ecology, geography, geology, marketing, meteorology, operations research, physics, remote sensing, statistics, and urban and regional planning (Novaes 2000).

The concept of Voronoi Diagram is very simple. Given a finite set of distinct, isolated points in a continuous space, we associate all locations in that space with the closest member of the point set. The result is a partitioning of the space in a set of regions where each region is related to only one of the points of the original set.

Since the 1970s, algorithms for computing Voronoi diagrams of geometric primitives have been developed in computational geometry and related areas. There are several ways to construct a Voronoi diagram. One of the most practical is the incremental method, described in Novaes (2000). This method is also one of the most powerful in the subject of numerical robustness. The total time complexity for this method is of $O(n^2)$. However, the average time complexity can be decreased to $O(n)$ by the use of special data structures as described in Novaes (2000), p. 264.

2 Concepts of Voronoi Diagram

In this section we will define the main concepts and some properties of Voronoi Diagrams. The concepts presented in this section were based in Okabe et al. (1992).

2.1 Definition of a Planar Ordinary Voronoi Diagram

Given a set of two or more but a finite number of distinct points in the Euclidian plane, we associate all locations in that space with the closest member(s) of the point set with respect to the Euclidean distance. The result is a tessellation off the plane into a set of regions associated with members off the point set (Okabe et al. 1992).

The mathematical definition is the following:

$$V(p_i) = \left\{ x \mid \|x - x_i\| \leq \|x - x_j\| \text{ for } j \neq i, j \in I_n \right\} \tag{1}$$

where V is the planar ordinary Voronoi diagram associated with p_i and the set given by:

$$V^o = \{V(p_1), \ldots, V(p_n)\} \tag{2}$$

An example of an ordinary Voronoi diagram is presented in Fig. 1.

2.2 Definition of a Multiplicatively Weighted Voronoi Diagram

In the ordinary Voronoi diagram, we assume that all generator points have the same weight. But, in many practical applications, we may have to assume that they have different weights in order to represent, for example the population of a city, or the level of hazardousness that an accident at a point can cause.

Voronoi diagrams for the weighted distance are more complicated to analyze. The sides of the polygons are no longer straight lines but are arcs of circles.

The multiplicative weighted Voronoi diagram is characterized by the weighted distance calculated by

Fig. 1 Example of ordinary
Voronoi diagram

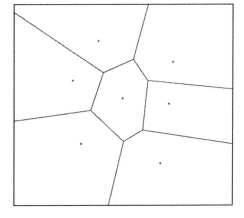

$$d_{mw}(p, p_i) = \frac{1}{w_i} ||x - x_i||, \quad w_i > 0 \tag{3}$$

where w_i is the weight associated with each point i. After a few steps of calculation, we obtain a bisector that is defined by

$$b(p_i, p_j) = \left\{ x \Big| \left\| x - \frac{w_i^2}{w_i^2 - w_j^2} x_j + \frac{w_j^2}{w_i^2 - w_j^2} x_i \right\| = \frac{w_i w_j}{w_i^2 - w_j^2} ||x_j - x_i|| \right\} \tag{4}$$

This bisector is the set of points that satisfy the condition that the distance from p to the point defined by

$$\frac{w_i^2 x_j}{w_i^2 - w_j^2} - \frac{w_j^2 x_i}{w_i^2 - w_j^2} \tag{5}$$

is constant. The bisector is a circle in $\mathbb{R}^2$. So, the dominance region of p_i over p_j with the weighted distance is written by:

$$\text{Dom}(p_i, p_j) = \left\{ x : \frac{1}{w_i} ||x - x_i|| \leq \frac{1}{w_j} ||x - x_j|| \right\}, \quad i \neq j \tag{6}$$

Figure 2 is an example of a multiplicatively weighted Voronoi Diagram with the coordinates of the points inside the parenthesis and the weights associated to them outside.

2.3 Definition of an Additively Weighted Voronoi Diagram

Similarly to the multiplicative weighted Voronoi diagram, the additively weighted Voronoi diagram (Fig. 3) is characterized by the weighted distance calculated by

Fig. 2 Example of a multiplicatively weighted Voronoi diagram

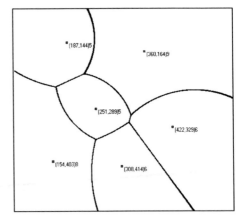

Fig. 3 Example of an additively weighted Voronoi diagram

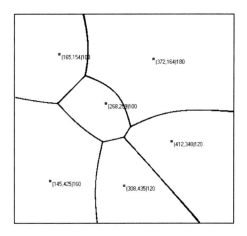

$$d_{mw}(p,p_i) = ||x - x_i|| - w_i \qquad (7)$$

2.4 Definition of a Compoundly Weighted Voronoi Diagram

Similarly to the multiplicative and additively weighted Voronoi diagram, the compoundly weighted Voronoi diagram is characterized by the weighted distance calculated by

$$d_{mw}(p,p_i) = \frac{1}{y_i}||x - x_i|| - w_i \qquad (8)$$

3 Definition of the Problem

This paper presents a model to optimize the bus-stop spacing of a bus line located in the city of São Paulo, Brazil. The city of São Paulo has approximately 10.5 million people and the public transportation system has about six million users daily (SPTrans—São Paulo Transportes S.A—http://www.sptrans.com.br).

The bus line that will be used in this paper is a new line that will be operational after the completion of the metro line number 4. That new metro line will be ready for use in 2010. The bus line was projected to transport passengers from the west region of the city to a metro station. At the station the passengers can switch to the metro line in order to get to their final destination. The purpose of the system is to find the optimal bus-stop spacing in order to minimize the total travel time of the passenger that goes from any point of the region to the metro station. The whole itinerary has 6.4 km.

Fig. 4 The bus line in a
Cartesian plane

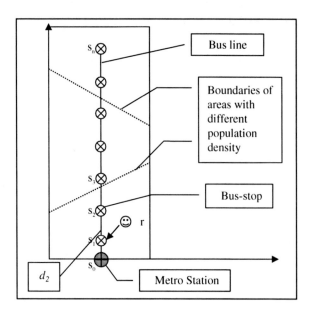

4 Formulation of the Model

Figure 4 shows the bus line in a Cartesian plan. The pattern on population density $\Phi(x, y)$ will be defined in relation with the variables 'x' and 'y'. The situation presented here is that the users in this area use bus lines to get to a bus terminal.

The area that this specific bus line can reach is called S. The ensemble of bus stops in the lines is represented by s_i where $i \in I_n$ and the number of bus stops is $n + 1$. There is also the distance d_i that represents de spacing between bus stops i and $i - 1$.

Considering that a user is at a specific point r (x_0, y_0) and wishes to go to bus stop s_1 (x_1, y_1), being $r, s_1 = S$. The travel distance D_a will be calculated as shown:

$$D_a = k * \|r - s_1\| = k * \sqrt{(x_1 - x_0)^2 + (y_1 - y_0)^2} \tag{9}$$

In this case, k is a correction factor to approximate the Euclidean distance to a walk distance. In this paper the value used is 1.3, as described in (Novaes 2000).

The time T_a that takes to go to the bus stop is calculated dividing the distance D_a by the user's speed on foot V_a. Which means:

$$T_a = \frac{D_a}{V_a} \tag{10}$$

Saka (2001) showed how to calculate how much time the bus takes to reach its destination (the terminal). The time is calculated as shown:

$$T_{\text{bus}} = T_{\text{ad}} + T_{\text{ed}} + T_c + T_o \tag{11}$$

In which:

a) T_{ad} = acceleration and deceleration time;
b) T_{ed} = passenger boarding and disembarkation time;
c) T_c = time delay due to traffic control (traffic lights, etc.);
d) T_o = travel time in normal traffic speed.

The total time that is lost in each bus stop can be represented as:

$$T_s = T_{ad} + T_{ed} \tag{12}$$

Adjusting to this particular case, as it is meant to calculate the ideal spacing between bus-stops, it is possible to eliminate the time delay due to traffic control (traffic lights, etc.) in order to concentrate the calculations in the time spent in the bus stops. So, we can say that the total travel time from stop i until the final destination is:

$$T_{bus} = (T_s * i) + \frac{D_i}{V_b} \tag{13}$$

where D_i is the distance from bus stop i to the terminal and V_b is the average speed of the bus on the route. And the total time is calculated as follows:

$$T_{tot} = T_a + T_{bus} \tag{14}$$

And the total time from bus stop i is

$$T_{tot} = k\frac{\|r - s_i\|}{V_a} + (T_s * i) + \frac{D_i}{V_b} \tag{15}$$

Assuming that every user will board in the bus-stop that minimizes his/her travel time, he/she will board the bus-stop that meets the following equation:

$$Min_i \left\{ k\frac{\|r - s_i\|}{V_a} + (T_s * i) + \frac{D_i}{V_b} \right\} \tag{16}$$

As a result of this factor, every stop will have its target area defined with:

$$V_i = \left\{ r | k\frac{\|r - s_i\|}{V_a} + (T_s * i) + \frac{D_i}{V_b} \leq k\frac{\|r - s_j\|}{V_a} + (T_s * j) + \frac{D_j}{V_b}; i \neq j; i, j \in I_n \right\} \tag{17}$$

which becomes, as described earlier, a additively weighted Voronoi region.

The total travel time from all passengers in the studied area will be calculated by:

$$T = \sum_{i=1}^{n} \int_{Vi} \left\{ k\frac{\|r(x, y) - s_i\|}{V_a} + (T_s * i) + \frac{D_i}{V_b} \right\} \phi(x, y) ds \tag{18}$$

This way, the optimization function is:

$$\min_{d_1,d_2,...,d_n} \sum_{i=1}^{n} \int_{Vi} \left\{ k\frac{\|r(x,y) - s_i\|}{V_a} + (T_s * i) + \frac{D_i}{V_b} \right\} \phi(x,y)ds \qquad (19)$$

That is a non restricted non-linear programming function.

5 The System

The system that was designed to solve that problem was written in Delphi 6.0 and utilizes some heuristics that are available in the literature.

For the non-linear programming problems we implemented three different heuristics: the Gradient method, the Conjugated Gradient method (Fletcher and Reeves 1964) and the Davidon-Fletcher-Powel method. Those methods are described by Luenberger (2005).

The system also can use three different rules to stop the line search: Armijo, Wolfe and Goldstein.

The gradient of the function is calculated using the Ridders' method of polynomial extrapolation described in Press et al. (2002).

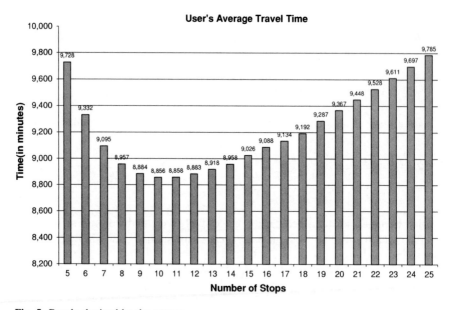

Fig. 5 Result obtained by the system

6 The Result

The system was used to find the optimal solution for the number of stops from $n = 5$ until $n = 25$. The solutions found indicates that the smallest total travel time is obtained with $n = 10$. Looking at the results in Fig. 5, we can see that if you increase the number of stops the user's average travel time will increase.

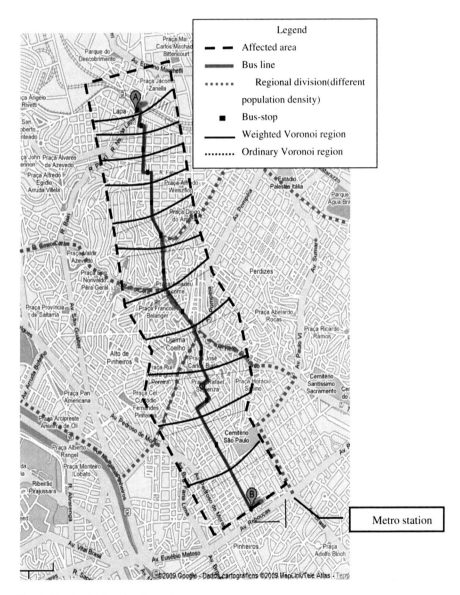

Fig. 6 Result obtained by the system

This result shows an average distance of 620 m from each bus stop. Using the usual bus-stop spacing of 250 m that is the most common in the city of São Paulo, we will have an increase of the travel time in more than 15%. This difference in a universe of six million passengers can be very significant.

Figure 6 shows a map with the location of the bus-stops obtained by the system. The figure shows the bus line and the area affected for it. Is this case we used 600 m for each side of the bus line to be the limit of the area affected by the bus line.

The map also shows the Voronoi regions (ordinary and weighted) associated to each bus-stop. The ordinary Voronoi region is the area where the user will find the nearest bus-stop from his location. The weighted Voronoi region is the area where the user will find the bus-stop that will take him to his destination in the smallest amount of time.

7 Conclusions

In this paper, we designed and implemented a system to find the optimum bus-stop spacing in order to minimize the total travel time of the passengers of a bus line that goes to a final destination. The system was developed using concepts of non-linear programming and Voronoi diagrams. The idea was to use both concepts together to find the optimal solution for the problem. The results showed that there is an optimal number of bus-stops but if this number is increased a little, it will not compromise the solution too much and it will make the user walk less. But if we compare the actual bus-stop spacing with the optimal one found by the system, we can observe that the travel time can be decreased in more than 15%.

The result can also be used as a parameter to be combined with others, like cost, number of vehicles, etc. in order to design a new line or to improve an existing one.

References

Ammons, D. N. Municipal benchmarks: Assessing local performance and establishing community standards. (2nd ed.). Thousand Oaks, CA: Sage Publications, 2001.

Fletcher, R. and Reeves, C. M. Function minimization by conjugate gradients. Computer Journal, 7 (149–154), 1964.

Frielander, A. Elementos de Programação Não Linear. Campinas, SP: Editora Unicamp, 1994.

Kehoe, O. V. Effects of Bus Stop Consolidation on Transit Speed and Reliability: a Test Case. A thesis submitted in partial fulfillment of the requirements for the degree of Master of Science in Civil Engineering University of Washington, 2004.

Luenberger, D. G., Linear and Nonlinear Programming. Second Edition, Springer Science + Business Media Inc., 2005.

Novaes, A. G. Logistics Districting With Multiplicatively Weighted Voronoi Diagrams. XI Congreso Panamericano de Ingeniería de Tránsito y Transporte, Gramado, RS. 19 al 24 de Noviembre del 2000.

Okabe, A.; Boots, B.and Sugihara, K. Spatial Tessellations Concepts and Applications of Voronoi Diagrams. Wiley, Chichester–New York–Brisbane–Toronto–Singapore, 1992.

Press, W. H., Teulosky, S. A., Vetterling, W. T. and Flannery, B. P. Numerical Recipes in C++. The Art of Scientific Computing. Second Edition, Cambridge University Press, 2002.

Reilly, J. M. (1997). Transit service design and operation practices in western european countries. Transportation Research Record, 1604, 3–8.

Saka, A. A. Model for determining optimum bus-stop spacing in urban areas. Journal of Transportation Engineering, n. 127 (3), pp. 195–199, USA, 2001.

Some Remarks on Stability and Robustness of Production Networks Based on Fluid Models

Bernd Scholz-Reiter, Fabian Wirth, Sergey Dashkovskiy, Michael Schönlein, Thomas Makuschewitz and Michael Kosmykov

1 Introduction

Modern production networks often consist of production facilities which are geographically distributed around the globe (Wiendahl and Lutz 2002). The generic structure of these networks leads to an increased complexity, which together with the dynamics of modern production networks poses a unique challenge for management and research (Scholz-Reiter et al. 2008). In particular perturbations or uncertainties of the market requirements may lead to an unstable behavior of a network (Scholz-Reiter et al. 2005) and the network may not be able to fulfil the customer demand. This raises the question, what type or size of perturbations is admissible without destroying the stability of a production network? This question can be answered using the stability radius of a network, which reflects the magnitude of the smallest possible perturbation that destabilizes the system (Hinrichsen and Pritchard 2005).

The dynamics of such complex production networks can be modelled by multiclass queueing networks. Dai (1995) and Dai and Jennings (2003) presented a new approach to investigate the stability of such networks using a fluid model, which is an analogue deterministic continuous model of the discrete stochastic model. The stability of a corresponding fluid limit model implies the stability of the original queueing network. In comparison to a queueing model the stability of

B. Scholz-Reiter and T. Makuschewitz
BIBA-Bremer Institut für Produktion und Logistik GmbH, University of Bremen, Bremen, Germany
e-mail: bsr@biba.uni-bremen.de

F. Wirth and M. Schönlein (✉)
Institute of Mathematics, University of Würzburg, Würzburg, Germany
e-mail: schoenlein@mathematik.uni-wuerzburg.de

S. Dashkovskiy and M. Kosmykov
Centre of Industrial Mathematics, University of Bremen, Bremen, Germany

H.-J. Kreowski et al. (eds.), *Dynamics in Logistics*,
DOI: 10.1007/978-3-642-11996-5_3, © Springer-Verlag Berlin Heidelberg 2011

a fluid model can be determined more easily. Based on the stability of a fluid model we present an approach to determine the stability radius. The results of this approach are validated using simulations of both the fluid model and the corresponding queueing network.

The outline of the paper is as follows. Section 2 briefly introduces the notation of a fluid model and the necessary and sufficient conditions of stability. This introduction is accompanied by the necessary adaptations for modelling a production network as a fluid model. In order to illustrate the characteristic dynamic behavior of a fluid and queueing network we introduce in Sect. 3 a test scenario. This scenario is used in Sect. 4 to investigate the robustness of the underlying production network. Section 5 closes with some conclusions and an outlook to future research.

2 Description of the Fluid Model

We follow the model description from Ye and Chen (2001). The considered network consists of locations S_j with $j \in \mathcal{J} = \{1, 2, \ldots, J\}$ and different types of products $\mathcal{P}_k$ with $k \in \mathcal{K} = \{1, 2, \ldots, K\}$. Every type of product is processed exclusively at one location. So there is a many-to-one mapping σ : $\{\mathcal{P}_1, \ldots, \mathcal{P}_K\} \to \{S_1, \ldots, S_J\}$. The mapping σ generates the so called constituency matrix C, with $c_{jk} = 1$ if $\sigma(\mathcal{P}_k) = S_j$ and $c_{jk} = 0$ otherwise. For every location the set $C(S_j) := \{\mathcal{P}_k \in \{\mathcal{P}_1, \ldots, \mathcal{P}_K\} : \sigma(\mathcal{P}_k) = S_j\}$ is assumed to be nonempty. Further every type of product $\mathcal{P}_k$ has an exogenous arrival rate α_k and a process rate μ_k of products per time unit. After a product of type $\mathcal{P}_k$ was processed at location $\sigma(\mathcal{P}_k)$ it has the possibility either to leave the network or to become a product of type $\mathcal{P}_l$. The transition matrix P sets the proportion of processed products that either turn to their next location or leave the network. To be precise, p_{kl} denotes the proportion of products of type $\mathcal{P}_k$ that become products of type $\mathcal{P}_l$ upon service completion. Hence $1 - \sum_{l=1}^{K} p_{kl}$ is the proportion that eventually leaves the network. It is assumed that the $K \times K$ matrix P has spectral radius strictly less than one, i.e. all products leave the network. The initial amount of products is represented through the K dimensional vector $Q(0)$. The model of the network is given by (α, μ, P, C) and $Q(0)$. The performance is described by the K dimensional product level process $\{Q(t): t \geq 0\}$ and the K dimensional allocation process $\{T(t): t \geq 0\}$, where $Q_k(t)$ denotes the amount of products $\mathcal{P}_k$ in the network at time t and $T_k(t)$ denotes the total amount of time in the interval $[0, t]$ that location $\sigma(\mathcal{P}_k)$ has devoted to processing products of type $\mathcal{P}_k$. For brevity we omit the calligraphic letters in the subscript. The next step is to fix a policy that predetermines the order in which the arriving products are processed at each location. We use the so called priority discipline, i.e. we consider a permutation π: $\{1,\ldots,K\} \to \{1,\ldots,K\}$. Given two types of products $\mathcal{P}_k$ and $\mathcal{P}_l$ that are processed at the same location $\sigma(\mathcal{P}_k) = \sigma(\mathcal{P}_l)$, we say that $\mathcal{P}_l$ has higher priority than $\mathcal{P}_k$ if

$\pi(l) < \pi(k)$. So products of type $\mathcal{P}_k$ are not processed as long as there are products of type $\mathcal{P}_l$. For $\mathcal{P}_k$ the set

$$H_k = \{\mathcal{P}_l : \mathcal{P}_l \in C(\sigma(\mathcal{P}_k)), \pi(l) \leq \pi(k)\}$$

denotes all products $\mathcal{P}_l$ that have at least priority $\pi(k)$ and are processed at the same location as $\mathcal{P}_k$. Finally the process of unused capacity $Y = \{Y(t):t \geq 0\}$ is introduced, where $Y_k(t)$ denotes the cumulative remaining capacity of location $\sigma(\mathcal{P}_k)$ for processing products of types that have strictly lower priority than products of type $\mathcal{P}_k$. The dynamics of the network under priority discipline are summarized as follows

$$Q(t) = Q(0) + \alpha t - (I - P^T)MT(t) \geq 0, \tag{1}$$

$$T(0) = 0 \text{ and } T(t) \text{ is nondecreasing}, \tag{2}$$

$$Y_k(t) = t - \sum_{l \in H_k} T_l(t) \text{ is nondecreasing, } k \in \mathcal{K}, \tag{3}$$

$$0 = \int_0^\infty Q_k(t) \, d\, Y_k(t), \ k \in \mathcal{K}, \tag{4}$$

where $M = diag(\mu_1,\ldots,\mu_k)$. Equation (4) describes the work-conserving property of the network and relation (1) is called the flow balance relation. The work-conserving property means that the idle time for a product of type $\mathcal{P}_k$ increases if and only if $Q_k(t) = 0$, i.e. there is no product of type $\mathcal{P}_k$ in the network waiting for being processed. The following theorem (Chen 1995) guarantees the existence of such a work-conserving allocation process.

Theorem 1 *For any fluid network (α, μ, P,C) with $Q(0)$ there is at least one work-conserving allocation T.*

The priority discipline is assumed to be preemptive. That is, if products of type $\mathcal{P}_m$ are processed at some time t, then this production is interrupted if the product level $Q_k(t)$ of any type $\mathcal{P}_k \in H_m$ with strict higher priority then $\mathcal{P}_m$ is positive. Any pair $(Q(t), T(t))$ that satisfies (1)–(4) is called a fluid solution of the work-conserving fluid network under priority discipline. The set of all feasible fluid level processes is denoted by

$$\Phi = \{Q(t) : \text{there is a } T(t) \text{ such that } (Q(t), T(t)) \text{ is a fluid solution}\}.$$

For $x \in \mathbb{R}^K$ we use $\|x\|_1 = \sum_{k=1}^K |x_k|$.

Definition 1 A fluid network is said to be stable, if there exists a finite time $\tau > 0$ such that $Q(\tau+\cdot) \equiv 0$ for any $Q(\cdot) \in \Phi$ with $\|Q(0)\|_1 = 1$.

We cite the following necessary condition for the stability of a fluid network (Chen 1995). Here $<$ has to be understood componentwise.

Theorem 2 *If a fluid network* (α, μ, P, C) *is stable, it holds that*

$$CM^{-1}(I - P^T)^{-1}\alpha < e, \qquad (5)$$

where $e = (1, \ldots, 1)^T$.

Remark 1 Common notations in the literature are $\lambda := (I - P^T)^{-1}a$ and $\rho := CM^{-1}\lambda$ which denote the total arrival rate and the nominal workload. Hence the condition (5) can be stated as $\rho_j < 1$ for all $j \in \mathcal{J}$.

Further in Chen (1995) there is also a sufficient condition for stability. Here a symmetric matrix A is called strictly copositive, if $x^T A x \geq 0$ for all $x \in \mathbb{R}^n$ with $x_i \geq 0$ and $x^T A x = 0$ only if $x = 0$. Here $x^- := \min\{0, x\}$.

Theorem 3 *A fluid network* (α, μ, P, C) *is stable if there exists a* $K \times K$ *symmetric strictly copositive matrix* $A = (a_{ik})$ *such that, for* $k = 1, \ldots, K$,

$$\sum_{i=1}^{K} \alpha_i a_{ik} - \min_{\mathcal{P}_i \in C(\mathcal{S}_k)} h_{ik} - \sum_{j=1, \mathcal{S}j \neq \sigma(\mathcal{P}_k)}^{J} \left(\min_{\mathcal{P}_i \in C(\mathcal{S}_j)} h_{ik} \right)^- < 0,$$

where $H = M(I - P)A$.

2.1 Adaption to Production Networks

The fluid model introduced in Sect. 2 can be used to model and to analyse the dynamics of a production network. To this end the production facilities of suppliers and OEM, warehouses and distribution centres are modelled as locations. These locations are numbered from 1 to J. The intermediate and final products of the network are classified into product types 1 to K that are processed or serviced by the locations. Thus the structure of the production network is given by the constituency and the transition matrix. The external and internal dynamics of the considered network are captured by the inflow rates of the product types and the processing rates of the product types at the assigned locations. The transition matrix of the fluid model contains the fragmentation of the material flows within the production network. The initial work in progress of the product types is given by $Q(0)$. If a location serves more than one product type a priority rule for the service needs to be determined.

3 Test Scenario

In this paper we simulate a network with three types of products and two locations. The parameters for the test scenario are given below.

$$\alpha = \begin{pmatrix} 0.15 \\ 0.15 \\ 0.10 \end{pmatrix}, \quad \mu = \begin{pmatrix} 0.6 \\ 0.9 \\ 0.5 \end{pmatrix}, \quad P = \begin{pmatrix} 0.25 & 0.15 & 0.20 \\ 0.05 & 0.25 & 0.15 \\ 0.20 & 0.25 & 0.10 \end{pmatrix}$$

$$Q(0) = \begin{pmatrix} 1 \\ 1 \\ 1 \end{pmatrix}, \quad C = \begin{pmatrix} 1 & 0 & 0 \\ 0 & 1 & 1 \end{pmatrix}$$

A schematic illustration is given in Fig. 1.

The priorities at the second location are chosen such that products of type 3 have higher priority than the products of type 2, i.e. $\pi = (\{1\}, \{3, 2\})$. Figure 2 shows

Fig. 1 Fluid network with two locations processing three types of products

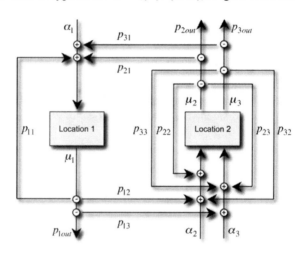

Fig. 2 WIP of the fluid model for the production network

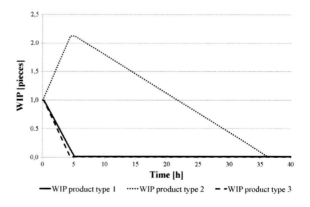

Fig. 3 WIP of the queueing
model for the production
network

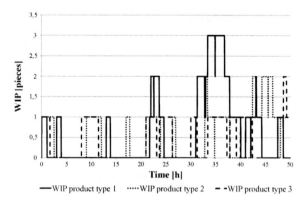

WIP product type 1   ⋯⋯WIP product type 2   ‒ ‒WIP product type 3

the behavior of the work in progress (WIP) under the preemptive priority discipline π. Here the WIP of type 2 products increases as long as the WIP of product
type 3 has not reached zero. At the time when $Q_3(t) = 0$ location 2 begins to
process type 2 products. Let $m \in (0, 1)$ be the proportion that location 2 allocates
to type 3 products such that the product level $Q_3(t)$ remains zero. Then the proportion that location 2 allocates to type 2 products is $1 - m$. Since the configuration is stable this allocation is able to empty the WIP of product type 2.
Consequently the WIP of all three product types reaches zero at some time τ and
the WIP stays zero beyond τ. Figure 3 shows the WIP of the associated queueing
network. Since the fluid model is stable the WIP of the queueing network becomes
zero from time to time.

4 An Approach to Describe Robustness of Production
Networks

In this section we focus on the question of quantifying the size of admissible
perturbations of the external arrival rate. The minimal size of a destabilizing
perturbation will be called the stability radius of the network. If the condition (5)
is violated Theorem 2 states that the network cannot be stable. The set of all
feasible arrival rates, i.e. all $\alpha \in \mathbb{R}_+^K$ such that that condition (5) is satisfied, forms
a polyhedron. So we increase the arrival rate by introducing a vector $\delta \in \mathbb{R}_+^K$
and consider a perturbation of the following form $\alpha \sim \alpha_\delta = \alpha + \delta$. Then we
consider the fluid network $(\alpha_\delta, \mu, P, C)$ and use the following notation
$\Delta = CM^{-1}(I - P^T)^{-1}$. The quantity of interest is the smallest $\|\delta\|_1$ such that
stability will no longer hold. Thus we give the following Definition.

Definition 2 The stability radius of the network (α, μ, P, C) is
$r(\alpha, \mu, P, C) = \inf\{\|\delta\|_1 : (\alpha_\delta, \mu, P, C) \text{ is not stable}\}$.

Fig. 4 Illustration of the
stability radius in the case
where the second arrival rate
is not perturbed

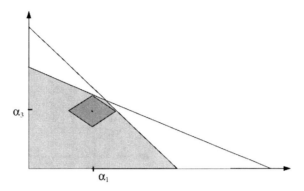

Remark 2 By Theorem 2 an upper bound for the stability radius is given by $\inf\{\|\delta\|_1 : \Delta(\alpha_\delta)\nleq e\}$.

In the following we give a geometric interpretation of the previous remark. In Fig. 4 the light grey area represents the set of all arrival rates that satisfy condition (5). Let α be an interior point.

The dark grey area is the largest open norm ball $B_r(\alpha)$ around α that lies completely in the light grey domain. The upper bound for the stability radius is calculated as follows, where we use $\Delta_j \cdot \alpha_\delta$ to denote the scalar product of the jth row with α_δ,

$$d_1 := \min\|\delta\|_1 \text{ such that } \Delta_1 \cdot \alpha_\delta \geq 1$$
$$d_2 := \min\|\delta\|_1 \text{ such that } \Delta_2 \cdot \alpha_\delta \geq 1 .$$
$$r(\alpha, \mu, P, C) \leq \min_{j=1,2} d_j$$

In regard to the test scenario Fig. 2 shows that the WIP of product type 1 reaches and remains zero long before the WIP of products type 2 and 3 do. Intuitively, one might expect that the network is able to cope with an additional arrival rate of type 1 products. A calculation for the test scenario using Remark 2 leads to $r(\alpha, \mu, P, C) \leq 0.0547781$ and the corresponding vector of perturbation is $\delta = (0, 0, 0.0547781)^T$. Since this is only an upper bound for the stability radius there may be arrival rates that are inside the dark grey area which lead to insta-bility. But this subset is very small as the following shows. For the perturbation $\delta = (0, 0, 0.05428493)^T$ the matrix

$$A = \begin{pmatrix} 10 & 0.374 & 0.53 \\ 0.374 & 0.645 & 0.916 \\ 0.53 & 0.916 & 1.3 \end{pmatrix}$$

satisfies the sufficient condition of Theorem 3 and the network is stable. The corresponding perturbed arrival rate is $\alpha_\delta = (0.15, 0.15, 0.15428493)^T$. In order to present the product level process and the queue length process clearly the

Fig. 5 WIP of the fluid
model with perturbed arrival
rate α_s

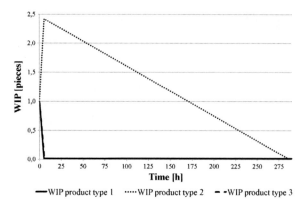

WIP product type 1   ·····WIP product type 2   − ·WIP product type 3

Fig. 6 WIP of the queueing
network with perturbed
arrival rate

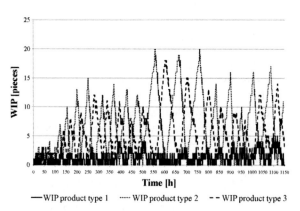

WIP product type 1   ····· WIP product type 2   − − WIP product type 3

following figures are generated for the perturbed arrival rate $\alpha_\delta = (0.15, 0.15, 0.153)^T$.

Figures 5 and 6 show the WIP of the fluid and queueing model for the perturbed production network. Further the pictures show that the time until the WIP of all types of products reaches zero has increased enormously as well as that the amplitude of the WIP rises by a factor of ten.

5 Conclusions and Outlook

We have introduced a fluid model that can be applied to a generic structure of locations in a production network. In particular, a scenario for two locations processing three types of products has been concerned. The parameter setting was chosen such that the fluid model is stable. Figure 3 illustrates the behavior of the corresponding queueing network. In the subsequent section the stability radius has been defined, which gives an upper bound to the additional arrival rates α_δ such that the production network is still able to handle the arrival rate. Again Fig. 6

illustrates this for the arrival rate. From Sect. 4 it can be seen that for the queueing network model corresponding to α_δ the amplitude of the queue length increases by a factor ten.

In future research we will also consider perturbations of the production rates as well as perturbations of the transition matrix. Moreover it is aspired to obtain analytical estimations of the queue length compared to increasing arrival rates or decreasing production rates.

Acknowledgments This research was funded by Volkswagen Foundation.

References

Chen, H. (1995): Fluid approximations and stability of multiclass queueing networks: work-conserving disciplines, Ann. Appl. Probab., 5, 637–655

Dai, J. (1995): On positive Harris recurrence of multiclass queueing networks: a unified approach via fluid limit models, Ann. Appl. Probab., 5, 49–77

Dai, J., Jennings, O. (2003): Stability of general processing networks, In: Stochastic modeling and optimization (Eds: D. Yao, H. Zhang, X. Zhou), Springer, New York

Hinrichsen, P., Pritchard, J. (2005): Mathematical Systems Theory I. Springer, Berlin

Scholz-Reiter, B., Wirth, F., Freitag, M., Dashkovskiy, S., Jagalski, T., de Beer, C., Rüffer, B. (2005): Some remarks on the stability of production networks, In: Proceedings of the International Scienific Annual Conference on Operations Research. Springer, Bremen, Germany, 91–96

Scholz-Reiter, B., Wirth, F., Dashkovskiy, S., Jagalski, T., Makuschewitz, T. (2008): Analyse der Dynamik großskaliger Netzwerke in der Logistik. Industrie Management, 3, 37–40

Ye, H.Q., Chen, H. (2001): Lyapunov method for the stability of fluid networks, Operations Research Letters 28, 125–136

Wiendahl, H-P., Lutz, S. (2002): Production in networks, Annals of the CIRP, 51(2), 1–14

Online Optimization with Discrete Lotsizing Production and Rolling Horizons

Wilhelm Dangelmaier and Bastian Degener

1 The Production System

We consider a production system for producing a certain set of products. The products are known in advance and can therefore be produced prior to a customer's order.

Orders are only given for a certain time horizon. In the course of time, new customer orders are revealed. No feasible plan can consider all customer orders, because only a certain, restricted part of the future is known in advance: the planning horizon. Although a scheduling strategy cannot provide an optimal solution due to a lack of complete knowledge about future customer orders, it should still avoid circumstances that will handicap future production processes. Hence, our basic rule: Independent of future events, the performance of the production, respectively the scheduling strategy, is not allowed to fall below a certain threshold. Of course, this threshold should depend on an optimal solution's value. Unfortunately, this threshold can only be determined as soon as all customer orders are known- and that is in retrospect. We are facing the challenge of how a company should produce today, even though only customer orders for the near future are known, or even worse, only the next customer order? Problems like this are handled by the framework of online optimizing (Dangelmaier 2008a, b; Danne et al. 2008; Krumke et al. 2005) Here, in contrast to classical offline optimization, the input is revealed piece by piece and hence the optimal solution is unknown in advance. Still certain bounds are guaranteed to be kept independent of future events.

W. Dangelmaier (✉) and B. Degener
Heinz Nixdorf Institut, Fürstenallee 11, 33102 Paderborn, Germany
e-mail: whd@hni.uni-paderborn.de

B. Degener
e-mail: degener@hni.uni-paderborn.de

H.-J. Kreowski et al. (eds.), *Dynamics in Logistics*,
DOI: 10.1007/978-3-642-11996-5_4, © Springer-Verlag Berlin Heidelberg 2011

The considered production system works taking into account the assumptions of the discrete lotsizing and scheduling problems (DLSP); the production of a product always covers complete time segments.

I^{PF}	Set of products respectively product indices, $i = \{1,...,n_{PF}\}$
T_P	Time model with the set of time segments respectively their indices $t = \{1,...,n'\}$ and the set of points of time for the end of a time segment (planning horizon) $T = \{0,...,n'\}$; $t = T$
b_{it}	Demand for product i in time segment t
a_t	Available capacity in time segment t
b_i	Production coefficient for product i
k_i^{set}	Setup costs for product i
k_{it}^{qty}	Cost per unit for product i in time segment t
k_i^{str}	Storage cost rate for product i per time segment t
B_{i0}	Initial inventory for product i
B_{iT}^{sht}	Minimum inventory level for product i at the end of time segment t
x_{it}	Lot size for product i in time segment t
B_{iT}	Demand for product i at the end of time segment t
δ_{it}^{set}	Setup costs for product t
δ_{it}^{pdn}	Production indicator for product i in time segment t
w_i	Maximum amount of product i in time segment t

The DLSP reduces the problem within a time segment to the decision whether to produce or not. Thus, the amount of production depends directly on the indicator variables and it holds $x_{it} = w_i \cdot \delta_{it}^{\text{pdn}}$. Since the DSLP basically considers only time segments of unit supply of capacity, w_i results as a time invariant ratio of the capacity per time segment a_t and the production coefficient b_i. This leads to the following formulation of the DLSP:

$$\text{Minimize} \quad \sum_{i,t} (k_i^{\text{set}} \cdot \max\{0, \delta_{it}^{\text{set}} - \delta_{i,t-1}^{\text{set}}\} + k_{it}^{\text{qty}} \cdot w_i \cdot \delta_{it}^{\text{set}} + k_i^{\text{str}} \cdot B_{iT})$$

under the restriction

$$\forall i \in I^{PF}, \forall t, T \in T_P : B_{iT} = B_{i,T-1} + w_i \cdot \delta_{it}^{\text{set}} - b_{it} \tag{DLSP, 1}$$

$$\forall t, T \in T_P : \sum_i \delta_{it}^{\text{set}} \leq 1 \tag{DLSP, 2}$$

$$\forall i \in I^{PF}, \forall t, T \in T_P : B_{iT} \geq B_{iT}^{\text{sht}} \tag{DLSP, 3}$$

$$\forall i \in I^{PF}, \forall t, T \in T_P : \delta_{it}^{\text{set}} \in \{0, 1\} \tag{DLSP, 4}$$

A schedule covers several periods with several time segments each (e.g. a horizon with 4 weeks and each week with 10 working shifts). It is updated periodically. At any one time the first period is considered to be the schedule implemented.

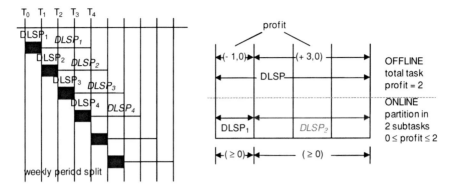

Fig. 1 Partition in two subtasks; an example where the DLSP solutions differ

The realization of a schedule is therefore the sustained implementation of respective first periods. Updates to the schedule only occur at proper points of time that are previously fixed for scheduling. Those are beyond the calendar (that is e.g. not during the week if a weekly planning cycle is considered). The scheduled demands (reduction in inventory) are already known "today" and can be postponed or changed in value until their final fixation (latest by the time they are scheduled in the first scheduling period).

2 Process Model for On-line Scheduling

Independent of the formal description of the production the following approach is chosen:

- Optimization of the 1 period + optimization of the planning horizon's remaining periods, considering the joint inventory at the end of the first period.[1]
- Weighting the two (partial) solutions.
- Dependent on the chosen weighting, a cost minimal solution for the corresponding planning horizon is chosen.

For instance, Fig. 1 displays a weekly continued planning, where always one week is fixed respectively. This week is divided for instance in days or shifts (4 week horizon, the schedule is repeated weekly, each week consists of 10 shifts).[2]

---

[1] Alternatively, one can construct an inventory at the end of the horizon that allows to carry forward the horizon without conflicts, if this is feasible in the given setting.

[2] Note that in the example, the profit for the first week of the complete DLSP is negative. Since there is rescheduling at the beginning of each week, it is possible that this value never changes to a positive value, although an off-line algorithm could make a profit over the entire period. A naive approach would therefore fail. We introduce a first possible solution.

At point of time T_0 the schedules DLSP_1 and $\textbf{\textit{DLSP}}_l$ are calculated. The schedule that maximizes (profit) respectively minimizes (costs) with $\alpha \cdot DLSP_1 + (1 - \alpha) \cdot DLSP_1 = \min$ with $0 \leq \alpha \leq 1$ is chosen. This way

- at the end of the first period disprofit is avoided, respectively a (minimum) profit is realized[3]
- a feasible schedule for the planning horizon is created
- due to the DLSP, a mathematical formulation of the problem is enabled

Of course, the solution for the DLSP over the (total-) horizon and the solutions of the two DSLP partial problems that are coupled by the inventory, need not to be identical, because each partial solution has to guarantee a positive value (refer to the example in Fig. 1).

3 The Formulation of the On-line Schedule as DLSP Model

The following notations are used in the approach:

T_P is the total horizon, T_T' and T_T' are the partial horizons at point of time T. k^{sales} is the sales revenue per production unit. A revenue can only be obtained in time segments with asset sale (demand) ($b_{it} > 0$).

B_{i0} initial inventory
δ_{i0}^{set} initial setup

Thus:

Maximize

$$\sum_{T_T' \in T_P} \left(\sum_{t,T \in T_T'} \left(b_{it} \cdot k_{it}^{\text{sales}} - (k_i^{\text{set}} \cdot \max\{0, \delta_{it}^{\text{set}} - \delta_{i,t-1}^{\text{set}}\} + k_{it}^{\text{qty}} \cdot w_i \cdot \delta_{it}^{\text{set}} + k_i^{\text{str}} \cdot B_{iT}) \right) \right)$$

under the following restrictions:

- Restrictions for partial problem 1 (partial horizon T_T'):

$$\forall i \in I^{PF}, \forall t, T \in T_T' : B_{iT} = B_{i,T-1} + w_i \cdot \delta_{it}^{\text{set}} - b_{it} \qquad \text{(DLSP, 1)}$$

$$\forall t, T \in T_T' : \sum_i \delta_{it}^{\text{set}} \leq 1 \qquad \text{(DLSP, 2)}$$

---

[3] We assume that an optimum plan is always connected with a profit.

$$\forall i \in I^{PF}, \forall t, T \in T'_T : B_{iT} \geq B^{sht}_{iT} \qquad \text{(DLSP, 3)}$$

$$\forall i \in I^{PF}, \forall t, T \in T'_T : \delta^{set}_{it} \in \{0, 1\} \qquad \text{(DLSP, 4)}$$

The initial inventory at the beginning of period T''_T is the inventory at the end of period T'_T of the previous planning cycle.

– Restrictions for partial problem 2 (partial horizon T''_T):

$$\forall i \in I^{PF}, \forall t, T \in T''_T : B_{iT} = B_{i,T-1} + w_i \cdot \delta^{set}_{it} - b_{it} \qquad \text{(DLSP, 1)}$$

$$\forall t, T \in T''_T : \sum_i \delta^{set}_{it} \leq 1 \qquad \text{(DLSP, 2)}$$

$$\forall i \in I^{PF}, \forall t, T \in T''_T : B_{iT} \geq B^{sht}_{iT} \qquad \text{(DLSP, 3)}$$

$$\forall i \in I^{PF}, \forall t, T \in T''_T : \delta^{set}_{it} \in \{0, 1\} \qquad \text{(DLSP, 4)}$$

– Joint constraints for both partial horizons T'_T and T''_T:

B_{iT} at the end of time horizon $T'_T = B_{iT}$ at the beginning of time horizon T''_T

With

$$K^{sales'}_T = \sum_{i,(t,T \in T'_T)} \left(b_{it} \cdot k^{sales}_{it} - \left(k^{set}_i \cdot \max\{0, \delta^{set}_{it} - \delta^{set}_{i,t-1}\} \right. \right.$$
$$\left. \left. + k^{qty}_{it} \cdot w_i \cdot \delta^{set}_{it} + k^{str}_i \cdot B_{iT} \right) \right) \quad \geq X \geq 0$$

and

$$K^{sales''}_T = \sum_{i,(t,T \in T''_T)} \left(b_{it} \cdot k^{sales}_{it} - \left(k^{set}_i \cdot \max\{0, \delta^{set}_{it} - \delta^{set}_{i,t-1}\} \right. \right.$$
$$\left. \left. + k^{qty}_{it} \cdot w_i \cdot \delta^{set}_{it} + k^{str}_i \cdot B_{iT} \right) \right) \geq Y$$

as well as $0 \leq \alpha \leq 1$ then yields $(\alpha \cdot K^{sales'}_T + (1 - \alpha) \cdot K^{sales'}_T) = \max \sum_T$.

4 Competitiveness of the Approach

The competitiveness is measured concerning the ratio of the total horizon of the off-line as well as the total horizon of the on-line approach. A schedule that is the solution to the DLSP-problem, can shift (net-) demands only towards presence, due to availability constraints. While the on-line solution can only use the planning horizon, the off-line solution is capable of using the entire horizon. However, note that the off-line solution still cannot shift the demands arbitrary far to the front in order to use the entire available capacity. The shifting is bounded, since resulting inventory costs dominate the costs for setup and production at some point of time. The best solution for the off-line algorithm (in order to beat the on-line algorithm as bad as possible according to the measure we introduced) is the period, respectively the time segment right before the planning horizon. Using this time slot can be iterated by the off-line algorithm, until the upper bound is reached that is set by the inventory cost. The number of those iterations can be set to [total horizon: (bound of offset + (bound of offset − planning horizon))].

Considering the formal description of the heuristic, it follows: a sales revenue is only possible if there are demands and those demands can be satisfied with costs that do not exceed the sales revenues. Thus, the heuristic tends to a just-in-time policy, but with a limited utilization of the capacity! If there are no demands present in a period, nothing can be shifted into this period. Otherwise the demand can always be satisfied by the maximal capacity (otherwise a shifting would be required; this might be for instance the 24-h capacity of the machinery).

• Production costs

We consider the worst case for the on-line algorithm by assuming a sequence of periods as follows: 1 period with maximal demand followed by a period with minimal demand. The minimal demand is [average demand − (maximal demand − average demand)]. This way, the off-line algorithm can level the needed capacity with minimal inventory: 1 period build up (maximal demand − average demand), 1 period reduction (maximal demand − average demand). If the off-line algorithm can balance over the entire planning horizon, for the on-line algorithm holds: 1 period with minimal load without shifting is followed by some periods with average load, is followed by 1 expensive load with maximal load. For the first case we have 1 period with minimal load + minimal balancing followed by 1 period with maximal load and minimal down balancing. Therefore the costs for the expensive capacity are maximal. In both cases the on-line algorithm is forced to choose the worse one.

• Setup costs
  We consider the worst case for the on-line algorithm:

    – The optimal lot size leads to an ordering cycle that is larger than the planning horizon. Thus, the off-line algorithm minimizes the sum of the setup and inventory costs.
    – If balancing is impossible (revenue = gross demand of the period) then there has to be disposed a product in each period (and this is also possible). The inventory is minimal, the setup costs are maximal.
    – If balancing is possible, but restricted by the minimal profit, this still holds (as an upper bound).

Fig. 2 Load model considering competitiveness

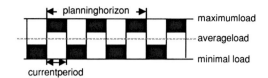

– Then we have to assume: number of products number of time segments/
  period.

• Formal deduction
  We consider two cases:

  – *First case:* shifting < planning horizon (Fig. 2).

The best case scenario for the off-line inventory, concerning a complete
balancing of capacity from one period into another, is the case where there are
z (e.g. 5) empty and cheap time segments at the end of a period, followed by a
period with z expensive time segments that are covered. Here an off-line
algorithm can completely balance its capacity, although at the price of additional inventory costs.

$$\text{costs}_{\text{off}} = \overbrace{1/2}^{a} \cdot \overbrace{\text{(number of producs in the expensive slot)}}^{b} \cdot \overbrace{\text{(inventory costs per period)}}^{c}$$

For the on-line algorithm:

$$\text{costs}_{\text{onl}} = (1 - \text{factor for the minimal profit}) \; \cdot a \cdot b \cdot c + (\text{factor for the minimal profit}) \cdot$$
$$b \cdot \text{additional costs for expensive periods}$$

In order to shift all expensive time segments to a previous period, the factor of
the minimal profit has to be set to "zero". Then the on-line algorithm aligns with
the off-line algorithm (for this case). If the factor of the minimal profit is set to
"1", no shifting to the previous period is possible and the entire costs for the
expensive time segments have to be applied.

Example:
  Factor of minimal profit $= \{0.1, 0.4, 0.8\}$
  10 normal time segments and 5 expensive time segments in one period
  Shifting to normal time segments in previous periods results in inventory
  costs for 5 segments
  Inventory costs per segment: $= 1/10$ (costs for producing in an expensive
  segment)
  (Costs for producing in expensive time segments): $= 2$(costs for producing in normal time segments)

Counting costs in units of production costs in normal segments.

$$\text{costs}_{\text{off}} = 10 \cdot \text{normal} + 1/2 \cdot 1/10 \cdot 5 \cdot 5 = 11.25$$
$$\text{costs}_{\text{onl}} = 10 + \text{normal} +$$
$$(1/2 \cdot 1/10 \cdot 5 \cdot 5) \cdot 0.9 + 0.1 \cdot 1 \cdot 5 = 11.625 \ (+0.5)$$
$$(1/2 \cdot 1/10 \cdot 5 \cdot 5) \cdot 0.6 + 0.4 \cdot 1 \cdot 5 = 12.75 \ (+2)$$
$$(1/2 \cdot 1/10 \cdot 5 \cdot 5) \cdot 0.2 + 0.8 \cdot 1 \cdot 5 = 14.25 \ (+4)$$

The capacity of the shifting is considered to be given—otherwise the off-line algorithm could not use them either. On the other hand, the minimal profit cannot be fixed. Here the minimal profit is in close relation to the expensive capacity. In fact it depends on the ratio of revenue and costs in a period by a certain part. In the worst case (consider an adversary) this can be "zero". The costs for the on-line algorithm are accordingly:

$$\text{costs}_{\text{onl}} = 10 \cdot \text{normal} + 5 \cdot \text{expensive} = 10 \cdot \text{normal} + 5 \cdot 2 \cdot \text{normal} = 10 + 10 = 20$$
$$\text{(if the costs of the expensive time segment are set to twice as expensive)}$$

The ratio on-line/off-line is therefore 20/11.25. This can hardly be considered as optimization in the classical sense. Thus, the minimal profit should be defined in a smarter way. An obvious approach would be accumulating the profits over the total horizon, starting with 0 at the beginning (e.g. the value of the last stock-taking) till the current period. It would be reasonable to take the current period into account for the cumulative. However, the following example (Fig. 3) illustrates that there might be no balancing at all.

The only requirement is an appropriate loss in the first (max-) period. Already in the 4th period one might expect balancing to be useful. Therefore, we choose the following approach: We use the "pull ahead" of the respective off-line solution in the current period, unless there is cumulated profit larger than zero. In this case only a certain fraction is used. That means: once there is a cumulated profit, it is not spend completely anymore. If there is no profit so far, we balance and try to keep the costs as low as possible:

– We shift to the front if the off-line algorithm was unable to make any profit in the past;
– We shift to the front if the off-line algorithm made a "large" profit in the past.

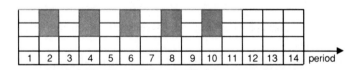

Fig. 3 Example for a load model

For that purpose an off-line solution for the entire past is calculated. The current period is integrated in this off-line solution and is accounted the average value of shifting forward in the past (average use of capacity in periods that where balanced). This is critical, if there is a minimal profit in the past. Then the off-line algorithm is able to balance, while the on-line algorithm has to use the expensive time slots. But the off-line algorithm will not further balance when the inventory costs exceed the savings. Then we have for the shifting forward of x periods:

$$\text{Savings} = x \cdot \text{expensive period} - \sum_{s=1}^{x} (2(s-1)+1) \cdot \text{inventory costs per period}$$

$$= x \cdot \text{expensive period} - x^2 \cdot \text{inventory costs per period}$$

From the first derivative to x

x/dx (savings) = (expensive period) $-$ ($2 \cdot$ inventory costs per period)
follows: The savings are maximal for
$x^* =$ (expensive period)/($2 \cdot$ inventory costs per period)
x^* is the best possible horizon for shifting to the front with maximal savings "savings*".

On the other hand, the on-line algorithm cannot avoid the expensive periods:

$\text{costs}_{\text{onl}} = x \cdot \text{normal} + x \cdot \text{expensive}$ (considering normal time slots as given)
$\text{costs}_{\text{off}} = \text{costs}_{\text{onl}} - \text{savings}^*$

This leads to the ratio ($\text{costs}_{\text{onl}}/\text{costs}_{\text{off}}$). In general holds:

- Best possible shifting $x^* = \left(k_p^t - k_p^n \right)/(2 \cdot k_B)$
- Maximal savings$^* = x^* \cdot \left(k_p^t - k_p^n \right) - (x^*)^2 \cdot k_B$
- Ratio on-line/off-line

$$\frac{k_{\text{onl}}}{k_{\text{off}}} = \frac{x^* \cdot (k_p^n) + x^* \cdot (k_p^t)}{x^* \cdot k_p^n + x^* \cdot k_p^t - x^*(k_p^t - k_p^n) + (x^*)^2 \cdot k_B} = \frac{x^*(k_p^t + k_p^n)}{2 \cdot k_p^n x^* + (x^*)^2 \cdot k_B}$$

- Example

(cost for producing in an expensive period):(inventory costs per period) = 10:1
normal period = $1/2$ expensive periods

$$\text{Savings} : x \cdot \text{expensive period} - \sum_{s=1}^{x} (2(s-1)+1) \cdot 1/10 \text{ expensive period}$$

$$k_p^t = 1; k_p^n = 0,5; k_B = 0,1; x^* = 2.5$$

$$\text{Savings*} = 1.25 - 0.625 = 0.625$$

$$\frac{k_{\text{onl}}}{k_{\text{off}}} = \frac{2.5 \cdot 1.5}{2 \cdot 0.5 \cdot 2.5 + 6.25 \cdot 0.1} = \frac{3.75}{3.125} = 1.2$$

	Inventory costs	Savings production costs	Savings Σ	
				$x * \dfrac{\Delta \text{expensive period}}{2 * \text{inventory costs period per}} = 2{,}5$
$x = 1$	1/10	$1 \cdot 0.5$	0.4	savings $= 2{,}5 \cdot 0{,}5 - 6{,}25 \cdot 0{,}1 = 0{,}625$
$x = 2$	3/10	$2 \cdot 0.5$	0.6	costs$_{\text{onl}} = 2{,}5$ normal $+ 2{,}5$ expensive $= 3{,}75$
$x = 3$	5/10	$3 \cdot 0.5$	0.6	costs$_{\text{off}} = 3{,}75 - 0{,}625 = 3{,}125$
$x = 4$	7/10	$4 \cdot 0.5$	0.4	costs$_{\text{onl}}$ / costs$_{\text{off}} = 3{,}75 / 3{,}125 = 1{,}2$
$x = 5$	9/10	$5 \cdot 0.5$	0	
$x = 6$	11/10	$6 \cdot 0.5$	−0.6	
$x = 7$	13/10	$7 \cdot 0.5$	−1.4	
$x = 8$	15/10	$8 \cdot 0.5$	−2.4	
$x = 9$	17/10	$9 \cdot 0.5$		

– *Second case: shifting > planning horizon* (Fig. 4).

We consider the time segment $2(V - H) + H$. In the second case holds for the shifting forward by x periods:

$$\text{Savings} = x \cdot \text{expensive period} - \sum_{s=1}^{x} (2(s - 1) + 1) \cdot \text{inventory costs per period}$$

$$= x \cdot \text{expensive period} - x^2 \text{inventory costs per period}$$

From the first derivative to x

$$d/dx(\text{savings}) = \Delta \text{ expensive period} - (H + 2x) \cdot \text{inventory cost per period}$$

follows: The savings are maximal for

$x^* = (\Delta \text{ expensive period} - H \cdot \text{ inventory costs per period})/(2 \cdot \text{ inventory costs per period})$

x^* is the best possible horizon for shifting to the front, thus the shifting horizon to use is $V = H + x$.

Fig. 4 Shifting planning horizon

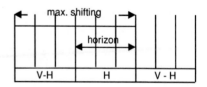

In general:

- Best possible shifting $x^* = \frac{(k_p^t - k_p^n) - d_H \cdot k_B}{2 \cdot k_B}$
- Maximal savings$^* = x^* \cdot (k_p^t - k_p^n) - (x^* \cdot d_H + (x)^2) \cdot k_B$
- Ratio on-line/off-line

$$\frac{k_{\text{onl}}}{k_{\text{off}}} = \frac{(d_h + x^*) \cdot k_p^n + x^* \cdot k_p^t}{(d_h + x^*) \cdot k_p^n + x^* \cdot k_p^t - x^*(k_p^t - k_p^n) + (x^* \cdot d_H + (x^*)^2) \cdot k_B}$$

$$= \frac{d_H \cdot k_p^n + x^*(k_p^t + k_p^n)}{d_H \cdot k_p^n + x^*(2 \cdot k_p^n + d_H \cdot k_B + x^* \cdot k_B)}$$

- Example

$$k_p^t = 1; \; k_p^n = 0.5; \; k_B = 0.1; \; d_H = 4; \; x^* = 0.5$$

$$\text{Savings*} = 0.25 - (0.5 \cdot 4 + 0.25) \cdot 0.1 = 0.25 - 0.225 = 0.025$$

$$\frac{k_{\text{onl}}}{k_{\text{off}}} = \frac{4 \cdot 0.5 + 0.5(1.5)}{4 \cdot 0.5 + 0.5(1 + 0.4 + 0.05)} = \frac{2.75}{2.725} = 1.01$$

	Inventory costs	Savings production costs	Savings Σ
$x = 1$	5/10	$1 \cdot 0.5$	0
$x = 2$	+7/10	$2 \cdot 0.5$	−0.2
$x = 3$	+9/10	$3 \cdot 0.5$	−0.6
$x = 4$	+11/10	$4 \cdot 0.5$	−1.2
$x = 5$	+13/10	$5 \cdot 0.5$	−2.0
$x = 6$	+15/10	$6 \cdot 0.5$	−3.0
$x = 7$	+17/10	$7 \cdot 0.5$	−4.2

$$x^* \frac{0.5 - 0.4}{2 \cdot 0.1} = 0.5$$

$\text{savings} = 0.25 - 0.225 = 0.025$

$\text{costs}_{\text{onl}} = 4.5 \cdot 0.5 + 0.5 \cdot 1 = 2.75$

$\text{costs}_{\text{off}} = 4.5 \cdot 0.5 + 0.5 \cdot 1 - 0.025 = 2.725$

$$\frac{\text{costs}_{\text{onl}}}{\text{costs}_{\text{off}}} = \frac{2.75}{2.725} = 1.01$$

5 Conclusion

The proposed approach offers for a special form of production (DLSP) an on-line optimization in the sense of a comparison to an optimal off-line solution. The approach is demonstrated in examples with production and inventory costs. An extension to other forms of production as well as other forms of costs (e.g.) setup costs should be possible.

Furthermore, the parameters for how costs for different time slots depend on each other as well as on the inventory costs are fixed in our example to give a first insight of how the model works. Future work will consider these parameters and dependency of the competitiveness. Note that the values chosen in this paper are reasonable for realistic scenarios.

References

Dangelmaier, W.: Online-Optimierungsansätze zur Steuerung der Produktion in der Serienfertigung. In: Schenk, M. (Hrsg.): Logistik—Effiziente und sichere Warenketten in Industrie und Handel. 11. IFF-Wissenschaftstage 25./26. Juni 2008. Tagungsband S. 145–154. Magdeburg: Fraunhofer-Institut für Fabrikbetrieb und -automatisierung 2008.

Dangelmaier, W.: PPS-Strategien und Online-Optimierung in der Serienfertigung bei kurzfristigem Horizont. In: Müller, W.; Spanner-Ulmer, B. (Hrsg.): Wandlungsfähige Produktionssysteme. S. 15–24. TBI'08. 13. Tage des Betriebs- und Systemingenieurs. II. Symposium Wissenschaft und Praxis. Wissenschaftliche Schriftenreihe des Instituts für Betriebswissenschaften und Fabriksysteme. Chemnitz, 13. November 2008.

Danne, Ch.; Blecken, A.; Dangelmaier, W.: Complexity-Induced uncertainty in Supply chains—A Framework and Case Studies. In: Pfohl, H.-Ch.; Wimmer, Th. (Hrsg.): Wissenschaft und Praxis im Dialog. Robuste und sichere Logistiksysteme. S. 71–88. 4. Wissenschaftssymposium Logistik. Hamburg: Deutscher Verkehrsverlag 2008.

Krumke, S. O.; Rambau, J.: Online Optimierung. Vorlesungsskript. Berlin: Technische Universität 2005.

Dynamic Vehicle Routing in Over Congested Urban Areas

Antonio G. N. Novaes, Enzo M. Frazzon and Paulo J. Burin

1 Introduction

Vehicle routing problems (VRPs) are usually related to efficiently assigning vehicles to tasks, such as picking up and delivering cargo, or accomplishing other services in a previously defined order so that tasks are completed within a certain time limit and vehicle capacities are not exceeded. Static approaches to vehicle routing problems, where all the characteristics of the routing process are known in advance, have been widely studied in the literature. When information on travel and service times gradually changes in the course of the system's operation, dynamic real-time techniques become increasingly important (Larsen et al. 2007; Flatberg et al. 2007; Zeimpekis et al. 2007). Many authors treat VRPs with a mathematical programming formulation. Other authors, instead of searching for specific optimal routes linking the servicing points, use approximate formulas to estimate the travelled distances and times in a continuous spatial format (Daganzo 1996; Langevin et al. 1996; Novaes et al. 2000). This is the approach we have adopted in this paper.

Travel and service times are essential information to solve VRPs. Most of the static models presented in the literature on VRPs assume pre-defined travel and servicing times. Travel times between relevant locations (depots and customers) are usually derived from shortest paths in a road network with known values on every link. Such an assumption may be far from reality in congested urban areas where traffic conditions change constantly (Fleischmann et al. 2004). Traffic information systems have been installed in large cities of the world with the

A. G. N. Novaes (✉) and P. J. Burin
Department of Industrial Engineering, Federal University of Santa Catarina,
Santa Catarina, Brazil
e-mail: novaes@deps.ufsc.br

E. M. Frazzon
IGS, University of Bremen, Bremen Germany
e-mail: fra@biba.uni-bremen.de

H.-J. Kreowski et al. (eds.), *Dynamics in Logistics*,
DOI: 10.1007/978-3-642-11996-5_5, © Springer-Verlag Berlin Heidelberg 2011

objective of reducing the negative effects of bottlenecks in street networks. These systems tend to increase the flow of vehicles by allowing higher vehicle speeds and by offering less-congested alternative routes to drivers. But in developing countries, the required large investments to install such systems often forbid their extensive use. In large cities such as São Paulo, Brazil, freight operators, that deliver or pickup cargo in congested urban areas, tend to assign larger number of visits to their vehicles in order to increase revenue. This often leads to non-performed tasks at the end of the daily cycle-time, impairing the logistics service level. This happens because, due to the extremely volatile traffic conditions and the great number of random variables along the route, the vehicle cycle-time usually shows great variability. As a consequence, the average cycle-time is low, and the operators try to increase vehicle usage by forcing larger loads. We intend to show in this paper how a simple dynamic vehicle routing problems (DVRP) model can improve this situation. No central navigational traffic system, as the one mentioned in Fleischmann et al. (2004), is assumed to exist. It is only necessary to have an on-board computer, a geo-referencing device and an electronic communication system linking the vehicle to nearby collaborative agents (other vehicles) and to the central depot.

The dynamic methodological procedure inserted into the DVRP presented in this paper is taken from Sequential Analysis (Wald 1947; Ghosh et al. 1997; Lai 2001). Although assuming a simple statistical hypothesis to be sequentially tested, the approach is sufficiently robust and can be extended to more complex situations.

2 The Proposed DVRP

The dynamic routing problem to be analysed in this paper comprises one depot and a homogeneous fleet of vehicles that serve an urban region $\Re$. Each vehicle is assigned to one district. The vehicle leaves the depot early in the morning, travels to the district, performs the tasks assigned to it, and returns to the depot when all services are finished or when the cycle-time exceeds a pre-defined level. Under the static approach, whenever the cycle-time exceeds the daily working limit, the unperformed services will be postponed to the next day, or even later. This situation has a serious logistics drawback but it is not uncommon in developing countries like Brazil. Conversely, under a dynamic policy scheme, part of the planned services will be transferred to other agents (vehicles) whenever the on-board system detects an abnormal traffic condition. This procedure will avoid unperformed tasks at the end of the working day, and must be accomplished within a suitable time advance in order to allow for the interchange of tasks among the agents. The objective of this paper is to compare these two strategies for the aforementioned problem, showing the advantages of the dynamic approach.

We assume that two distinct urban traffic congestion patterns can occur: (a) situation H_0, which corresponds to a typical working day, showing a standard congestion pattern; (b) situation H_1, which represents a "hectic" congested pattern due to severe accidents, unpredictable public transport strikes, heavy rain, etc. Under the static approach, the dispatcher will define the district size and contour in such a way as to keep the unperformed tasks within a certain limit. Under the dynamic situation the dispatcher assumes that the optimistic scenario H_0 will occur and set up the districts and the routes accordingly. The tour will be periodically monitored and if the system detects significant deviation that indicates the occurrence of hypothesis H_1, part of the remaining tasks will be transferred to other agents (other vehicles).

3 Sequential Estimation

In classic hypothesis testing a sample is extracted from the data and examined. Following its statistical analysis, one of two possible actions is taken: accept the null hypothesis H_0, or accept the alternative hypothesis H_1. In sequential tests there is a third possible course of action when the evidence is ambiguous: take more observations until the evidence strongly favours one of the two hypotheses. Thus, sequential analysis follows a dynamic sequence of observations in such a way that the decision to terminate or not the experiment depends, at each stage, on the previous test results. In the case of simple hypothesis, the strength of the evidence for H_1 is given by the ratio of the probability of the data under H_1 to the probability of the data under H_0 (Wald 1947; Lai 2001). Denote this likelihood ratio by Δ. The Neyman-Pearson lemma implies that for a given amount of information the likelihood ratio test is the most powerful test (Wald 1947). Thus, one accepts hypothesis H_1 if Δ is big enough, and decides to accept H_0 otherwise. How big Δ must be to lead to the decision H_1 depends on its sampling distribution under H_0 and H_1, and on the errors of first and second kind (Wald 1947; Ghosh et al. 1997). An error of the first kind is committed if one rejects hypothesis H_0 when it is true. Conversely, an error of the second kind is committed if one accepts H_0 when H_1 is true. We shall denote α the probability of an error of the first kind, and β the probability of an error of the second kind.

A simple case of sequential estimation, but of particular interest, arises when only one unknown parameter μ is required to define the distribution of the random variable x under analysis. Let $f(x, \mu)$ denote the probability density function of x, where x is continuous. Conversely, if x is discrete, $f(x, \mu)$ represents its probability. Suppose that μ can take only two values, say μ_0 and μ_1. Let $x_1, \dots, x_m$ be a set of m independent observations on x. Due to the statistical independence, the joint probability density function is

$$f(x_1, \mu)f(x_2, \mu) \dots f(x_m, \mu) \tag{1}$$

Let H_0 be the hypothesis that $\mu = \mu_0$, and H_1 the hypothesis that $\mu = \mu_1$. To apply the sequential probability ratio test (SPRT) for testing $H_0 : \mu = \mu_0$ against

$H_1{:}\mu = \mu_1$, two positive constants A and B $(B < A)$ are chosen. At each stage m of the experiment the ratio

$$\pi_m = \frac{f(x_1, \mu_1) f(x_2, \mu_1) \ldots f(x_m, \mu_1)}{f(x_1, \mu_0) f(x_2, \mu_0) \ldots f(x_m, \mu_0)} \tag{2}$$

is defined. If $B < \pi_m < A$ the experiment continues by taking an additional observation. If $\pi_m \geq A$ the process is terminated with the rejection of H_0, and if $\pi_m \leq B$, the process is terminated with the acceptance of H_0 (Wald 1947; Lai 2001). The values of A and B for which the test has a required strength (α, β) are

$$A = (1 - \beta)/\alpha \quad \text{and} \quad B = \beta/(1 - \alpha) \tag{3}$$

For purposes of practical computation it is more convenient to compute the logarithm of the ratio π_m. Let

$$z_i = \ln\left(\frac{f(x_i, \mu_1)}{f(x_i, \mu_0)}\right) \tag{4}$$

Define

$$\pi_m^* = \ln(\pi_m) = z_1 + \cdots + z_m \tag{5}$$

The experiment continues if $\ln B < \pi_m^* < \ln A$ by taking an additional observation; the process is terminated with the rejection of H_0 if $\pi_m^* \geq \ln A$; and the process terminates with the acceptance of H_0 if $\pi_m^* \leq \ln B$. Suppose $f(x, \mu)$ is distributed according to an Erlang distribution. Its probability density function is

$$f(x, \mu) = \frac{\mu^\lambda}{(\lambda - 1)!} x^{\lambda-1} e^{-\mu x}, \quad \text{with} \quad \lambda = 1, 2, \ldots, \infty, \mu > 0, x \geq 0, \tag{6}$$

with $E[x] = \lambda/\mu$ and $var[x] = \lambda/\mu^2$. Substituting (6) into (4) and (5), one has

$$\pi_m^* = \lambda m \ln\left(\frac{\mu_1}{\mu_0}\right) - (\mu_1 - \mu_0) \sum_{i=1}^m x_i \tag{7}$$

The same sequential estimation procedure can be used with regard to the more general, one-sided composite hypothesis testing, namely $H_0{:}\mu \leq \mu_0$ against $H_1{:}\mu \geq \mu_1$, with $\mu_1 > \mu_0$ (Wald 1947; Lai 2001).

In this application, the variable that commands the success or failure of the adopted dynamic strategy is time. Since the servicing points are fixed for a given route and since the stopping times to serve the clients do not depend on traffic conditions, the variable that controls the time process is the vehicle average speed. The speed is assumed to take two distinct values under hypotheses H_0 and H_1, namely $v_z^{(0)}$ and $v_z^{(1)}$, with $v_z^{(0)} > v_z^{(1)}$. Let d be the route distance between two consecutive clients, and let $x_i^{(0)} = d/v_z^{(0)}$ and $x_i^{(1)} = d/v_z^{(1)}$ be the vehicle displacement times under hypotheses H_0 and H_1 respectively. We will show in Sect. 4 that d follows an Erlang distribution with $\lambda = 3$, and since $v_z^{(0)}$ and $v_z^{(1)}$ are

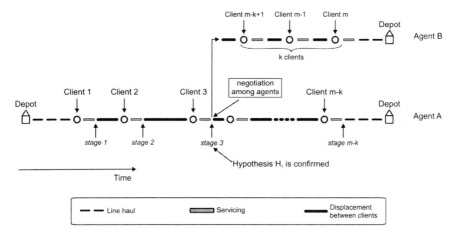

Fig. 1 The vehicle routing sequence and the decision stages

constants, the variables $x_i^{(0)}$ and $x_i^{(1)}$ are also represented by the same kind of distribution. From the above definition of the Erlang distribution, one has $\mu_0 = \lambda/x_i^{(0)}$ and $\mu_1 = \lambda/x_i^{(1)}$. Then, after completing a visiting task, the on-board system applies the sequential test with relation (7), in order to infer which hypothesis is binding, or if it is necessary to proceed with the test further. Figure 1 shows a schematic representation of the vehicle routing sequence and the decision stages where the SPRTs are performed. Assume the vehicle (agent A, in Fig. 1) left the depot with an assignment of m visits. Suppose the sequential test indicates the acceptance of hypothesis H_1 at stage 3, as shown in Fig. 1. At this point, the on-board computer checks how many of the remaining visits will be transferred to another agent B. Let k be the number of visits to be transferred. Upon negotiation, agent B agrees to perform the k tasks. Of course, if there are two or more visits to be transferred, more than one agent can be involved in the transference.

4 The Routing Model

Approximate formulas are used to estimate the vehicle travelled distances and times along the route (Daganzo 1996; Novaes et al. 2000). In this paper we assume that the density δ of visiting points over the region is constant. Irregular density patterns and analysis of districting methodologies are presented in Novaes et al. (2000), Galvão et al. (2006), and Novaes et al. (2009).

Let n be the number of visits to be performed in a tour. For varying values of n an ad hoc model generates the corresponding n servicing points randomly scattered over the district and applies a 3-opt heuristic to solve the corresponding travelling salesman problem (TSP), thus obtaining a number of simulated results. We assume, as mentioned, that the displacement time between two clients within the

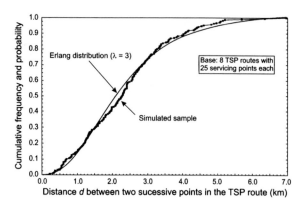

Fig. 2 Fitting an Erlang distribution to the distance d

district suffers heavily under traffic congestions. According to Novaes and Graciolli (1999), the distance d between two successive visiting points in the tour can be estimated as

$$d = k_0 \, \delta^{-1/2}. \tag{8}$$

A district of 4.5 km^2 in the city of São Paulo, Brazil, was taken as a data basis for our example. A total of 25 points were randomly generated on it. Applying a 3-opt heuristic, the optimal travelling salesman tour was generated. The distances d between two successive points in the resulting TSP was then computed. Repeating the simulation 8 times, a sample of $8 \times 25 = 200$ values of d was generated. An Erlang distribution of order $\lambda = 3$ was fitted to the data (Fig. 2).

The total vehicle cycle-time TC is the sum of the line-haul time (both ways), the local travel time (i.e., the vehicle displacement times within the district) and the servicing times. The random variables that compose the cycle time are assumed to be independently distributed, and since n is usually large ($n > 15$), the cycle time can be assumed to be normally distributed in the simulation. Let TC_i^* be the maximum expected value of the cycle time in district i. Assuming a 0.02 confidence level, the expected value of TC_i^* is

$$TC_i^* = E[TC_i] + 2.06 \quad \text{var}[TC_i]^{1/2}, \quad \text{with} \quad TC_i^* \leq HC, \tag{9}$$

where HC is the maximum daily effective working time.

Since the problem involves a great number of random variables, many events and diverse decision instants, and since we are not focussing the spatial routing process itself but only an approximate representation, we have opted to use simulation in order to analyze the static and dynamic versions of the problem. Of course, for further developments of this research, more detailed modelling will have to be done, as suggested in Sect. 6. Numerical values of the variables were assumed in this work with the sole purpose of illustrating the possibilities of the proposed approach.

5 Static and Dynamic Simulation of the DVRP

Figure 3 shows the simulation diagram of the dynamic version of the routing problem. The simulation scheme of the equivalent static situation is obtained by cutting off the transference routine and registering the non-performed visits in the tour. Daily tours were simulated for different values of n and assuming that situations H_0 and H_1 occur with probability 0.7 and 0.3, respectively. We assume that the district has a fixed number n of clients, but the probability that a client will order a service on a specific day is 0.8. Thus, the number n_e of effective visits in a daily tour is represented by a binomial distribution with success rate 0.8.

The DVRP model is applied next, and at each stage of the process (Sect. 3) the sequential probability ratio test (SPRT) is applied to variable x_i. The values $\alpha = 0.01$ and $\beta = 0.01$ were assumed, and applying (3), it leads to $A = 99$ and $B = 10.101 \times 10^{-3}$. The process of transferring visits to other agents (vehicles) follows a simple scheme in this example. Whenever the SPRT signals the occurrence of hypothesis H_1, the model estimates the cycle-time necessary to accomplish the remaining tasks. If the expected total cycle-time exceeds the HC limit (9), the model will extract visits from the tour until $TC^* \leq HC$ is observed. The extracting process starts from the last planned visit of the tour and proceeds backwards, thus leaving more time for the effective transference of tasks. The vehicle cycle-time usually shows a great variability. As a consequence, the

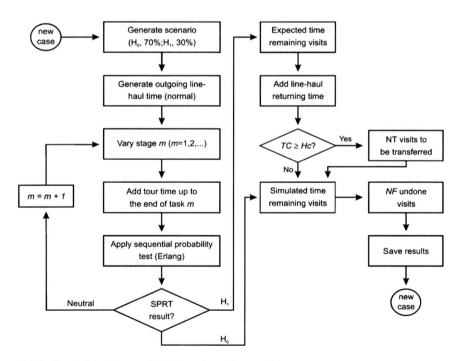

Fig. 3 Simulation diagram of the dynamic routing problem

Fig. 4 Simulation results of the static routing problem

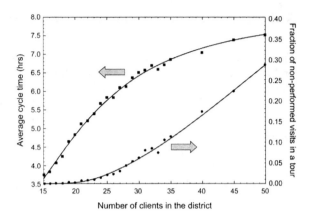

simulation must involve a large number of cases in order to guarantee meaningful results. A total of 20,000 daily tours were simulated for each value of n.

Figure 4 shows the simulation results for the static routing problem. Two outcomes are depicted in Fig. 4: (a) the average vehicle cycle-time, and (b) the fraction of non-performed visits as referred to the total assigned visits. Suppose the operator must keep a low percentage of non-performed visits, say 1% of the planned tasks. From Fig. 4 it can be seen that the district must contain about 23 clients, leading to an expected number of 18.4 visits per tour (an 80% rate), about 0.18 non-performed visits per tour, and an expected daily cycle-time of 5.4 h per tour. One must recall that the non-performed visits will be accomplished next day or later.

Figure 5 shows the simulation results of the dynamic routing problem. Assuming the logistics operator sets up a district containing 28 clients, with an expected number of 22.4 visits per tour, the DVRP simulation shows that an average of 1.5 visits per tour would have to be transferred to other vehicles (6.7%).

Fig. 5 Simulation results of the dynamic routing problem

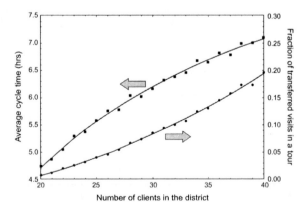

In this case, the average cycle time is 6.0 h. As an example, in a simulation run with 28 clients, and 23 effective visits in the tour, the SPRT signalled situation H_1 at stage 16, that happened at instant 5 h:2 min after departure. The last two planned visits were transferred to another agent. Thereafter, the vehicle performed 5 additional visits, leading to a total of 21 tasks and getting back to the depot at time 6 h:43 min.

With this DVRP procedure the logistics operator would achieve, in general, the following results: (a) no task is postponed to the next day; (b) the vehicle would be able to make $0.8 \times 28 - 1.5 = 20.9$ visits per tour, as against $0.8 \times 23 = 18.4$ visits under the static scheme, a 13.6% increase; (c) the cycle-time is 11.1% greater, reflecting the better usage of the fleet; (d) an average of 1.5 visits per tour would be transferred to other agents in the dynamic case.

These results are not optimal since only one vehicle (agent) has been considered in the application. Additionally, the present formulation assumes that the tasks to be transferred to other agents are the last ones in the routing sequence, which is a non-optimal assumption. Also, algorithms to choose the tasks to be removed, insertion methods, routing rearrangements, and other optimising procedures (Goel 2008), are not investigated here. Thus, this application must be understood as a preliminary demonstration of the practical possibilities of the method.

6 Conclusions and Research Prospects

In this paper we described an approximate dynamic model for the real-time management of a vehicle performing visits on an over congested urban area. One feature of the model is to use sequential estimation procedures to detect significant changes in traffic conditions, hence transferring part of the remaining tasks to other agents.

With this approach no service visits will be postponed to the next day or later, thus improving the service level. The system requires an on-board computer, a geo-referencing device and a communication system linking the vehicle to other agents and to the central depot. The present model formulation assumes that the tasks to be transferred to other agents are the last ones in the routing sequence. Algorithms to decide which tasks are to be removed from the route, considering time restrictions, routing rearrangement, and other optimising procedures (Goel 2008), will be investigated further. Conversely, insertion procedures and evaluation criteria will be developed in order to analyse multiple-vehicle situations. In addition, the model presented here will be extended to a set of districts where the vehicles (agents) work in conjunction, with the simulation covering real-time task exchanges, and leading to an improved districting pattern when compared to the equivalent static solution. With regard to cargo delivery cases, an additional restriction occurs, namely the need to physically transfer the products to other vehicles, with the necessary place and time matching.

References

Daganzo C.F. (1996) Logistics systems analysis, Springer, Berlin.

Flatberg, T., Hasle, G., Kloster, O., Nilssen, E., Riise, A. (2007) Dynamic and stochastic vehicle routing in practice. In: Zeimpekis, V., Tarantilis, C.D., Giaglis, G., Minis, I. Dynamic fleet management. Springer, New York.

Fleischmann, B., Gnutzmann, S., Sandvoß, E. (2004), Dynamic vehicle routing based on on-line traffic information. Transportation Science, 38 (4), pp. 420–433.

Galvão, L.C., Novaes, A.G., Souza de Cursi, J.E., and Souza, J.C. (2006), A Multiplicatively-weighted Voronoi diagram approach to logistics districting, Computers & Operations Research 33, pp. 93–114.

Ghosh, M., Mukhopadhyay, N. and Sen, P.K. (1997), Sequential estimation, Wiley, New York.

Goel, A. (2008) Fleet telematics, Springer, New York.

Lai, T.L. (2001), Sequential analysis: some classical problems and new challenges, Statistica Sinica 11, pp. 303–408.

Langevin A., Mbaraga P., Campbell J.F. (1996), Continuous approximation models in freight distribution: an overview, Transportation Research – B, V 30, pp. 163–188.

Larsen, A., Madsen, O., Salomon, M. (2007) Classification of dynamic vehicle routing systems. In: Zeimpekis, V., Tarantilis, C.D., Giaglis, G., Minis, I. Dynamic fleet management. Springer, New York.

Novaes A.G., Graciolli O.D. (1999), Designing multi-vehicle tours in a grid-cell format. European Journal of Operational Research, V 119, pp. 613–634.

Novaes, A.G., Souza de Cursi J.E., and Graciolli, O.D. (2000), A continuous approach to the design of physical distribution systems, Computers & Operations Research 27, pp. 877–893.

Novaes A.G., Souza de Cursi J.E., da Silva A.C.L. and Souza, J.C. (2009), Solving continuous location-districting problems with Voronoi diagrams, Computers & Operations Research 36, pp. 40–59.

Wald, A. (1947), Sequential analysis, Wiley, New York.

Zeimpekis, V., Minis, I., Mamassis, K., Giaglis, G. (2007) Dynamic management of a delayed delivery vehicle in a city logistics environment. In: Zeimpekis, V., Tarantilis, C.D., Giaglis, G., Minis, I. Dynamic fleet management. Springer, New York.

Serving Multiple Urban Areas
with Stochastic Customer Requests

Stephan Meisel, Uli Suppa and Dirk Mattfeld

1 Introduction

Recent economical developments provide new challenges to vehicle routing. For example, new internet-based business models lead to an increasing number of shipments operated by express service providers and less-than-truckload companies. Further, a growing importance of field service can be noticed in many companies. In addition, growing resource prices strengthen the need for more efficient vehicle routing operations.

In this context, efficient vehicle routing requires explicit anticipation of customer requests appearing while a vehicle is en route. In particular, the routing of a vehicle must be adjusted dynamically to the behavior of customers possibly requesting for service. In case service cannot be provided directly a requesting customer must be rejected explicitly.

In order to meet these requirements of dynamic vehicle routing a planner must rely on technological support. Modern data processing and analysis technologies provide a basis for anticipation in terms of information about customer behavior derived from historical data. Positioning systems make it possible to include the up to date geographical location of a vehicle into planning. Finally, mobile communication devices allow for online transmission of routing adjustments and other planning results to drivers. The approach presented in this contribution takes the availability of such technologies for granted.

S. Meisel (✉), U. Suppa and D. Mattfeld
Decision Support Group, Carl-Friedrich Gauss-Department, University of Braunschweig,
38106, Braunschweig, Germany
e-mail: stephan.meisel@tu-bs.de

U. Suppa
e-mail: u.suppa@tu-bs.de

D. Mattfeld
e-mail: d.mattfeld@tu-bs.de

H.-J. Kreowski et al. (eds.), *Dynamics in Logistics*,
DOI: 10.1007/978-3-642-11996-5_6, © Springer-Verlag Berlin Heidelberg 2011

We treat a dynamic vehicle routing problem with one vehicle and a number of known customer locations. Customers are divided into the two distinct sets of early requesting customers and late requesting customers. Early requesting customers are known to definitely request for service, whereas late requests appear randomly over time according to individual request probabilities. Late requests must be either confirmed or rejected directly after becoming known. The goal is serving the maximum number of requesting customers within a fixed period of time. Throughout this period of time the vehicle must move from a specified start depot to a specified end depot.

This problem setting closely resembles the situation of a service provider (e.g. a less-than-truckload carrier) having one vehicle assigned to a geographical region in order to serve the different customers there. We consider the case of a number of customer agglomerations within this region in detail. This corresponds to the situation of a vehicle covering a number of neighboring small and medium size urban areas.

Our focus is on a generic solution approach to the problem by approximate dynamic programming (ADP). We present computational results for the case of multiple urban areas and compare these results to solutions gained from state-of-the-art waiting strategies applied to the problem.

The structure of the remainder of the paper is as follows. In Sect. 2 we put our contribution in the context of related work. In Sect. 3 we derive the optimality equations for our problem and present an approximate dynamic programming approach for solution. Section 4 explains our experimental setup in detail and gives computational results as well as an analysis of the performance of the approximate dynamic programming approach. Section 5 covers conclusions and future work.

2 Related Work

The high relevance of dynamic vehicle routing is reflected by the increasing number of publications within the field over the past two decades. The first general discussions of dynamic vehicle problems are due to Psaraftis (1988, 1995). Subsequently, survey papers have been published by Gendreau and Potvin (1998), Bianchi (2000) as well as Larsen et al. (2007, 2008).

Two major categories of traditional approaches can be distinguished for dynamic vehicle routing problems with stochastic customer requests. On the one hand a huge amount of publications discusses so called re-optimization procedures based on a rolling time horizon. On the other hand, many contributions rely on heuristics in terms of dispatching rules for the assignment of customer requests to vehicles.

Gendreau et al. (1999) represent a typical example of a re-optimization approach. They propose tabu search optimization procedures for a dynamic vehicle routing problem with time windows. The routes of a number of vehicles are continuously improved by tabu search. If a new request becomes known, it is integrated heuristically into the present solution before optimization continues.

In case a request is satisfied, optimization is interrupted again and a new request is assigned to each idle vehicle according to the current solution.

Planning by dispatching rules implies assignment of service requests to a vehicle without search procedures. Early works on the use of dispatching rules are due to Bertsimas and van Ryzin. They rely on queueing theory for analysis of different dispatching rules applied to the traveling repairman problem. In Bertsimas and van Ryzin (1991) they derive an optimal rule for the case of a single vehicle and a small number of service requests. Papastavrou (1996) develops a dispatching rule for the same problem and provides bounds for its deviation from the optimum. Similar approaches are discussed by Bertsimas and van Ryzin (1993) for the case of multiple vehicles with limited capacity.

Both traditional re-optimization procedures and dispatching rules do not explicitly realize anticipation of future requests. Powell (1986, 1996) was among the first to show explicit anticipation of future request to be necessary. More recent approaches propose rules for including waiting times into the route of a vehicle. Thus requests of customers located close to the current position of the vehicle can be anticipated. Examples are provided by the works of Mitrovic-Minic and Laporte (2004) as well as Branke et al. (2005). Thomas and White (2004) cover anticipatory route selection for the case of new customer requests appearing while the vehicle is en route. Thomas (2007) proposes anticipation by waiting heuristics for a problem similar to the problem treated in this contribution.

Up to now, only very few works consider the use of approximate dynamic programming for anticipation in dynamic vehicle routing. Topaloglu (2007) investigates application of ADP in the context of dynamic fleet management with random travel times. Secomandi (2000) provides an ADP approach for the case of a single vehicle with limited capacity and a priori known customers requests with stochastic request volumes. A different ADP approach to the same problem is presented by Novoa and Storer (2009).

In Sect. 3 we develop an ADP approach for the case of a single vehicle without capacity constraints, serving stochastic customer requests subject to confirmations or rejections.

3 Model and Solution Method

The formulation of dynamic programming optimality equations (Bellman, 1957) is required as a precondition for an ADP approach. In Sect. 3.1 we develop such a formulation for our problem. Based on this our ADP method is described in Sect. 3.2.

3.1 Optimality Equations

Our problem setting comprises a set C of potential customers with known geographic locations. Each customer $c_i \in C$ is assigned a probability p_i, indicating the probability of c_i requesting for service within the given time period T of vehicle operation. In case of an early requesting customer $c_i \in C_E$ we set $p_i = 1$. The sets of early requesting customers and late requesting customers are complementary, i.e. $C = C_E \cup C_L$ and $C_E \cap C_L = \emptyset$.

3.1.1 State Space Definition

At each point in time the state s_{it} of a customer c_i is represented as follows:

- $s_{it} = 0$ if c_i has not requested service yet
- $s_{it} = 1$ if c_i has launched a request but has not been confirmed or rejected yet
- $s_{it} = 2$ if the request of c_i has been confirmed
- $s_{it} = 3$ if the request of c_i has been either rejected or served.

Further, at each point in time $t \in [0, T]$ the vehicle is located at a position n_t. At $t = 0$ the vehicle is located at the start depot c_S and at $t = T$ it is located at the end depot c_E, i.e. $n_0 = c_S$ and $n_T = c_E$ with $c_S \neq c_E$. The overall state S_t of the system at time t is represented as $S_t = (n_t, s_{1t}, s_{2t}, .., s_{|C|t}, s_{Et})$.

3.1.2 Action Space Definition

A decision d_t consists of both a confirmation (respectively rejection) operation d_c and a vehicle move operation d_m, i.e. $d_t = (d_c, d_m)$. Further, a decision must be taken each time the vehicle arrives at a customer location as well as each time the vehicle has waited for one time unit at a customer's location or at the start depot. New customer requests occurring between two subsequent decision times t_0 and t_1 become known in t_1. The first decision must be taken in $t = 0$.

Concerning d_c, each $c_i \in R_t = \{c_i | s_{it} = 1\}$ must be either confirmed or rejected. Let $\wp(R_t)$ be the powerset of new requesting customers at time t and $\overline{\wp}(R_t)$ the set of all the subsets of $\wp(R_t)$ that can be confirmed without forcing a violation of the time horizon T. Thus, $d_c \in \overline{\wp}(R_t)$. A set d_c determines the customers making a state transition from $s = 1$ to $s = 2$ (in case of rejection to $s = 3$).

Concerning d_m, either waiting or moving onto the next confirmed customer is possible. Thus, the next vehicle position $d_m \in M_t = n_t \cup \{c_i | s_{it} = 2\}$. Note that $M_t = M_t(d_c)$.

3.1.3 State Dynamics and Goal

Obviously the state dynamics of the system are determined by both the customer behavior in terms of the request probabilities p_i and the decisions d_t. For a detailed derivation of state transition probability matrices based on p_i we refer to the work of Thomas (2007).

Our goal is maximization of the total number of confirmed late requests within the time period $[0, T]$, i.e. max $\sum_{t \in [0,T]} |d_c(t)|$. Let S_t^d be the system state immediately after decision d_t has been taken. Then, the value $V_t(S_t^d)$ of the post-decision state S_t^d can be specified in terms of the optimality equations:

$$V_t(S_t^d) = E[\max_{d_{t+1}}(|d_c(t+1)| + V_{t+1}(S_{t+1}^d))|S_t^d]$$

For a detailed discussion of the formulation of the optimality equations around the post-decision state variables we refer to Powell (2007). Note that formulation of the equations around the post-decision state variables leads to a significant reduction of the size of our state space. There are no customers with $s_{it} = 1$ in a post-decision state, reducing the size of our state space from $|C| * |T| * 4^{|C|}$ down to $|C| * |T| * 3^{|C|}$. Nevertheless the size of our state space still makes the use of traditional methods of dynamic programming prohibitive. Thus we rely on the approximate dynamic programming approach described in the next subsection.

3.2 Approximate Dynamic Programming Approach

We develop an ADP value iteration algorithm for our problem. To this end, we first specify an approximation $\tilde{V}_t(S)$ of the value function $V_t(S)$. Subsequently we explain how the action space is approximated before we give an outline of the resulting value iteration algorithm.

3.2.1 Value Function Approximation

For each point in time $t \in [0, T]$ the value function is approximated by a linear approximation architecture based on one feature f extracted from the state of the system, i.e. $\tilde{V}_t(S) = \gamma_t f(S)$. Determination of the feature value for a state S_{t_0} requires the following definitions:

- Slack: The slack is defined as the surplus of time achieved if the customers with $s_{it_0} = 2$ would be served without waiting and without considering further late requests. Hence, the slack is $T - t_{finish}$, where t_{finish} is the arrival time at c_E. Note that $(t_0 < t_{finish} \leq T)$.
- Weighted deviation: The deviation associated with a customer c_i with $s_{it_0} = 0$ is the additional amount of travel time $t_{add}(c_i)$ resulting from a confirmation of c_i. The weighted deviation of c_i is defined as $p_i * t_{add}(c_i)$.

The feature is defined as $f(S_t) = \dfrac{T - t_{finish}}{\sum_{c_i|s_{it}=0} p_i * t_{add}(c_i)}$ and we act on the hypothesis of a linear relation between $f(S_t)$ and $V_t(S_t)$.

3.2.2 Action Space Approximation

Determination of the set of feasible decisions is another major challenge we face in each decision time step t. Both identification of $\overline{\wp}(R_t)$ and identification of M_t for each $d_c \in \overline{\wp}(R_t)$ may be very time consuming operations. Hence, we introduce an approximation of the action space as follows.

At $t = 0$ a feasible tour including each $c_i \in C_E$ is determined. Subsequently confirmed late request customers are inserted into this tour using an insertion heuristic. In a decision time step a subset of feasible decisions $d_t = (d_c, d_m)$ is determined by iteration over the elements r of $\wp(R_t)$. For each r

- A new tour is constructed by inserting the elements of r successively into the current tour.
- $d = (r, n_t)$ is considered feasible if $t_{finish} < T$ with respect to the new tour.
- $d = (r, c_{next})$ is considered feasible if $t_{finish} \leq T$ with respect to the new tour, where c_{next} represents the customer next to n_t with respect to the new tour.

Concerning the vehicle movement operation, this approximation implies a tradeoff between anticipating late requests by waiting at the current location and moving onto the next confirmed customer. A distinction of alternative move operations is not made and the move operation considered for execution depends on the insertion procedure used. Note that feasible sets r may also be considered infeasible due to construction of routes by mere insertion.

3.2.3 Learning Algorithm

Learning the parameters γ_t is done by a double pass algorithm. For a general discussion of double pass value iteration we refer to Powell (2007). The algorithm is based on simulation of a number of trajectories. For each trajectory the following steps are executed:

1. For each decision time $t = 0, 1, 2, \ldots T$:
   1.1 Determine $d_t^* = \text{argmax}_{d_t} \left[|d_c(t)| + \gamma_t f(S_t^d) \right]$
   1.2 Execute a state transition according to d_t^*

2. For each decision time $t = T, T - 1, \ldots, 1$:
   2.1 Calculate $v_t = |d_c(t)| + v_{t+1}$
   2.2 Update the value of γ_{t-1}: $\gamma_{t-1} \leftarrow U(\gamma_{t-1}, S_{t-1}^d, v_t)$

4 The Case of Serving Multiple Urban Areas

The methodology described in Sect. 3 is applied to the case of a single vehicle serving multiple urban areas. We describe our experimental setup for this case in Sect. 4.1. In Sect. 4.2 we show our method to outperform state-of-the-art waiting heuristics from literature.

4.1 Experimental Setup

In order to reflect the situation of a vehicle serving multiple urban areas, we generated our problem instances based on Solomon's RC101 instance (Solomon 1987). This instance consists of 49 customers geographically grouped into 5 clusters of approximately 10. In addition a start and an end depot are located in-between the clusters. We randomly generated a request probability between 0.1 and 0.99 for each of the 49 customers. A problem instance is then created by randomly selecting a number of early request customers from the three clusters offering the highest total sum of individual request probabilities. We set our instances to result in a ratio of approximately 50% of the total number of customers requesting for service. Among these we assume about 40% early request customers. The length of the time horizon is held constant for each of the instances. It corresponds to the double average length of the tour containing only early request customers. 30 problem instances were generated for testing purposes.

The ADP algorithm is configured with a stochastic gradient method as update procedure. Stepsizes are reduced according to a generalized harmonic sequence. The learning phase comprises 10,000 trajectories for each instance. Decisions are taken following a pure exploitation sampling policy. The initial tour containing the early request customers was created by application of a greedy randomized adaptive search procedure (GRASP). Confirmed late request customers were inserted into this tour according to the insert savings criterion.

4.2 Computational Results

We compared our algorithm for each of the 30 test instances to a number of waiting heuristics developed by Thomas (2007). Thomas proposes 5 different heuristics and concludes the "distribute available waiting time (DW)" heuristic to outperform the others. The basic idea of DW is to equally distribute available waiting time among the locations of customers currently being in state $s = 2$. We tested DW as well as ADP on a set of 250 randomly generated test trajectories and compared the average number of confirmations of the two. ADP outperformed DW for 28 out of 30 problem instances. Figure 1 illustrates the relative difference of solution quality.

The comparison of corresponding test trajectories showed two characteristic features of ADP. ADP tends to allocate waiting time at the position of the last requesting customer so far within a cluster. Further ADP tends to avoid certain clusters by rejecting all the requests from this cluster even though confirmation would have been feasible. The former allows for additional confirmations of late requests within the cluster currently visited. The latter saves time by reducing long

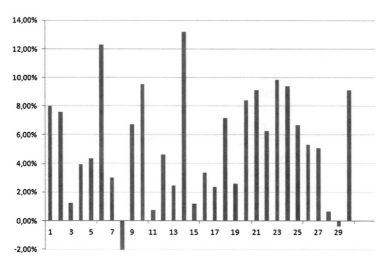

Fig. 1 Relative difference of solution quality of ADP compared to DW over the 30 problem instances considered

time distance moves between clusters. Both characteristics are responsible for the good performance of ADP.

Figure 2 shows a typical evolution of the mean solution quality of ADP throughout the learning phase. Starting from 9.2 confirmations of late requesting customers in iteration 1 we observe a strong increase of solution within the first 250 iterations up to about 10.15 confirmations. However, the further improvement up to 10.4 confirmations takes about 3,500 iterations. This quick increase at the beginning in combination with a relatively slow increase throughout the rest of the learning phase evolution are known to be typical for the pure exploitation sampling strategy.

5 Conclusions

We treated a dynamic vehicle routing problem with one vehicle, stochastic customer requests and confirmations by approximate dynamic programming. The approach is based on a formulation of the optimality equations around the post-decision state variables. A linear architecture with one feature is used for approximation of the value function. A strategy for action space approximation is presented introducing a tradeoff between anticipation by waiting and moving onto the next customer. A double pass learning algorithm is applied to problem instances representing the case of the vehicle serving a multiple urban areas. Computational results show this approach to outperform state-of-the-art waiting heuristics.

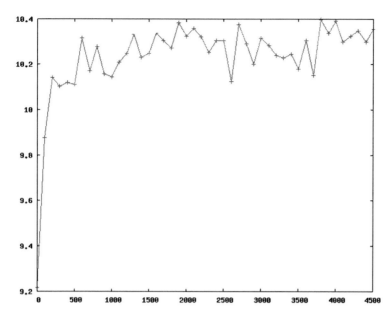

Fig. 2 Average number of confirmations by ADP over learntrajectories for problem instance 1

Future work comprises the adaptation of the approach to other types of problem instances as well as improvement of the crucial algorithmic components.

References

Bellman R (1957) Dynamic Programming. Princeton University Press

Bertsimas DJ, van Ryzin G (1991) A stochastic and dynamic vehicle routing problem in the Euclidean plane Operations Research 39:601–615

Bertsimas DJ, van Ryzin G (1993) Stochastic and dynamic vehicle routing in the Euclidean plane with multiple capacitated vehicles. Operations Research 41:60–76

Bianchi L (2000) Notes on dynamic vehicle routing – The state of the art. Technical Report IDSIA-05-01, Idsia, Manno-Lugano (CH)

Branke J, Middendorf M, Nöth G, Dessouky M (2005) Waiting strategies for dynamic vehicle routing. Transportation Science 39:298–312

Gendreau M, Potvin JY (1998) Dynamic vehicle routing and dispatching. In: Crainic T, Laporte G (eds) Fleet Management and Logistics, Kluwer, London

Gendreau M, Guertin F, Potvin JY, Taillard E (1999) Parallel tabu search for real-time vehicle routing and dispatching. Transportation Science 33:381–390

Larsen A, Madsen OBG, Solomon MM (2007) Classification of dynamic vehicle routing systems. In Zeimpekis V et al. (eds) Dynamic Fleet Management, Springer, New York

Larsen A, Madsen OBG, Solomon MM (2008) Recent developments in dynamic vehicle routing systems. In Golden B et al. (eds) The Vehicle Routing Problem: Latest Advances and New Challenges, Springer, New York

Mitrovic-Minic S, Laporte G. (2004) Waiting Strategies for the dynamic pickup and delivery problem with time windows. Transportation Research B 38:635–655

Novoa C, Storer R (2009) An approximate dynamic programming approach for the vehicle routing problem with stochastic demands. European Journal of Operational Research 196:509–515

Papastavrou JD (1996) A stochastic and dynamic routing policy using branching processes with state dependent immigration. European Journal of Operational Research 95:167–177

Powell W (1986) A stochastic model of the dynamic vehicle allocation problem. Transportation Science 20:117–129

Powell W (1996) A stochastic formulation of the dynamic assignment problem with an application to truckload motor carriers. Transportation Science 30:195–219

Powell W (2007) approximate dynamic programming. Wiley, Hoboken (NJ)

Psaraftis H (1988) Dynamic vehicle routing problems. In: Goldan B, Assad A (eds.) Vehicle Routing: Methods and Studies. North-Holland, Amsterdam

Psaraftis H (1995) Dynamic vehicle routing: Status and prospects. Annals of Operations Research 61:143–164

Secomandi N (2000) Comparing neuro-dynamic programming algorithms for the vehicle routing problem with stochastic demands. Computers & OR 27:1201–1225

Solomon MM (1987) Algorithms for the Vehicle Routing and Scheduling Problem with Time Windows. Operations Research 35:254–265

Thomas B (2007) Waiting Strategies for Anticipating Service Requests at Known Customer Locations. Transportation Science 43:319–331

Thomas B, White C (2004) Anticipatory route selection. Transportation Science 38:473–484

Topaloglu H (2007) A parallelizable and approximate dynamic programming-based dynamic fleet management model with random travel times and multiple vehicle types. In Zeimpekis V et al. (eds) Dynamic Fleet Management, Springer, New York

Stability Analysis of Large Scale Networks of Autonomous Work Systems with Delays

H. R. Karimi, S. Dashkovskiy and N. A. Duffie

1 Introduction

Production networks are emerging as a new type of cooperation between and within companies, requiring new techniques and methods for their operation and management (Wiendahl and Lutz 2002). Coordination of resource use is a key challenge in achieving short delivery times and delivery time reliability. These networks can exhibit unfavourable dynamic behaviour as individual organizations respond to variations in orders in the absence of sufficient communication and collaboration, leading to recommendations that supply chains should be globally rather than locally controlled and that information sharing should be extensive (Helo 2000; Huang et al. 2003). However, the dynamic and structural complexity of these emerging networks inhibits collection of the information necessary for centralized planning and control, and decentralized coordination must be provided by logistic processes with autonomous capabilities (Monostori et al. 2004).

A production network with several autonomous work systems is depicted in Fig. 1. The behaviour of such a network is affected by external and internal order flows, planning, internal disturbances, and the control laws used locally in the work systems to adjust resources for processing orders (Duffie et al. 2008). In prior work, sharing of capacity information between work systems has been modelled (Kim and Duffie 2006) along with the benefits of alternative control laws and

H. R. Karimi (✉)
Department of Engineering, Faculty of Engineering and Science,
University of Agder, 4898, Grimstad, Norway
e-mail: hamid.r.karimi@uia.no

S. Dashkovskiy
Centre for Industrial Mathematics, University of Bremen, Bremen, Germany

N. A. Duffie
Department of Mechanical Engineering, University of Wisconsin-Madison,
Madison, WI, USA

H.-J. Kreowski et al. (eds.), *Dynamics in Logistics*,
DOI: 10.1007/978-3-642-11996-5_7, © Springer-Verlag Berlin Heidelberg 2011

Fig. 1 Production network consisting of a group of autonomous work systems

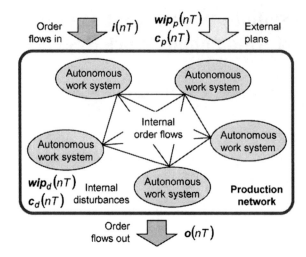

reducing delay in capacity changes (Kim and Duffie 2005; Nyhuis et al. 2005). Several authors have described both linear and nonlinear dynamical models for control of variables such as inventory levels and work in progress (WIP), including the use of pipeline flow concepts to represent lead times and production delays (John et al. 1994; Bai and Gershwin 1994). Delivery reliability and delivery time have established themselves as equivalent buying criteria alongside product quality and price (see Wiendahl and Lutz 2002; Towill et al. 1997). High delivery reliability and short delivery times for companies demand high schedule reliability and short throughput times in production. In order to manufacture economically under such conditions, it is necessary to minimise WIP levels in production and utilise operational resources in the best possible way.

Production Planning and Control (PPC) has become more challenging as manufacturing companies adapt to a fast changing market (Ratering and Duffie 2003; Scholz-Reiter et al. 2005; Wiendahl and Breithaupt 2000). Current PPC methods often do not deal with unplanned orders and other types of turbulence in a satisfactory manner (Kim and Duffie 2004). Assumptions such as infinite capacity and fixed lead time are often made, leading to a static view of the production system may not be valid because WIP affects lead time and performance, while capacity is finite and varies both according to plan and due to unplanned disturbances such as equipment breakdowns, worker illness, market changes, etc. Understanding the dynamic nature of production systems requires new approaches for the design of PPC based on company's logistics (Wiendahl et al. 2002). The controllers implicitly interact to adjust capacity to eliminate backlog as the system maintains its planned WIP level (Kim and Duffie 2004). A discrete closed-loop PPC model was developed and analyzed by Duffie and Falu (2002) in which two discrete controllers, one for backlog and one for WIP, with different periods between adjustments of work input and capacity, respectively, were selected and evaluated using transfer function analysis and time-response simulation. A second architecture for continuous WIP control and discrete backlog control, with delay capacity adjustment, was developed and analyzed by (Ratering and Duffie 2003) for cases of high and low WIP.

On the other hand, delay-differential systems are assuming an increasingly important role in many disciplines like economic, mathematics, science, and engineering. For instance, in economic systems, delays appear in a natural way since decisions and effects are separated by some time interval. The delay effects problem on the stability of systems is a problem of recurring interest since the delay presence may induce complex and undesired behaviors (oscillation, instability, bad performance) for the schemes (Han 2002; Fridman 2001; He et al. 2007; Park 1999; Karimi 2008). Over the past few decades, discrete-time systems with time-delay have received little attention compared with its continuous-time counterpart (Boukas and Liu 2001; Wang et al. 1999; Gao et al. 2005, 2007). The stability of time-delay systems is a fundamental problem because of its importance in the analysis of such systems. With regard to the stability analysis issue, (Verriest and Ivanov 1995) studied the sufficient conditions for the asymptotic stability of the discrete-time state-delayed systems by using an algebraic matrix inequality approach. The basic method for stability analysis is the direct Lyapunov method, and by this method, strong results have been obtained. But finding Lyapunov functions for nonautonomous delay-difference systems is usually a difficult task. In contrast, many methods different from Lyapunov functions have been successfully applied to establish stability results for difference equations with delay, for example, (Kipnis and Komissarova 2006; Levitskaya 2005; Kipnis and Nigmatulin 2004). Recently, in (Karimi 2006) a computational method was presented using Haar wavelets to determine the piecewise constant feedback controls for a finite-time linear optimal control problem of a time-varying state-delayed system.

In this paper, we contribute to the problem of stability analysis for a class of production networks of autonomous work systems with delays in the capacity changes. The system under consideration does not share information between work systems and the work systems adjust capacity with the objective of maintaining a desired amount of WIP. Attention is focused to derive explicit sufficient delay-dependent stability conditions for the network using properties of matrix norm. Finally, numerical results are provided to demonstrate the proposed approach.

2 Model of Autonomous Work Systems

A linear discrete-time dynamic approach for modeling the flow of orders into, out of, and between work systems was chosen because it promotes straightforward calculation of fundamental dynamic properties such as characteristic times and damping. Assume that there are N work systems in a production network, as shown in Fig. 1, and that vector $i(nT)$ is the rate at which orders are input to the N work systems from sources external to the production network, which is constant over time $nT \leq t < (n + 1)T$ where $n = 0, 1, 2, \ldots$ and T is a time period between capacity adjustments (for example, 1 shop-calendar day [scd]). The total orders that have been input to the work systems up to time $(k + 1)T$ then can be represented as the vector (Duffie et al. 2008).

$$w_i((n + 1)T) = w_i(nT) + T(i(nT) + R(nT)^T c_a(nT)) \tag{1a}$$

where vector $c_a(nT)$ is the rate at which orders are output from the N work systems during time $nT \leq t < (n + 1)T$ (the actual capacity of each work system) and R is a matrix in which element approximates the fraction of the flow out of work system j that flows into work system k.

The total number of orders that have been output by the work systems up to time $nT \leq t < (n + 1)T$ can be represented by the vector

$$w_o((n + 1)T) = w_o(nT) + T c_a(nT) \tag{1b}$$

while the rate at which orders are output from the network during time $nT \leq t < (n + 1)T$ is

$$o(nT) = R_o(nT) c_a(nT) \tag{1c}$$

where $R_o(nT)$ is a diagonal matrix in which non-zero diagonal elements represents the fraction of orders flowing out of work systems that flow out of the network during time $nT \leq t < (n + 1)T$. $R_o(nT)$ is assumed to be constant during this period, and

$$R_{o_{ii}}(nT) + \sum_{\substack{j=1 \\ j \neq i}}^{N} R_{o_{ij}}(nT) = 1 \tag{1d}$$

$R(nT)$ and $R_o(nT)$ represent the structure of order flow in the network. The WIP in the work systems is

$$wip_a(nT) = w_i(nT) - w_o(nT) + w_d(nT) \tag{1e}$$

where $w_d(nT)$ represents local work disturbance, such as rush order, that affect the work system. Furthermore, the actual capacity of each work system depends on three components as follows:

$$c_a(nT) = c_p(nT) + c_m((n - d)T) - c_d(nT) \tag{1f}$$

where $c_d(nT)$ represents local capacity disturbances such as equipment failures, $c_p(nT)$ denotes planned capacities of the work systems and $c_m(nT)$ represents local capacity adjustments to maintain the WIP in each work system in the vicinity of the planned levels $wip_p(nT)$ using gain k_c and is described in the form of

$$c_m(nT) = k_c(wip_a(nT) - wip_p(nT)) \tag{1g}$$

It is assumed that a delay dT exists in the capacity changes $c_m(nT)$ for logistic reasons such as operator work rules. In this network, the work systems do not share information regarding the expected physical flow of orders between them.

A capacity plan is required for each work system. For constants $R(nT)$ and $R_o(nT)$, the transfer functions relating $wip_a(z)$ and $c_a(z)$ to the inputs $i(z), w_d(z), wip_p(z)$, $c_p(z)$ and $c_d(z)$ are

$$wip_a(z) = \left((I - z^{-1})I + k_c T(I - R^T)z^{-(d+1)} \right)^{-1} \left(Tz^{-1}i(z) + (I - z^{-1})w_d(z) \right.$$
$$+ k_c T(I - R^T)z^{-(d+1)}wip_p(z) - T(I - R^T)z^{-1}c_p(z) + T(I - R^T)z^{-1}c_d(z) \Big)$$
$$(2a)$$

$$wip_a(z) = \left((I - z^{-1})I + k_c T(I - R^T)z^{-(d+1)} \right)^{-1} \left(Tz^{-1}i(z) + (I - z^{-1})w_d(z) \right.$$
$$+ k_c T(I - R^T)z^{-(d+1)}wip_p(z) - T(I - R^T)z^{-1}c_p(z) + T(I - R^T)z^{-1}c_d(z) \Big)$$
$$(2b)$$

Our purpose is to investigate the stability of the network (1) respect to the delay parameter and the controller gain which is characterize by the roots of

$$\det((I - z^{-1})I - Az^{-(d+1)}) = 0 \qquad (3)$$

with $A := -k_c T(I - R^T)$.

3 Stability Analysis

In this section, sufficient conditions for the stability of the network (1) respect to the delay parameter and the controller gain are proposed using characteristic equation.

The characteristic Eq. 3 can be represented in the form of

$$\det(A + Iz^d - Iz^{(d+1)}) = 0 \qquad (4)$$

and (4) is corresponding to the characteristic equation of the following system

$$x_n = x_{n-1} + Ax_{n-d-1} \qquad (5)$$

Levitskaya in (Levitskaya 2005) established that (5) is asymptotically stable if and only if any eigenvalue of the matrix A lies inside the oval of the complex plane bounded by a curve

$$\Gamma = \left\{ z \in C : z = 2i \sin \frac{\varphi}{2d + 1} e^{i\varphi}, |\varphi| \leq \frac{\pi}{2} \right\}. \qquad (6)$$

Remark 1 Let λ_i be eigenvalues of the matrix $A = -k_c T(I - R^T)$. Eq. 5 is asymptotically stable if and only if

$$|\lambda_i| < 2 \sin \frac{\pi}{2(2d+1)}$$

Theorem 1 *If the system (5) is asymptotically stable, then all eigenvalues of A lie inside the unit disk.*

Proof It is sufficient to consider the stability ovals (6) and to remark that $|2 \sin(\pi/2(2d/1))| \le 1$ for $k > 1$.

In the sequel, we will obtain the necessary and sufficient condition in terms of the eigenvalues location of the matrix A for the asymptotic stability of Eq. 5.

Lemma 1 (Kipnis and Komissarova 2006) *If* $\sum_{i=1}^{k} \|A_i\| < 1$, *then the linear system* $x_n = \sum_{i=1}^{k} A_i x_{n-i}$ *is asymptotically stable.*

Theorem 2 *If*

$$\|A + I\| + d\|A\|^2 < 1 \tag{7}$$

then (5) is asymptotically stable.

Proof Eq. 5 is rewritten as

$$
\begin{aligned}
x_n &= (A + I)x_{n-1} - A(x_{n-1} - x_{n-d-1}) \\
&= (A + I)x_{n-1} - A\sum_{i=1}^{d}(x_{n-i} - x_{n-i-1}) \\
&= (A + I)x_{n-1} - A\sum_{i=1}^{d} Ax_{n-i-d-1}
\end{aligned}
\tag{8}
$$

According to Lemma 1, from (8) we conclude (7).

Now, we introduce an additional stability condition for (5) depending on whether the delay d is odd or even.

Theorem 3 *If*

$$\left\| I + (-1)^d A \right\| + d\|A\|(2 + \|A\|) < 1 \tag{9}$$

then (5) is asymptotically stable.

Proof If d is even Eq. 5 is rewritten as

$$
\begin{aligned}
x_n &= (A + I)x_{n-1} - A(x_{n-1} - x_{n-d-1}) \\
&= (A + I)x_{n-1} - A\sum_{i=1}^{d}(-1)^{i+1}(2I x_{n-i-1} + Ax_{n-k-i})
\end{aligned}
\tag{10}
$$

and if d is odd we have

$$x_n = (I - A)x_{n-1} + A(x_{n-1} - x_{n-d-1})$$

$$= (I - A)x_{n-1} + A\sum_{i=1}^{d}(-1)^{i+1}(2I\,x_{n-i-1} + Ax_{n-k-i}) \tag{11}$$

Similar to the proof of Theorem 2, the inequality (9) is concluded.

4 Numerical Results

Consider the case of a supplier of components to the automotive industry and for which production data documents orders flowing between five work systems over a 162-day period. These work systems and the order-flow structure over this period is illustrated in Fig. 2. In this network, all order flows are unidirectional; therefore, the fundamental dynamic properties of capacity adjustment in the individual work systems are independent. Then, the internal flow of orders is approximated using the following matrix (Duffie et al. 2008),

$$R = \begin{bmatrix} 0 & 106/341 & 235/341 & 0 & 0 \\ 0 & 0 & 0 & 188/401 & 204/401 \\ 0 & 0 & 0 & 100/236 & 129/236 \\ 0 & 0 & 0 & 0 & 268/295 \\ 0 & 0 & 0 & 0 & 0 \end{bmatrix}$$

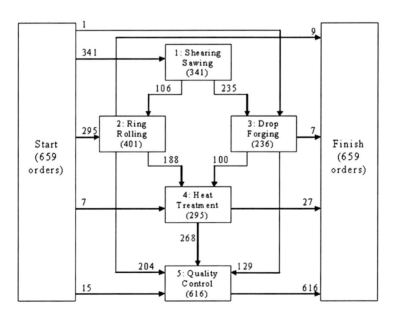

Fig. 2 A production network consisting of five work systems

in which element R_{ij} is the total number of orders that went from work system i to work system j divided by the total number of orders that left work system i.

Consider the sampling time $T = 1$ scd. It is clear that the condition in Lemma 1 cannot be applied. Applying all of the *Theorems* derived, the conditions of maximum controller gain for the asymptotic stability of the network are shown in Table 1. The result from Table 1 guarantees the asymptotic stability of system under consideration.

5 Conclusion

The problem of stability analysis for a class of production networks of autonomous work systems with delays in the capacity changes was investigated in this paper. The system under consideration does not share information between work systems and the work systems adjust capacity with the objective of maintaining a desired amount of local work in progress (WIP). In terms of properties of matrix norm some explicit sufficient delay-dependent stability conditions were derived for the network. Finally, numerical results were provided to demonstrate the proposed approach.

Acknowledgments This research has been funded by the German Research Foundation (DFG) as part of the Collaborative Research Center 637 'Autonomous Cooperating Logistic Processes: A Paradigm Shift and its Limitations' (SFB 637).

6 Appendix

$\|.\|$ is any matrix norm which satisfies the following conditions:
   (I) $\|A \geq 0\|$, and $\|A = 0\|$ if and only if $A = 0$,
   (II) for each $c \in R$, $\|cA\| = |c|\|A\|$,
   (III) $\|A + B\| \leq \|A\| + \|B\|$,
   (IV) $\|AB\| \leq \|A\| \cdot \|B\|$ for all $(m \times m)$ matrices A, B.
   In addition, matrix norm should be concordant with the vector norm $\|.\|_*$, that is,

$$\|Ax\|_* \leq \|A\| \cdot \|x\|_*$$

Table 1 Controller gain k_c w.r.t. d

	Theorem 1	Theorem 2	Theorem 3
$d = 1$	1.0000	0.8500	0.8650
$d = 2$	0.6180	0.6250	0.6850
$d = 3$	0.4450	0.4750	0.4875
$d = 4$	0.3473	0.3845	0.3950

for all $x \in \Re^m$ and any $(m \times m)$ matrix A. For real $(m \times m)$ matrix A, we define, as usual, $\|A\|_1 = \max_{1 \leq j \leq m} \sum_{i=1}^{m} |a_{ij}|$ and $\|A\|_\infty = \max_{1 \leq i \leq m} \sum_{j=1}^{m} |a_{ij}|$

References

Bai S.X., Gershwin S.B., 'Scheduling manufacturing systems with work-in-process inventory control: Multi-part-type systems.' *Int. J. Production Research*, vol. 32, no. 2, pp. 365–385, 1994.

Boukas E.K., Liu Z.K., 'Robust H_∞ control of discrete-time Markovian jump linear systems with mode-dependent time-delays.' *IEEE Transactions on Automatic Control*, vol. 46, pp. 1918–1924, 2001.

Duffie N., Falu I., 'Control-theoretic analysis of a closed-loop PPC system.' *Annals of the CIRP*, vol. 5111, pp. 379–382, 2002.

Duffie N.A., Roy D., Shi L., 'Dynamic modelling of production networks of autonomous work systems with local capacity control.' *CIRP Annals-Manufacturing Technology*, vol. 57, pp. 463–466, 2008.

Fridman E., 'New Lyapunov-Krasovskii functionals for stability of linear retarded and neutral type systems.' *Systems and Control Letters*, vol. 43, pp. 309–319, 2001.

Gao H., Lam J., Xie L.H., Wang C.H., 'New approach to mixed H_2/H_∞ filtering for polytopic discrete-time systems' *IEEE Transactions on Signal Processing*, vol. 53, no. 8, pp. 3183–3192, 2005.

Gao H., Lam J., Wang Z., 'Discrete bilinear stochastic systems with time-varying delay: Stability analysis and control synthesis.' *Chaos, Solitons and Fractals*, vol. 34, no. 2, pp. 394–404, 2007.

Han Q.L., 'Robust stability of uncertain delay-differential systems of neutral type.' *Automatica*, vol. 38, no. 4, pp. 719–723, 2002.

He Y., Wang Q.G., Lin C., Wu M., 'Delay-range-dependent stability for systems with time-varying delay.' *Automatica*, vol. 43, pp. 371–376, 2007.

Helo P., 'Dynamic modelling of surge effect and capacity limitation in supply chains.' *Int. J. Production Research*, vol. 38, no. 17, pp. 4521–4533, 2000.

Huang G.Q., Lau J.S.K., Mak K.L., 'The impacts of sharing production information on supply chain dynamics: A review of the literature.' *Int. J. Production Research*, vol. 41, no. 7, pp. 1483–1517, 2003.

John S., Naim M.M., Towill D.R., 'Dynamic analysis of a WIP compensated decision support system.' *Int. J. Manufacturing System Design*, vol. 1, no. 4, pp. 283–297, 1994.

Karimi H.R., 'A computational method for optimal control problem of time-varying state-delayed systems by Haar wavelets.' *Int. J. Computer Mathematics*, vol. 83, no. 2, pp. 235–246, 2006.

Karimi H.R., 'Observer-based mixed H_2/H_∞ control design for linear systems with time-varying delays: An LMI approach.' *Int. J. Control, Automation, and Systems*, vol. 6, no. 1, pp. 1–14, 2008.

Kim J.-H., Duffie N.A., 'Backlog control for a closed loop PPC system.' *CIRP Annals*, vol. 53, no. 1, pp. 357–360, 2004.

Kim J.-H., Duffie N., 'Design and analysis of closed-loop capacity control for a multi-workstation production system.' *Annals of the CIRP*, vol. 54, no. 1, pp. 455–458, 2005.

Kim J.-H., Duffie N., 'Performance of coupled closed-loop workstation capacity controls in a multi-workstation production system.' *Annals of the CIRP*, vol. 55, no. 1, pp. 449–452, 2006.

Kipnis M., Komissarova D., 'Stability of a delay difference system.' *Advances in Difference Equations*, vol. 2006, Article ID 31409, pp. 1–9.

Kipnis M., Nigmatulin R.M., 'Stability of trinomial linear difference equations with two delays,' *Automation and Remote Control*, vol. 65, no. 11, pp. 1710–1723, 2004.

Levitskaya I.S., 'A note on the stability oval for $x_{n+1} = x_n + Ax_{n-k}$.' *Journal of Difference Equations and Applications*, vol. 11, no. 8, pp. 701–705, 2005.

Monostori L., Csaji B.C., Kadar B., 'Adaptation and learning in distributed production control.' *Annals of the CIRP*, vol. 53, no. 1, pp. 349–352, 2004.

Nyhuis P., Cieminski G., Fischer A., 'Applying simulation and analytical models for logistic performance prediction.' *Annals of the CIRP*, vol. 54, no. 1, pp. 417–422, 2005.

Park P., 'A delay-dependent stability criterion for systems with uncertain time-invariant delays.' *IEEE Transactions on Automatic Control*, vol. 44, pp. 876–877, 1999.

Ratering A., Duffie N., 'Design and analysis of a closed-loop single-workstation PPC system.' *Annals of CIRP*, vol. 52, no. 1, pp. 355–358, 2003.

Ratering A., Duffie N., 'Design and analysis of a closed-loop single-workstation PPC system.' *Annals of the CIRP*, vol. 5211, pp. 355–358, 2003.

Scholz-Reiter B., Freitag M., de Beer C., Jagalski T., 'Modelling dynamics of autonomous logistic processes: Discrete-event versus continuous approaches.' *CIRP Annals—Manufacturing Technology*, vol. 54, no. 1, pp. 413–416, 2005.

Towill D.R., Evans G.N., Cheema P., 'Analysis and design of an adaptive minimum reasonable inventory control system.' *Journal of Production Planning and Control*, vol. 8, no. 6, pp. 545–577, 1997.

Verriest E., Ivanov A.F., 'Robust stability of delay-difference equations.' *Proceedings of 34th Conference on Decision and Control*, pp. 386–391, New Orleans, USA, 1995.

Wang Z., Huang B., Unbehauen H., 'Robust H_∞ observer design of linear state delayed systems with parametric uncertainty: the discrete-time case.' *Automatica*, vol. 35, pp. 1161–1167, 1999.

Wiendahl H.-P., Breithaupt J.-W., 'Automatic production control applying control theory.' *Int. J. Production Economics*, vol. 63, no. 1, pp. 33–46, 2000.

Wiendahl H.-P., Lutz S., 'Production in networks.' *Annals of the CIRP*, vol. 52, no. 2, pp. 573–586, 2002.

Wiendahl H.-H., Roth N., Westkamper E., 'Logistical positioning in a turbulent environment,' *Annals of the CIRP*, vol. 5111, pp. 383–386, 2002.

Local Input-to-State Stability
of Production Networks

Sergey Dashkovskiy, Michael Görges and Lars Naujok

1 Introduction

Production and supply networks or other modern logistic structures are typical examples of complex systems with a nonlinear and sometimes chaotic behavior. Their dynamics subject to many different perturbations due to changes on market, changes in customer behavior, information and transport congestions, unreliable elements of the network, etc.

One approach to handle such complex systems is to shift from centralized to decentralized or autonomous control, i.e., to allow the entities of a network to make their own decisions based on some given rules and available local information. However a system emerging in this way may become unstable and hence be not effective in performance. Typical examples of unstable behaviour are unbounded growth of unsatisfied orders or unbounded growth of amount of workload to be processed by a machine and causes high inventory costs or loss of customers. To avoid instability it is worth to investigate its behavior in advance.

Mathematical methods can help to handle complex systems. In particular mathematical modelling and analysis provide helpful tools for investigation of such objects and can be used for design, optimization and control of such networks and for deeper understanding of their dynamical properties.

This paper focuses on the stability analysis of a production network, in order to identify stable parameter constellation. In particular cases stability means that the number of unsatisfied orders or/and amount of workload to be processed by a machine remain bounded over time in spite of disturbances. By application of the

S. Dashkovskiy and L. Naujok (✉)
Centre of Industrial Mathematics, University of Bremen, Bremen, Germany
e-mail: larsnaujok@math.uni-bremen.de

M. Görges
Planning and Control of Production Systems (PSPS), BIBA, Bremer Institut für
Produktion und Logistik GmbH at the University of Bremen, Bremen, Germany

H.-J. Kreowski et al. (eds.), *Dynamics in Logistics*,
DOI: 10.1007/978-3-642-11996-5_8, © Springer-Verlag Berlin Heidelberg 2011

stability analysis to a logistical network we can draw conclusions of its behaviour and derive conditions to guarantee stability, which avoid negative outcomes mentioned above. The results of this analysis can be used to design logistical networks in order to have good properties to achieve economic goals. Obviously stability is decisive for the performance and vitality of a network.

In this paper we propose a model for a production logistic scenario comprising several autonomous production plants connected through transport routs. This network is modelled by ordinary differential equations. We show how its stability can be analyzed with help of small gain theorems recently developed for general type of dynamic networks. Explicit conditions of the production rates will be derived by application of mathematical systems theory of interconnected systems.

In Sect. 2, we describe the given production network with its conditions and model it mathematically by differential equations. A mathematical background is given in Sect. 3, which is used in Sect. 4 to derive stability conditions of the production network. In Sect. 5 some simulation results and their interpretations are given. Conclusions and outlines can be found in Sect. 6.

2 Model Description

In this section, we describe the given production network, which we model and analyze the system in view of stability with help of differential equations.

The production network in Fig. 1 consists of six geographically distributed production locations, which are connected. In logistic there are many flows, e.g. material, information or worth flows. In Fig. 1 the material flow is described by arrows and the information flow by dashed arrows. The *state* of each production location is denoted by $x_i(t) \in \mathbb{R}$ for $i = 1,...,6$, where $t \in \mathbb{R}_+$ can be interpreted as time and $\mathbb{R}_+$ denotes all positive real values. In the rest of this paper we write *subsystem i* for the ith production location. All six subsystems form the production network, which we name simply (whole) system.

We describe the production network by the information flow and interpret the state of the ith subsystem as the number of unsatisfied orders within ith production location. Subsystem 6 gets some orders of its product from the customers, denoted by $d(t) \in \mathbb{R}_+$. While processing the orders, subsystem 6 orders components, which it needs for production from subsystem 4 and 5. These two subsystems send orders for components, which they need to subsystem 2 and 3. Their orders will be sent to subsystem 1, which gets instantly its raw material from an external source.

Fig. 1 The production network

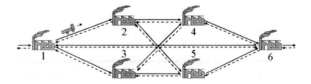

The orders from subsystem 1 to subsystem 6 are interpreted as a kind of payment or the demand for its production of subsystem 1 of the final product of the given production network from subsystem 6.

We suppose all subsystems are autonomously controlled, it means the ability to adjust the production rate of the production location. This can be achieved by varying work times of the workers, transportation times of the products or the number of used machines for production. $\alpha_i \in \mathbb{R}_+$ denotes the (constant) maximum production rate of subsystem i. The actual production rate of subsystem i $(\tilde{f}_i)$ converges to α_i, if the state of subsystem $x_i(t)$ is large and $\tilde{f}_i$ tends to zero, if the state of subsystem $x_i(t)$ tends to zero. This means, if there are many orders, the actual production rate is near to the maximum production rate and if there are no orders nothing will be produced. Therefor the actual production rate of each subsystem at time t is given by

$$\tilde{f}_i(x_i(t)) := \alpha_i(1 - \exp(-x_i(t))), \quad i = 1, \ldots, 6.$$

With these considerations we can model the system presented in Fig. 1 by differential equations for each subsystem, which are nothing but a description of changes of the state $x_i(t)$ of subsystem i along time $t \in \mathbb{R}_+$:

$$\begin{aligned}
\dot{x}_1(t) &= c_{12}\tilde{f}_2(x_2(t)) + c_{13}\tilde{f}_3(x_3(t)) - \tilde{f}_1(x_1(t)), \\
\dot{x}_2(t) &= c_{24}\tilde{f}_4(x_4(t)) + c_{25}\tilde{f}_5(x_5(t)) - \tilde{f}_2(x_2(t)), \\
\dot{x}_3(t) &= c_{34}\tilde{f}_4(x_4(t)) + c_{35}\tilde{f}_5(x_5(t)) - \tilde{f}_3(x_3(t)), \\
\dot{x}_4(t) &= c_{46}\tilde{f}_6(x_6(t)) - \tilde{f}_4(x_4(t)), \\
\dot{x}_5(t) &= c_{56}\tilde{f}_6(x_6(t)) - \tilde{f}_5(x_5(t)), \\
\dot{x}_6(t) &= d(t) + c_{61}\tilde{f}_1(x_1(t)) - \tilde{f}_6(x_6(t)),
\end{aligned} \tag{1}$$

where the constants $c_{ij} \in \mathbb{R}_+$ can be interpreted as the number of orders of components to subsystem i from subsystem j.

By definition of $f_i(x, d) := \dot{x}_i(t), i = 1, \ldots, 6, x := (x_1, \ldots, x_6)^T$ and $f(x, u) := (f_1(x, d), \ldots, f_6(x, d))^T$ we can write the whole system as

$$\dot{x}_i(t) = f(x(t), d(t)), \quad t \in \mathbb{R}_+. \tag{2}$$

Now the question arises, under which conditions the subsystems are stable, which means that the states of all subsystems will not increase to infinity. In other words, under which conditions all states of the subsystem and therefore of the whole system are bounded, which means stability of the production network?

3 Mathematical Background

For investigation of the stability of system (1) and (2), respectively, we need some mathematical results. We present a stability property and a tool how to check, weather the system has the stability property.

We consider nonlinear dynamical system of the form

$$\dot{x}(t) = f(x(t), u(t)), \tag{3}$$

where $t \in \mathbb{R}_+$ is the time, $\dot{x}(t)$ the derivate of the state $x(t) \in \mathbb{R}^N$ with the initial value x_0, input $u(t) \in \mathbb{R}^m$, which is an essentially bounded measurable function and $f : \mathbb{R}^{N+m} \to \mathbb{R}^N$ nonlinear. To have existence and uniqueness of a solution of (3), function f has to be continuous and locally Lipschitz in x uniformly in u. The solution is denoted by $x(t; x_0, u)$ or $x(t)$ in short.

To describe the given production network we generalize (3) and consider $n \in \mathbb{N}$ interconnected systems. These are in general nonlinear dynamical systems of the form

$$\dot{x}(t) = f_i(x_1(t), \ldots, x_n(t), u_i(t)), \quad i = 1, \ldots, n, \tag{4}$$

where $t \in \mathbb{R}_+$, $x_i(t) \in \mathbb{R}^{N_i}$, $u_i(t) \in \mathbb{R}^{M_i}$, which are essentially bounded measurable functions, $f_i : \mathbb{R}^{\sum_j N_j + M_i} \to \mathbb{R}^{N_i}, i = 1, \ldots, n$, where f_i are continuous and locally Lipschitz in $x = (x_1^T, \ldots, x_n^T)^T$ uniformly in u_i. We consider x_j as internal input and u_i as external input of the ith subsystem $i, j = 1, \ldots, n, i \neq j$. The solution is denoted by $x(t; x_i, x_j : j \neq i, u_i)$ or $x(t)$ in short.

If we define $N := \sum_{i=1}^n N_i, m := \sum_{i=1}^n M_i, x = (x_1^T, \ldots, x_n^T)^T$, and $f := (f_1^T, \ldots, f_n^T)$, then (4) becomes

$$\dot{x}(t) = f(x(t), u(t)), \quad t \in \mathbb{R}_+. \tag{5}$$

We denote the standard euclidian norm in $\mathbb{R}^n$ by $\|\cdot\|$ and the essential supremum norm for essentially bounded functions u in $\mathbb{R}_+$ by $\|u\|_\infty$. We need some classes of functions to define the stability property, which we will use. A function $f : \mathbb{R}^n \to \mathbb{R}_+$ is said to be *positive definite*, if $f(0) = 0$ and $f(x) > 0, \forall x \in \mathbb{R}^n$ holds. A class K function $\gamma : \mathbb{R}_+ \to \mathbb{R}_+$ is continuous, $\gamma(0) = 0$ and strictly increasing. If it is additionally unbounded then it is of class K_∞. We call a function $\beta : \mathbb{R}_+ \times \mathbb{R}_+ \to \mathbb{R}_+$ of class KL if β is continuous, $\beta(\cdot, t) \in K$ and $\beta(r, \cdot)$ strictly decreasing with $\lim_{t \to \infty} \beta(r, t) = 0, \forall t, r > 0$.

Now we define local input-to-state stability (LISS) and input-to-state stability (ISS), respectively, for each subsystem of (4). For system (3) the definition of LISS and ISS, respectively, can be found for example in Dashkovskiy et al. (2007b) and Sontag (1989), respectively.

Definition 1 The ith subsystem of (4) is called LISS, if there exist constants $\rho_i, \rho_j^i, \rho_i^u > 0, \gamma_{ij}, \gamma_i \in K_\infty$ and $\beta_i \in KL$, such that for all initial values $\|x_i^0\| \leq \rho_i, \|x_i\|_\infty \leq \rho_j^i$ and all inputs $\|u_i\|_\infty \leq \rho_i^u$ the inequality

$$\left\| x_i(t; x_i^0, x_j : j \neq i, u_i) \right\| \leq \max\left\{ \beta_i\left(\left\| x_i^0 \right\|, t \right), \max_{j \neq i} \gamma_{ij}\left(\left\| x_j \right\|_\infty \right), \gamma_i\left(\left\| u_i \right\|_\infty \right) \right\} \tag{6}$$

is satisfied $\forall t \in \mathbb{R}_+$. γ_{ij} and γ_i are called (nonlinear) gains.

Note that, if $\rho_i, \rho_j^i, \rho_i^u = \infty$ then the ith subsystem is ISS (see Dashkovskiy et al. 2007a). LISS and ISS, respectively, mean that the norm of the trajectories of each subsystem is bounded.

Furthermore we define the *gain matrix* $\Gamma := (\gamma_{ij}), i, j = 1, \ldots, n, \gamma_{ii} = 0$, which defines a map $\Gamma : \mathbb{R}_+^n \to \mathbb{R}_+^n$ by

$$\Gamma(s) := \left(\max_j \gamma_{1j}(s_j), \ldots, \max_j \gamma_{nj}(s_j) \right)^T, \quad s \in \mathbb{R}_+^n. \tag{7}$$

Previous investigations of two interconnected systems established a small gain condition to guarantee stability (see Jiang et al. 1994, 1996). In Dashkovskiy et al. (2007a) an ISS small gain theorem for general networks was proved, where the small gain condition is of the form

$$\Gamma(s) \not\geq s, \quad \forall s \in \mathbb{R}_+^n \setminus \{0\}. \tag{8}$$

Notation $\not\geq$ means that there is at least one component $i \in \{1, \ldots, n\}$ such that $\Gamma(s)_i < s_i$. Here we recall a local version of the small gain condition:

Definition 2 Γ satisfies the *local small gain condition* (LSGC) on $[0, w^*]$, provided that

$$\Gamma(w^*) < w^* \quad \text{and} \quad \Gamma(s) \not\geq s, \quad \forall s \in [0, w^*], s \neq 0. \tag{9}$$

Further details of (9) can be found in Dashkovskiy et al. (2007b). The small gain condition is equivalent to the compliance of the cycle condition (see Ruffer 2007, Lemma 2.3.14 for details). We quote the local version of the small gain theorem:

Theorem 1 *Let all subsystems of (4) satisfy (6). Suppose Γ satisfies LSGC. Then there exist constants $\rho, \rho^u > 0, \beta \in KL$ and $\gamma \in K_\infty$, such that the whole system (5) is LISS.*

The proof can be found in Dashkovskiy et al. (2007b), Theorem 4.2. An important tool to verify LISS and ISS, respectively, are Lyapunov functions. For systems of the form (3) one can find the definition of Lyapunov functions for example in Jiang et al. (1996) and Dashkovskiy et al. (2007b).

Definition 3 A smooth function $V_i : \mathbb{R}^{N_i} \to \mathbb{R}_+$ is called LISS Lyapunov function of the ith subsystem of system (4), if it satisfies the following two conditions:

There exist functions $\psi_{1i}, \psi_{2i} \in K_\infty$ such that

$$\psi_{1i}(\|x_i\|) \leq V_i(x_i) \leq \psi_{2i}(\|x_i\|), \quad \forall x_i \in \mathbb{R}^{N_i}. \tag{10}$$

There exist $\chi_{ij}, \chi_i \in K_\infty$, a positive function μ_i and constants $\rho_i^0, \rho_j^i, \rho_i^u > 0$ such that

$$V_i(x_i) \geq \max\left\{\max_j \chi_{ij}(V_j(x_j)), \chi_i(||u_i||)\right\} \Rightarrow \nabla V_i(x_i) \cdot f_i(x, u) \leq -\mu_i(V_i(x_i))$$

$$(11)$$

for all $x_i \in \mathbb{R}^{N_i}, ||x_i^0|| \leq \rho_i^0, ||x_j|| \leq \rho_j^i, u_i \in \mathbb{R}^{M_i}, ||u_i|| \leq \rho_i^u, \chi_{ii} = 0$, where ∇ denotes the gradient of V_i. Functions χ_{ij} and χ_i are called LISS Lyapunov gains.

Note that, if $\rho_0^i, \rho_i^u = \infty$ then the LISS Lyapunov function of the ith subsystem becomes an ISS Lyapunov function of the ith subsystem (see Dashkovskiy et al. 2010).

To check if the whole system of the form (5) has the LISS or ISS property one can use LISS or ISS Lyapunov functions, respectively. If there exists a LISS or ISS Lyapunov function for a subsystem of (4) then the subsystem has the LISS or ISS property, respectively. Furthermore, if all subsystems have a LISS of ISS Lyapunov function and the LISS or ISS Lyapunov gains satisfy the small gain condition, then the whole system of the form (5) is LISS or ISS, respectively (see Dashkovskiy and Ruffer 2006, Dashkovskiy et al. 2007b or Dashkovskiy et al. 2010).

With this mathematical theory we can derive conditions, for which the subsystems and the whole system are stable. This will be presented in the next section.

4 Stability of the Model

In this section we investigate all six subsystems of (1) to check if they have the LISS or ISS property, respectively. Therefore we choose a Lyapunov function candidate for each subsystem and check, whether conditions (10) and (11) are satisfied.

Remark 1 It can be shown that for any non-negative initial condition all subsystems of (1) are non-negative, since the term $f_i(x_i)$ is zero for $x_i = 0$ and $d = 0, i = 1, \ldots, n$.

We choose $V_i(x_i) = x_i$ as Lyapunov function candidate for $i = 1, \ldots, 6$. V_i satisfies condition (10). For the investigation of the first subsystem we define

$$\chi_{1j}(x_j) := -\ln\left(1 - \frac{c_{12}\alpha_2 + c_{13}\alpha_3}{(1 - \varepsilon_{1j})\alpha_1}\left(1 - \exp(-x_j)\right)\right) \leq x_1 = V_1(x_1),$$

$j = 2, 3, 1 > \varepsilon_{1j} > 0$, which implies

$$c_{1j}\alpha_j\left(1 - \exp(-x_j)\right) \leq \frac{c_{1j}\alpha_j}{c_{12}\alpha_2 + c_{13}\alpha_3}(1 - \varepsilon_{1j})\alpha_1(1 - \exp(-x_1)).$$

To guarantee that χ_{1j} is well defined the condition

$$c_{12}\alpha_2 + c_{13}\alpha_3 < \alpha_1(1 - \varepsilon_{1j}) < \alpha_1 \qquad (12)$$

has to be satisfied. With this consideration it follows

$$
\begin{aligned}
&\nabla V_1(x_1(t))f_1(x_1(t),\ldots,x_6(t),d(t)) \\
&= c_{12}\alpha_2(1 - \exp(-x_2)) + c_{13}\alpha_3(1 - \exp(-x_3)) - \alpha_1(1 - \exp(-x_1)) \\
&\leq \left(\frac{(1 - \varepsilon_{12})\alpha_1 c_{12}\alpha_2}{c_{12}\alpha_2 + c_{13}\alpha_3} + \frac{(1 - \varepsilon_{13})\alpha_1 c_{13}\alpha_3}{c_{12}\alpha_2 + c_{13}\alpha_3} - \alpha_1 \right)(1 - \exp(-x_1)) \\
&\leq -\varepsilon_1\alpha_1(1 - \exp(-x_1)) = -\mu_1(V_1(x_1(t))),
\end{aligned}
$$

where $\varepsilon_1 := \min\{\varepsilon_{12}, \varepsilon_{13}\}$ and $\mu_1(r) := \varepsilon_1\alpha_1(1 - \exp(-r))$ is a positive definite function.

The reason of the introduction of the constant value ε_{1j} is to guarantee that μ_1 is positive definite. V_1 satisfies condition (11) and is the ISS Lyapunov function of the first subsystem from which we know that the first subsystem has the ISS property for all $x_j \in \mathbb{R}_+, j = 1, 2, 3$, if condition (12) holds.

For subsystem 2 to 5 we do similar calculations and get the gains

$$\chi_{2j}(x_j) := -\ln\left(1 - \frac{c_{24}\alpha_4 + c_{25}\alpha_5}{(1 - \varepsilon_{2j})\alpha_2}(1 - \exp(-x_j))\right), \qquad 1 > \varepsilon_{2j} > 0, \quad j = 4, 5,$$

$$\chi_{3j}(x_j) := -\ln\left(1 - \frac{c_{34}\alpha_4 + c_{35}\alpha_5}{(1 - \varepsilon_{3j})\alpha_3}(1 - \exp(-x_j))\right), \qquad 1 > \varepsilon_{3j} > 0, \quad j = 4, 5,$$

$$\chi_{j6}(x_6) := -\ln\left(1 - \frac{c_{j6}\alpha_6}{(1 - \varepsilon_{j6})\alpha_j}(1 - \exp(-x_6))\right), \qquad 1 > \varepsilon_{j6} > 0, \quad j = 4, 5$$

and conditions

$$\alpha_2 > c_{24}\alpha_4 + c_{25}\alpha_5, \quad \alpha_3 > c_{34}\alpha_4 + c_{35}\alpha_5, \quad \alpha_4 > c_{46}\alpha_6, \quad \alpha_5 > c_{56}\alpha_6 \qquad (13)$$

for which the subsystems 2 to 5 have the ISS property.

For subsystem 6 from

$$
\begin{aligned}
\chi_6(d(t)) &:= -\ln\left(1 - \frac{d(t)(\|d\|_\infty + c_{61}\alpha_1)}{\|d\|_\infty(1 - \varepsilon_{6d})\alpha_6}\right) \leq x_6 = V_6(x_6), \\
\chi_{61}(x_1) &:= -\ln\left(1 - \frac{\|d\|_\infty + c_{61}\alpha_1}{(1 - \varepsilon_{61})\alpha_6}(1 - \exp(-x_1))\right) \leq x_6 = V_6(x_6),
\end{aligned}
\qquad (14)
$$

with $0 < \varepsilon_{61}, \varepsilon_{6d} < 1$ we get

$$
\begin{aligned}
&\nabla V_6(x_6(t))f_6(x_1(t),\ldots,x_6(t),d(t)) \\
&= d(t) - \alpha_6(1 - \exp(-x_6(t))) + c_{61}\alpha_1(1 - \exp(-x_1(t))) \\
&\leq -\varepsilon_6\alpha_6(1 - \exp(-x_6(t))) = -\mu_6(V_6(x_6(t))),
\end{aligned}
$$

where $\varepsilon_6 := \min\{\varepsilon_{61}, \varepsilon_{6d}\}$ and $\mu_6(r) := \varepsilon_6\alpha_6(1 - \exp(-r))$ is positive definite, if

$$\alpha_6 > ||d||_\infty + c_{61}\alpha_1 \tag{15}$$

holds true to guarantee that χ_6 and χ_{61} are well defined. Function χ_6 as defined in (14) is $\in K$, but we can find a continuation of χ_6 such that the composed function is K_∞. Hence V_6 satisfies condition (11) and from Sect. 3 we know that subsystem six has the LISS property for all $x_6^0 \in \mathbb{R}_+$ and $||d||_\infty < \alpha_6 - c_{61}\alpha_1 =: \rho''$.

With $\exp(-r) < 1, r > 0 \Leftrightarrow (1-a)\exp(-r) < (1-a), 0 < a < 1 \Leftrightarrow \exp(-r) < 1 - a + a\exp(-r) \Leftrightarrow -\ln(1 - a + a\exp(-r)) < r$ it follows

$$\chi_{12} \circ \chi_{24} \circ \chi_{46} \circ \chi_{61}(r)$$
$$= -\ln\left(1 - \frac{c_{12}\alpha_2 + c_{13}\alpha_3}{(1-\varepsilon_{12})\alpha_1}\frac{c_{24}\alpha_4 + c_{25}\alpha_5}{(1-\varepsilon_{24})\alpha_2}\frac{c_{46}\alpha_6}{(1-\varepsilon_{46})\alpha_4}\frac{||d||_\infty + c_{61}\alpha_1}{(1-\varepsilon_{61})\alpha_6}(1 - \exp(-r))\right)$$
$$< r, \quad r > 0.$$

By similar calculations the following holds

$$\begin{array}{ll}
\chi_{12} \circ \chi_{24} \circ \chi_{46} \circ \chi_{61}(r) < r, & \chi_{13} \circ \chi_{34} \circ \chi_{46} \circ \chi_{61}(r) < r, \\
\chi_{12} \circ \chi_{25} \circ \chi_{56} \circ \chi_{61}(r) < r, & \chi_{13} \circ \chi_{35} \circ \chi_{56} \circ \chi_{61}(r) < r,
\end{array} \tag{16}$$

for $r > 0$, such that the cycle condition and therefor the small gain condition is satisfied. We conclude that all subsystems are LISS or ISS, respectively, and we can apply Theorem 1, such that the whole system is LISS for all $x, x_0 \in \mathbb{R}_+^6$ and $||d||_\infty < \rho''$ with additional conditions (12), (13) and (15).

5 Simulation Results

To verify and demonstrate the results of the previous section we simulate all subsystems with help of Matlab.

At first we choose values for the parameters $c_{ij} : c_{61} = 0.0001, c_{12} = 4, c_{13} = 3, c_{24} = 4, c_{34} = 9, c_{25} = 6, c_{35} = 2, c_{46} = 8, c_{56} = 4$. Consider constant orders $d \equiv 20$. Then the stability conditions (SC) (12), (13) and (15) become

$$\begin{array}{lll}
\alpha_1 > 4\alpha_2 + 3\alpha_3, & \alpha_2 > 4\alpha_4 + 6\alpha_5, & \alpha_3 > 9\alpha_4 + 2\alpha_5, \\
\alpha_4 > 8\alpha_6, & \alpha_5 > 4\alpha_6, & \alpha_6 > 20 + 0.0001\alpha_1.
\end{array}$$

By solving this system of linear inequalities we get the condition

$$\alpha := (\alpha_1, \alpha_2, \alpha_3, \alpha_4, \alpha_5, \alpha_6)^T > (9731.55, 1174.5, 1677.86, 167.79, 83.9, 20.98)^T.$$

With the choice $\alpha = (9750, 1180, 1680, 169, 85, 21)^T$ and $x_0 = (1, 1, 1, 1, 1, 1)^T$ the simulation results are presented in Fig. 2, where the number of orders (No^i) of each subsystem for time t is displayed. We see, that all trajectories of the subsystems are bounded.

Fig. 2 No^i, if (SC) are satisfied

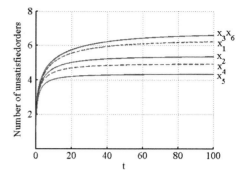

Fig. 3 No^i, if (SC) are not satisfied

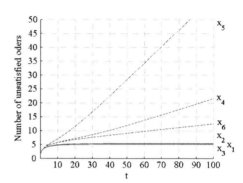

Now we choose the maximum production rates only a bit smaller:

$$\alpha = (9730, 1174, 1677, 167, 83, 20.9)^T.$$

The simulation results are displayed in Fig. 3. We see that the trajectories of the subsystems 1 to 3 are bounded, but the trajectories of the subsystems 4 to 6 are unbounded, which means that the whole system is not stable.

By further simulations of the system we discover that for other inputs where $\|d\|_\infty < \rho^u$ is not satisfied, the system can be stable. We consider all values c_{ij} as before, choose the maximum production rates $\alpha = (9750, 1180, 1680, 169, 85, 21)^T$ such that conditions (12) and (13) are satisfied and replace d by $d(t) = 20(\sin(t) + 1)$. It is $\|d\|_\infty = 40 > \rho^u$, but by simulation results, which are presented in Figs. 4 and 5, all subsystems and therefor the whole system are stable.

This result is caused by the usage of the "worst case" within the mathematical theory, namely the supremum norm $\|\cdot\|_\infty$. In particular for oscillating inputs (e.g. seasonal changes of demand) the maximum value is used for all the time to derive stability conditions, such that lower inputs will not be considered over the time. Whereas in the Matlab simulation the actual input for time t is used, which is not the maximum value for all the time for an oscillating input and therefore lower stability conditions can be obtained. By mathematical theory used in this paper it is not possible to cover all inputs for which the system is stable, in particular

Fig. 4 Simulation results for x_1 to x_5 with $d(t) = 20(\sin(t) + 1)$

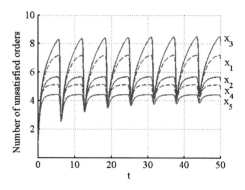

Fig. 5 Simulation results for x_6 with $d(t) = 20(\sin(t) + 1)$

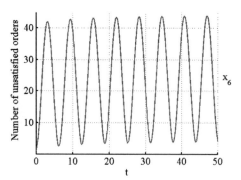

oscillating inputs. This is an actual mathematical problem to find the domain of stability as large as possible.

6 Conclusions and Outline

In this paper, we have described a model for networks of autonomous production plants. This model was investigated on stability. In particular necessary conditions for its stable behavior were provided. This paper illustrates an approach for modelling and analysis of autonomous logistic systems, which can be transferred to other more complex logistical networks equivalently. By application of the stability analysis as presented here one can derive stability conditions to guarantee stability of the network and they help to design the network to avoid negative outcomes and to achieve economic goals.

For validation of the provided methods a comparison of the obtained results with simulations provided by discrete event simulation is of interest and is planned for the future research.

Acknowledgments This research is funded by the German Research Foundation (DFG) as part of the Collaborative Research Centre 637 "Autonomous Cooperating Logistic Processes: A Paradigm Shift and its Limitations" (SFB 637).

References

Dashkovskiy, Sergey; Rüffer, Björn S.; Wirth, Fabian R.; *An ISS small gain theorem for general networks*, Math. Control Signals Systems 19 (2007), no. 2, pp 93–122.

Dashkovskiy, Sergey; Rüffer, Björn S.; Wirth, Fabian R.; *On the construction of ISS Lyapunov functions for networks of ISS systems*, Proceedings of the 17th Int. Symposium on Mathematical Theory of Networks and Systems, Kyoto, Japan, July 24–28, 2006, pp 77–82.

Dashkovskiy, Sergey; Rüffer, Björn S.; Wirth, Fabian R.; *Numerical verification of local input-to-state stability for large networks*, Proceedings of the 46th IEEE Conference on Decision and Control, New Orleans, LA, USA, December 12–14, 2007, pp 4471–4476.

Dashkovskiy, Sergey; Rüffer, Björn S.; Wirth, Fabian R.; *Small gain theorems for large scale systems and construction of ISS Lyapunov functions*, SIAM J. Control Optim. 48 (2010), no. 6, pp 4089–4118. http://arxiv.org/pdf/0901.1842.

Jiang, Z.-P.; Teel, A. R.; Praly, L.; *Small-gain theorem for ISS systems and applications*, Math. Control Signals Systems 7 (1994), no. 2, pp 95–120.

Jiang, Z.-P.; Mareels, I.M.Y.; Wang, Y.; A Lyapunov Formulation of Nonlinear Small Gain Theorem for Interconnected ISS Systems, Automatica 32 (1996), no. 9, pp 1211–1215.

Rüffer, B. S.; *Monotone dynamical systems, graphs, and stability of large-scale interconnected systems*, PhD Thesis, Universität Bremen, Germany, 2007.

Sontag, Eduardo D.; Smooth stabilization implies coprime factorization, IEEE Trans. Automat. Control. 34 (1989) no. 4, pp 435–443.

An Approach to Model Reduction of Logistic Networks Based on Ranking

Bernd Scholz-Reiter, Fabian Wirth, Sergey Dashkovskiy,
Michael Kosmykov, Thomas Makuschewitz and Michael Schönlein

1 Introduction

Performance and competitiveness of a large-scale logistic network depend on the capability of the network to meet the expectations of the customers (Vahrenkamp 2007). This capability is strongly connected to an effective management of the material flow within the network. The material flow is subject to the complex and often global structure of the network as well as to the dynamics of production and transportation processes (Scholz-Reiter et al. 2008). In order to support the management, a better understanding of the dynamics related to the material flow and their consequences for the performance of the logistic network is required. In the literature several methods exist to analyse the material flow.

Three different methods can be utilized for the investigation of the material flow of a logistic network (Tutsch 2006). First, the material flow of the real-world network can be measured. Second, simulations can be carried out in order to analyse changes in the structure and dynamics of the logistic network for different scenarios. Third, mathematical methods can be applied in order to obtain a more precise understanding of the involved processes. Both, simulation and mathematical methods require usually a model of the real-world logistic network.

Model development faces two major challenges. First, the model should exhibit almost the same properties compared to the real-world logistic network and

B. Scholz-Reiter and T. Makuschewitz
BIBA-Bremer Institut für Produktion und Logistik GmbH, University of Bremen,
Bremen, Germany
e-mail: bsr@biba.uni-bremen.de

F. Wirth and M. Schönlein
Institute of Mathematics, University of Würzburg, Wurzburg, Germany

S. Dashkovskiy and M. Kosmykov (✉)
Centre of Industrial Mathematics, University of Bremen, Bremen, Germany
e-mail: kosmykov@math.uni-bremen.de

H.-J. Kreowski et al., *Dynamics in Logistics*,
DOI: 10.1007/978-3-642-11996-5_9, © Springer-Verlag Berlin Heidelberg 2011

second, it should be tailored to the applied methods in order to enable consolidated findings. The size of a model is often crucial for a successful application of a certain analysis method. A model of lower size facilitates simulations or the application of mathematical methods. Since, logistic networks often consist of a large number of locations and transportation connections between them a representative model of lower size is desired.

Our approach to model reduction is based on a ranking scheme of the locations that takes the material flows within the network and the structure of the connections between the locations into account. For this purpose the PageRank algorithm (Page et al. 1999), which has been a core component of Google Internet search engine in its early days, is used. The original ranking algorithm is extended by the results of a material flow analysis. A material flow analysis provides valuable information about the importance of the connections between the locations by analysing the quantities of material flow between the locations (Arnold and Furmans 2007). These quantities can be incorporated into the ranking algorithm in order to enhance the ranking. The adapted ranking algorithm provides in terms of its application to a logistic network a reasonable ranking (Scholz-Reiter et al. 2009; Langville and Meyer 2006). In order to derive a model of lower size we propose to focus on locations with a low importance for the network. According to their connections to other locations of the network we investigate three different approaches for model reduction. These approaches involve the exclusion and aggregation of individual locations as well as the exclusion of subparts of the network. The paper shows that by applying these approaches for model reduction representative models can be derived.

The outline of the paper is as follows. In Sect. 2 our proposed adaptation of the PageRank algorithm to logistic networks is presented. Three different approaches to model reduction based on the structural properties of the locations are introduced in Sect. 3 and illustrated by examples. Section 4 summarise the findings of this paper and provides an outlook to future research.

2 Adaptation of the PageRank Algorithm to Logistic Networks

Before we introduce a notion of importance rank of a logistic location in a network we describe the network itself as a model.

We model a logistic network as a directed graph in the following way. Let the logistic locations be numbered by $1, \ldots, n$ and each of them be a node of this graph. There is an edge from node i to node j if there is a material, information or monetary flow from the ith to jth location. In our approach only the aggregated quantity of a material flow between the locations over a certain period of time is considered. Let a_{ij} be a number quantifying this flow. In particular, if there is no flow from location i to location j then $a_{ij} = 0$. The matrix $\mathbf{A} = (a_{ij})_{i,j=1,\ldots,n}$

Fig. 1 Weighted directed
graph of a logistic network.
The numbers in the circles are
the numbers of nodes and the
numbers near the edges are
their weights

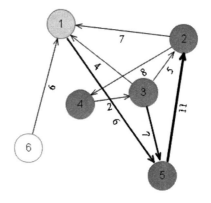

describes the interconnection structure and the flows of a given network and is
called weighted adjacency matrix of the graph. For example the network in Fig. 1
has the following weighted adjacency matrix:

$$A = \begin{pmatrix} 0 & 0 & 0 & 0 & 6 & 0 \\ 7 & 0 & 0 & 8 & 0 & 0 \\ 4 & 5 & 0 & 0 & 7 & 0 \\ 0 & 0 & 2 & 0 & 0 & 0 \\ 0 & 11 & 0 & 0 & 0 & 0 \\ 6 & 0 & 0 & 0 & 0 & 0 \end{pmatrix}$$

The following types of matrices will be used in this paper:
A matrix $\mathbf{A}$ is called *column-normalized* if for all $i = 1,\ldots, n$

$$\sum_{j=1}^{n} a_{ij} = \begin{cases} 1, & \text{if there exists j such that } a_{ij} \neq 0 \\ 0, & \text{if for all j } a_{ij} = 0. \end{cases}$$

A matrix $\mathbf{A}$ is *column-stochastic* if $\sum_{j=1}^{n} a_{ij} = 1$ for all $i = 1,\ldots, n$.

It is called *primitive* if there exists a positive integer k such that the matrix $\mathbf{A}^k$
has only positive elements.

We call an adjacency matrix $\mathbf{A}$ *irreducible* if the corresponding graph is
strongly connected, i.e., for every nodes i and j of the graph there exists a sequence
of directed edges connecting i to j. Note that any primitive matrix is irreducible.

Now we are ready to introduce the notion of importance of logistic locations in
a network. In the sequel we call it rank of a node or location. We say that the rank
of a certain location depends on the network structure, its position within the
network and flows in this network. As the weighted adjacency matrix of a network
contains the information about its structure and flows between its nodes we use it
to define the importance of the nodes. To define the rank of logistic locations we
use the idea of the PageRank (Page et al. 1999), which was originally designed for
the ranking web pages in the Internet. This idea and the algorithm for its calculation can be adapted in the following way to logistic networks. We say that the

rank of a location i depends on the flows a_{ij}, from this node to other locations $j = 1,\ldots, n$ as well as ranks of these locations The more important locations receive material from a given location the more important it is. In comparison with the original PageRank locations do not share their rank equally between their suppliers but rather proportionally to the flows from their suppliers described by a_{ij}. I.e., the proportions $\tilde{a}_{ij}$ are $\tilde{a}_{ij} = \frac{a_{ij}}{\sum_{k \in r_j^d} a_{kj}}$. Such proportions were proposed in Baeza-Yates and Davis (2004) as a modification of PageRank. Thus the rank NR_i of the node i should be calculated by

$$NR_i = \sum_{j=1}^{n} \tilde{a}_{ij} NR_j, \quad i = 1,\ldots,n. \tag{1}$$

However we introduce a parameter α, $0 < \alpha \leq 1$ and for a given graph with n nodes we define the rank NR_i of a location i by

$$NR_i = \alpha \sum_{j=1}^{n} \tilde{a}_{ij} NR_j + (1 - \alpha) \frac{1}{n}, \quad i = 1,\ldots,n, \tag{2}$$

i.e., the rank of location i is the sum of the proportion of ranks that locations j contribute and a small positive term, where n is usually a large number. Note that for $\alpha = 1$ Eq. 2 coincides with (1). Thus the value $1 - \alpha$, is usually taken close to zero in order to preserve the information about the real structure of the network. It is interpreted as a probability that a location supplies material to locations with which it has no direct link (partnership). This positive term is important for the existence and uniqueness of a solution of (2).

Now we see that the ranks NR_i can be calculated by solving the linear system of algebraic equations (2). However this problem can be nontrivial in the case of large number of nodes n. To solve Eq. 2 we transform it into an eigenvalue problem that can be solved numerically (Langville and Meyer 2006). The needed transformation steps are:

1. Column-normalization:

$$H = (H_{ij}), H_{ij} = \frac{a_{ij}}{\sum_{k=1}^{n} a_{kj}} = \tilde{a}_{ij}, \quad \text{if} \quad \sum_{k=1}^{n} a_{kj}^T \neq 0 \quad \text{and} \quad H_{ij} = 0, \text{ otherwise.} \tag{3}$$

2. Make the matrix stochastic:

$$S = H + \frac{eb^T}{n}, \quad e = (1,\ldots,1)^T \in R^n, \quad b_i = 1, \quad \text{if} \sum_{j=1}^{n} H_{ji} = 0 \quad \text{and} \tag{4}$$
$$b_i = 0, \text{ otherwise.}$$

Table 1 Ranks of the
original network in Fig. 1

Location	Rank
1	0.1040 (5)
2	0.2455 (1)
3	0.2016 (3)
4	0.2056 (2)
5	0.1778 (4)
6	0.0655 (6)

3. Making the matrix primitive:

$$G = \alpha S + (1 - \alpha)E, \quad E = \frac{ee^T}{n} \tag{5}$$

Let $p = (p_1, \ldots, p_n)^T$ be a normalized and nonnegative vector. Multiplying the right side of matrix G by vector p we obtain the right-hand side of Eq. 2 with $NR_j = p_j$ and the problem (2) is equivalent to the problem of finding the right-eigenvector $p = Gp$ of G that corresponds to the eigenvalue 1. Since the matrix G is stochastic and irreducible the Perron-Frobenius theorem guarantees that 1 is an eigenvalue of G and all other eigenvalues have absolute value less then 1, i.e., 1 is the spectral radius of G (Meyer 2000). Since G is primitive and hence irreducible from the Perron-Frobenius theorem in Meyer (2000) it follows that the corresponding eigenvector is unique up to a scalar factor.

Applying our method to the network given in Fig. 1 with $\alpha = 0.9$ we obtain the following ranks which are given in Table 1. In parentheses we number the locations by their importance.

The group of the most important locations is highlighted in dark grey. Light grey is used for the group of locations with average importance and white is used for the background of the group of the least important locations. We see that in this example the most important location is location 2. This can be explained by the reason that it delivers large amounts of material to other locations. Location 6 delivers material to only one location with an average rank and therefore has the lowest rank and hence it is the least important location.

3 Approaches for Network Model Reduction

For the analysis of large logistic networks the reduction of the size of their model is often needed. Our main idea for reduction of a model is to exclude or aggregate the nodes of the lowest ranks, i.e., the less important locations. We propose three different rules how to do this. These rules are chosen in a way to conserve the main structure of the network. In the following we describe these rules and show their implementation on some simple examples. In these examples we will observe the changes of the ranks and especially of the order of the nodes by their ranks.

3.1 Exclusion of Low-Ranked Locations Connected to Only One Location

Locations with low rank connected to only one location do not describe the major structure of the network. Thus such locations are excluded from the graph as well as their links.

Example 1 Consider the location 6 in Fig. 1 that is the least important location and is connected only to the location 1. Applying the given rule it is excluded from the graph. The reduced model is shown in Fig. 2. The ranks of locations of the reduced graph are given in Table 2. Note that the ranks of the locations in the reduced model have the same order as in the original network.

3.2 Aggregation of Low-Ranked Locations

Locations with low ranks appearing in a parallel or sequential connection in the network are aggregated and considered as one location. The weights of incoming and outgoing links of the aggregated location are defined as the sums of the corresponding weights of the former individual links. In this case the information about the original material flow through these locations is kept in the reduced network. In Altman and Tennenholtz (2005) a theoretical analysis of similar rules is presented but it was applied to the original PageRank algorithm without weighted edges of the considered graph.

Fig. 2 Reduced network

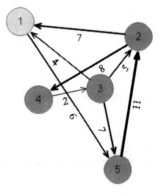

Table 2 Ranks of the
reduced network

Location	Rank
1	0.1039 (5)
2	0.2707 (1)
3	0.2202 (2)
4	0.2171 (3)
5	0.1882 (4)

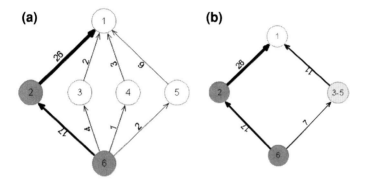

Fig. 3 a Original network. **b** Reduced network

To illustrate this rule consider the following two examples.

Example 2 (parallel connection) Figure 3a illustrates a logistic network, which consists of only one customer (location 1), who orders products from four OEM's (locations 2–5). The OEM's are connected to one supplier (location 6).

Locations 3, 4 and 5 have low ranks, see Table 3a, and are connected in a parallel way to the same locations 1 and 6. Hence, they are aggregated to one location called '3–5' and a reduced model with 4 locations is obtained, see Fig. 3b. The new incoming link to '3–5' corresponds to the sum of the former three individual incoming links. The same applies to the outgoing links. Note that the

Table 3a Ranks of the
original network

(a)

Location	Rank
1	0.0911 (6)
2	0.1456 (2)
3	0.0953 (5)
4	0.0974 (4)
5	0.1037 (3)
6	0.4668 (1)

Table 3b Ranks of the
reduced network

Location	Rank
1	0.1375 (4)
2	0.2196 (2)
3-5	0.1723 (3)
6	0.4706 (1)

Fig. 4 a Original network.
b Reduced network

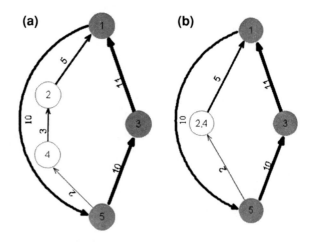

Table 4a Ranks of the
original network

Location	Rank
1	0.2828 (2)
2	0.1051 (5)
3	0.1952 (3)
4	0.1194 (4)
5	0.2975 (1)

Table 4b Ranks of the
reduced network

Location	Rank
1	0.3202 (2)
2,4	0.1226 (4)
3	0.2247 (3)
5	0.3326 (1)

Table 5 Ranks of the
original network

Location	Rank	Location	Rank
1	0.2946 (01)	10	0.0319 (04)
2	0.2189 (03)	11	0.0165 (11)
3	0.2525 (02)	12	0.0088 (15)
4	0.0123 (14)	13	0.0088 (15)
5	0.0135 (13)	14	0.0162 (12)
6	0.0188 (08)	15	0.0169 (09)
7	0.0088 (15)	16	0.0231 (05)
8	0.0190 (07)	17	0.0168 (10)
9	0.0226 (06)	-	-

order of the locations by their ranks in the reduced model remains unchanged, cf. the Tables 3a, 3b.

Example 3 (sequential connection) In the network in Fig. 4a the locations 2 and 4 have low rank (see Table 4a) and they are connected sequentially. Thus they are aggregated to one location that is denoted by '2, 4'. The weight of the incoming link is the weight of the former incoming link of location 4 and the weight of the outgoing link is the weight of the former outgoing link of location 2. Table 4b shows that the ranks in the reduced graph have the same order as in the original graph.

We see that in the last three examples the rules of model reduction do not change the order of importance between the unchanged locations.

3.3 Exclusion of Sub-networks of Low-ranked Locations Connected to Only One Important Location

A sub-network of locations that have low ranks is excluded from the graph if it is connected to one important location only. The material flows between the remaining locations are kept unchanged and therefore the main structure of the original network is preserved.

We illustrate this rule by the following example.

Example 4 The logistic network in Fig. 5 contains three locations with high ranks, see Table 5. There are three groups of low rank locations. Each of them has connections to one of the high rank nodes only. By the third rule all three groups are excluded to obtain the reduced graph, see Fig. 6, which consists of 3 locations. The order of the three most important locations is again unchanged, cf. Table 6.

Note that each of the rules defined above can be applied simultaneously to different nodes. Thus the network can be significantly reduced to the desired size. The next example illustrates the application of all three rules to one network.

Fig. 5 Network with 3 sub-networks

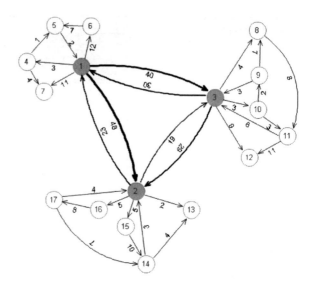

Fig. 6 Reduced network

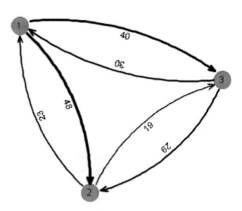

Table 6 Ranks of the reduced network

Location	Rank
1	0.3890 (1)
2	0.2832 (3)
3	0.3278 (2)

Example 5 The graph in Fig. 7 consists of 33 locations. Their ranks are shown in Table 7. The most important locations are locations 2, 5 and 14. Locations 1, 3, 6, 8, 27, 28 will be treated as average rank nodes. The rest of the nodes are identified as low rank locations.

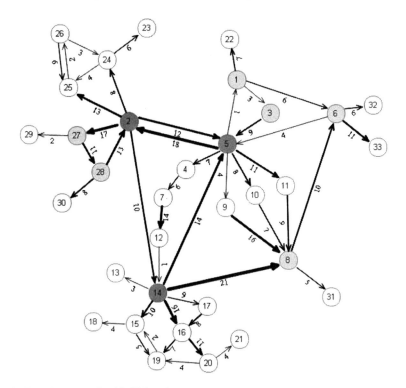

Fig. 7 Complex network with 33 locations

Locations 22, 29, 30, 31, 32, 33 are deleted by applying the rule from Sect. 3.1. The locations 4, 7, 12 are aggregated using the rule of Sect. 3.2 for sequential connection and locations 9–11 are aggregated using the rule for parallel connections from Sect. 3.2. Sub-network of nodes 13, 15–21 and sub-network of nodes 23–26 are excluded according to the rule from Sect. 3.3. The reduced model consists of 11 locations and is shown in Fig. 8. The corresponding ranks of the locations are given in Table 8. Locations 2 and 5 are still the most important ones. However the rank of location 14 is of the order of the average rank nodes.

4 Conclusions and Future Research

We have proposed three rules for the reduction of the size of the model of a given large-scale logistic network. For this purpose a method for identifying the nodes to be excluded or aggregated was developed that takes the structure of the network and material flows between its locations into account. This method extends the idea of the well known PageRank algorithm. Examples given in this paper illustrate how this reduction rules can be applied to obtain reasonable models of lower size to describe large logistic networks.

Table 7 Ranks of the
original network

Location	Rank	Location	Rank
1	0.0567 (04)	18	0.0066 (23)
2	0.1847 (02)	19	0.0087 (22)
3	0.0401 (06)	20	0.0140 (17)
4	0.0230 (10)	21	0.0066 (23)
5	0.1969 (01)	22	0.0066 (23)
6	0.0327 (08)	23	0.0066 (23)
7	0.0193 (12)	24	0.0122 (19)
8	0.0296 (09)	25	0.0189 (13)
9	0.0142 (16)	26	0.0145 (15)
10	0.0099 (21)	27	0.0416 (05)
11	0.0066 (23)	28	0.0346 (07)
12	0.0149 (14)	29	0.0066 (23)
13	0.0066 (23)	30	0.0066 (23)
14	0.1077 (03)	31	0.0109 (20)
15	0.0145 (15)	32	0.0066 (23)
16	0.0218 (11)	33	0.0066 (23)
17	0.0128 (18)	-	-

Fig. 8 Reduced network
with 11 locations

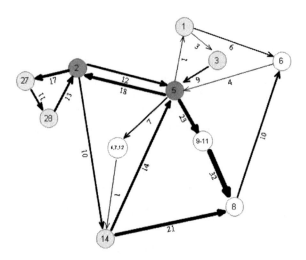

The observed numerical results raise theoretical questions related to the stability of the order of ranks with respect to the application of the proposed reduction rules. These questions have to be investigated analytically and are a matter of the further research. The development of further reduction rules and analysis of their properties are also planed for the future work.

Table 8 Ranks of the
reduced network

Location	Rank	Location	Rank
1	0.0762 (06)	14	0.0975 (03)
2	0.2262 (02)	8	0.0320 (09)
3	0.0607 (07)	9-11	0.0301 (10)
4,7,12	0.0212 (11)	27	0.0937 (05)
5	0.2336 (01)	28	0.0943 (04)
6	0.0345 (08)	-	-

Acknowledgment B. Scholz-Reiter, F. Wirth, M. Kosmykov, T. Makuschewitz and M. Schönlein are supported by the Volkswagen Foundation (Project Nr. I/82684 "Dynamic Large-Scale Logistics Networks"). S. Dashkovskiy is partially supported by the DFG as a part of Collaborative Research Center 637 "Autonomous Cooperating Logistic Processes—A Paradigm Shift and its Limitations".

References

Altman, A., Tennenholtz, M. (2005): Ranking systems: the PageRank axioms. Proceedings of the 6th ACM conference on Electronic commerce, Vancouver, BC, Canada, 1–8.

Arnold, D., Furmans, K. (2007): Materialfluss in Logistiksystemen. 5. Auflage, Springer, Berlin.

Baeza-Yates, R., Davis, E. (2004): Web Page Ranking Using Link Attributes. Archive Proceedings of the 13th international World Wide Web conference, Alternate Track Papers & Posters, New York, NY, USA, 328–329.

Langville, A. N., Meyer, C. D. (2006): Google's PageRank and Beyond: The Science of Search Engine Rankings. Princeton University Press, Princeton, NJ.

Meyer, C. D. (2000): Matrix Analysis and Applied Linear Algebra. SIAM, Philadelphia.

Page, L., Brin, S., Motwani, R., Winograd, T. (1999): The PageRank citation ranking: Bringing order to the web. Technical report, Stanford University.

Scholz-Reiter, B., Wirth, F., Dashkovskiy, S., Jagalski, T., Makuschewitz, T. (2008): Analyse der Dynamik großskaliger Netzwerke in der Logistik. In: Industrie Management, 3, 37–40.

Scholz-Reiter, B., Wirth, F., Dashkovskiy, S., Makuschewitz, T., Kosmykov, M., Schönlein, M. (2009): Application of the PageRank algorithm for ranking locations of a production network. Proceedings of 42nd CIRP Conference on Manufacturing Systems.

Tutsch, D. (2006): Performance Analysis of Network Architectures. Springer, Berlin, Heidelberg, New York.

Vahrenkamp, R. (2007): Logistik: Management und Strategien. 5. Auflage, Oldenbourg, München.

Optimization of Spare Parts Lot Size for Supply of Equipment's Park

Iryna Morozova, Mykhaylo Postan and Lyudmyla Shyryaeva

1 Description of Manufacturing System

It is well-known that redundancy of unreliable equipment or its parts is an effective way of operational reliability of production process increasing. One of the main tasks of so-called engineering logistics is supply the equipment's park by spare parts. In particular, this includes the determination of optimal lot size of spare parts. Not also that organization of optimal supply of spare parts is an important element of Materials Requirements Planning conception or Logistics Requirement Planning system (Orlisky 1975; Ballou 1993). From the scientific point of view, this problem can be solved with the help of queueing, inventory control, and mathematical reliability theories (Prabhu 1998; Rolsky et al. 1998; Gnedenko and Kovalenko 2005). During the last decades for mathematical modeling the manufacturing systems functioning under uncertainty the apparatus of so-called Markovian drift processes (with sign-variable drift) is widely used (Gnedenko and Kovalenko 2005; Postan 2006, 2008).

The aim of our paper is modeling, by means of Markov drift processes, the manufacturing system "equipment's park (production line)—warehouse for final product" and formulation of optimization problem concerning the determination of optimal lot size o spare parts for renewal of unreliable units.

Consider a manufacturing system which includes the production line without intermediate buffers, consisting of N units (machines, equipment), and warehouse for final product storage. Production rate of line is W (it is assumed that arbitrary unit has the same production rate).With the rate W a final product comes to warehouse with unlimited capacity (i.e. we ignore the possibility to be filled up for the warehouse). From the warehouse product removes with the constant rate

I. Morozova, M. Postan (✉) and L. Shyryaeva
Odessa National Maritime University, 34 Mechnikov Str., 65029, Odessa, Ukraine
e-mail: postan@ukr.net

H.-J. Kreowski et al. (eds.), *Dynamics in Logistics*,
DOI: 10.1007/978-3-642-11996-5_10, © Springer-Verlag Berlin Heidelberg 2011

$U < W$ (if it isn't empty). In other words, the demand is fixed and given by the value U.

Any unit may be subjected to breakdowns and renewals. It is assumed that in small interval $(t, t + \Delta t)$ with probability $\lambda_i \Delta t + o(\Delta t)$ the i-th unit fails independently from others units and with probability $1 - \lambda_i t + o(\Delta t)$ a failure doesn't occur. Immediately after failure of i-th unit its repair occurs by substitution of failed spare part (nonrenewable) with a new one taken from the warehouse for reserved spare parts (the repair time is ignored). The inventory level of spare parts is governed by the following rule:

(a) an order point coincides with the moment of time when the last spare part is taken from warehouse;
(b) a lot size of a new party of spare parts is fixed and equals to n.

The time of order execution is random variable with distribution function $A(t)$.

Thus if arbitrary unit fails and there are no spare parts at warehouse at this moment, then production line is stopped until a new party of spare parts will arrive. During this period another units don't fail.

2 Construction of Mathematical Model and Its Analysis

For mathematical modelling many manufacturing systems functioning under uncertainty it is very convenient to apply a special subclass of Markovian processes, namely, so-called Markov drift processes (with sign-variable drift) (Poston 2006, 2008).

Below, we shall assume that

$$A(t) = 1 - e^{-at} \sum_{i=0}^{r-1} \frac{(at)^i}{i!},$$

i.e. executive time has the Erlang distribution of r-th order. Note that Erlang distribution is the simplest particular case of so-called phase type distributions (Neuts 1994). It is well-known that this distribution describes a big variety of random flows: from Poisson to regular.

Let us introduce the following designations:

$v(t)$ is the amount of reserved spare parts at warehouse at moment t;
$\alpha(t)$ is the current number of phase of Erlang distribution at moment t;
$\xi(t)$ is the amount of final product at warehouse at moment t.

The process $v(t)$ takes the values from the set $\{-1, 0, 1, ..., n\}$, where value -1 means that production line is failed and there is no one spare part in reserve. The possible values of process $\alpha(t)$ belongs to the set $\{0, 1, ..., r - 1\}$. The process $\xi(t)$ is the random walk on semi-interval $[0, \infty)$ with sticky bound in 0.

Consider the stochastic process $(Z(t), \xi(t))$, where

$$Z(t) = \begin{cases} v(t) & \text{if } v(t) = 1, 2, \ldots, n, \\ (v(t), \alpha(t)) & \text{if } v(t) = -1 \text{ or } 0. \end{cases}$$

The phase space of process $(Z(t), \xi(t))$ is $D \times [0, \infty)$, where

$$D = \{k : k = 1, 2, \ldots, n\} \cup \{(k, i) : k = -1, 0; \ i = 0, 1, \ldots, r - 1\}.$$

So defined it will be the Markov drift process for which:

(a)

$$D^- = \{(-1, i) : i = 0, 1, \ldots, r - 1\}, \ D^+ = D/D^-, \ D = D^- \cup D^+;$$

(b) process $\xi(t)$ increases with the velocity $V = W - U$ when $Z(t) \in D^+$ and decreases with the velocity U when $Z(t) \in D^-$.

Denote

$$\Pr\{v(t) = k, x < \xi(t) < x + dx\} = q_k(x, t)dx, \ k = 1, 2, \ldots, n; x \geq 0,$$

$$\Pr\{v(t) = k, \alpha(t) = i, x < \xi(t) < x + dx\} = q_{ki}(x, t)dx, \ k = -1, 0; i = 0, 1, \ldots, r - 1; x \geq 0,$$

$$\Pr\{v(t) = -1, \alpha(t) = i, \xi(t) = 0\} = p_i^-(t), \ i = 0, 1, \ldots, r - 1.$$

Suppose that there exists the limit probabilistic distribution of process $(Z(t), \xi(t))$

$$q_k(x) = \lim_{t \to \infty} q_k(x, t), \quad k = 1, 2, \ldots, n; \quad x \geq 0,$$

$$q_{ki}(x) = \lim_{t \to \infty} q_{ki}(x, t), \quad k = -1, 0; \quad i = 0, 1, \ldots, r - 1; \quad x \geq 0,$$

$$p_i^- = \lim_{t \to \infty} p_i^-(t), \quad i = 0, 1, \ldots, r - 1.$$

For this limit distribution finding the following system of ordinary differential equations (ODE) and corresponding boundary conditions are valid:

$$\begin{aligned} -Uq'_{-1,0}(x) &= -aq_{-1,0}(x) + \lambda q_{00}(x), \\ -Uq'_{-1,i}(x) &= -aq_{-1,i}(x) + \lambda q_{0i}(x) + aq_{-1,i-1}(x), \quad i = 1, 2, \ldots, r - 1, \quad (1) \\ Vq'_{00}(x) &= -(\lambda + a)q_{00}(x) + \lambda q_1(x), \end{aligned}$$

$$\begin{aligned} Vq'_{0i}(x) &= -(\lambda + a)q_{0i}(x) + aq_{0,i-1}(x), i = 1, 2, \ldots, r - 1, \\ Vq'_k(x) &= -\lambda q_k(x) + \lambda q_{k+1}(x), k = 1, 2, \ldots, n - 2, \\ Vq'_{n-1}(x) &= -\lambda q_{n-1}(x) + \lambda q_n(x) + aq_{-1,r-1}(x), \\ Vq'_n(x) &= -\lambda q_n(x) + aq_{0,r-1}(x), x > 0, \end{aligned} \quad (2)$$

$$-Uq_{-1,i}(0) = -ap_i^- + ap_i^-, \quad i = 1, 2, \ldots, r-1,$$
$$Uq_{-1,0}(0) = ap_0^-,$$
$$q_{00}(0) = q_{01}(0) = \ldots = q_{0,r-1}(0) = 0, \tag{3}$$
$$q_1(0) = \ldots = q_{n-2}(0) = q_n(0) = 0,$$
$$Vq_{n-1}(0) = ap_{r-1}^-,$$

where

$$\lambda = \sum_{m=1}^{N} \lambda_m.$$

The normalizing condition is given by

$$\int_0^\infty \left[\sum_{k=1}^{n} q_k(x) + \sum_{i=0}^{r-1} (q_{-1,i}(x) + q_{0,i}(x)) \right] dx + \sum_{i=0}^{r-1} p_i^- = 1. \tag{4}$$

Let us denote

$$p_k = \int_0^\infty q_k(x)dx, \quad k = 1, 2, \ldots, n,$$

$$p_{ki} = \delta_{k0}p_i^- + \int_0^\infty q_{ki}(x)dx, \quad k = -1, 0; \quad i = 0, 1, \ldots, r-1, \tag{5}$$

where δ_{k0} is the Kroneker's delta.

Then after integration of both sides of Eqs. (1), (2) from $x = 0$ to $x = \infty$, taking into account the conditions (3), (4) and relations (5) we obtain the following system of Kolmogorov equations for determination of stationary state-probabilities of Markov process $Z(t)$:

$$-ap_{-1,0} + \lambda p_{00} = 0,$$
$$-ap_{-1,i} + \lambda p_{0i} + ap_{-1,i-1} = 0, \quad i = 1, 2, \ldots, r-1,$$
$$-(a + \lambda)p_{00} + \lambda p_1 = 0,$$
$$-(a + \lambda)p_{0i} + ap_{0,i-1} = 0, \quad i = 1, 2, \ldots, r-1,$$
$$-\lambda p_k + \lambda p_{k+1} = 0, \quad k = 1, 2, \ldots, n-2, \tag{6}$$
$$-\lambda p_{n-1} + ap_{-1,r-1} + \lambda p_n = 0,$$
$$-\lambda p_n + ap_{0,r-1} = 0,$$

$$\sum_{i=0}^{r-1} (p_{-1,i} + p_{0i}) + \sum_{k=1}^{n} p_k = 1.$$

The system (6) may easily be solved and its solution is given

$$p_{0i} = \left(\frac{a}{a+\lambda}\right)^i p_{00}, \quad i = 1, 2, \ldots, r-1,$$

$$p_{-1,i} = \left(1+\frac{a}{\lambda}\right)\left[1 - \left(\frac{a}{a+\lambda}\right)^{i+1}\right] p_{00}, \quad i = 0, 1, \ldots, r-1,$$

$$p_1 = p_2 = \cdots = p_{n-1} = p_{00}, \tag{7}$$

$$p_n = \frac{a}{\lambda}\left(\frac{a}{a+\lambda}\right)^{r-1} p_{00},$$

$$p_{00} = \left[r\left(1+\frac{a}{\lambda}\right) + (n-1)\left(1+\frac{a}{\lambda}\right) + \frac{a}{\lambda}\left(\frac{a}{a+\lambda}\right)^{r-1}\right]^{-1}.$$

From (4) and (7), it follows the relation

$$\sum_{i=0}^{r-1} p_i^- = \sum_{i=0}^{r-1} p_{-1,i} - \frac{V}{U}\left(\sum_{i=0}^{r-1} p_{0i} + \sum_{k=0}^{n} p_k\right)$$

or after some calculations (see (7))

$$\sum_{i=0}^{r-1} p_i^- = \left(1+\frac{a}{\lambda}\right)\left[\frac{\lambda}{a}r + \left(\frac{a}{a+\lambda}\right)^r - \frac{nV}{U} - 1\right] p_{00}. \tag{8}$$

Since the right-hand side of equality (8) must be strictly positive, from (8) it follows the necessary condition that manufacturing system under consideration is stable

$$\frac{\lambda}{a}r + \left(\frac{a}{a+\lambda}\right)^r > 1 + \frac{nV}{U}. \tag{9}$$

The solution to the system of Eqs. 1–4 in the terms of Laplace-transform is given by the following formulas:

$$q_{0i}^*(s) = \left(\frac{a}{a+\lambda+Vs}\right)^i q_{00}^*(s), \quad i = 0, 1, \ldots, r-1,$$

$$q_{-1,i}^*(s) = \left(\frac{a}{a-Us}\right)^{i+1}\left[\frac{\lambda}{a}q_{00}^*(s)\sum_{j=0}^{i}\left(\frac{a-Us}{a+\lambda+Vs}\right)^j - p_0^-\right.$$

$$\left. + \sum_{j=1}^{i}(p_{j-1}^- - p_j^-)\left(\frac{a-Us}{a}\right)^j\right], \quad i = 0, 1, \ldots, r-1, \tag{10}$$

$$q_k^*(s) = \left(\frac{\lambda+Vs}{\lambda}\right)^{k-1}\frac{a+\lambda+Vs}{\lambda}q_{00}^*(s), \quad k = 1, 2, \ldots, n-1,$$

$$q_n^*(s) = \frac{a}{\lambda+Vs}\left(\frac{a}{a+\lambda+Vs}\right)^{r-1} q_{00}^*(s),$$

$$q_{00}^*(s) = \frac{a}{a + \lambda + Vs}$$

$$\times \left[\sum_{j=1}^{r-1} \left(p_{j-1}^- - p_j^- \right) \left(\frac{a}{a - Us} \right)^{r-j} + p_{r-1}^- - \left(\frac{a}{a - Us} \right)^r p_0^- \right]$$

$$\times \left\{ \left(\frac{\lambda + Vs}{\lambda} \right)^{n-1} - \frac{\lambda a^r}{\lambda + Ws} \left[\frac{1}{(a - Us)^r} + \frac{Us}{(\lambda + Vs)(a + \lambda + Vs)^r} \right] \right\}^{-1},$$

$$(11)$$

where

$$q_k^*(s) = \int_0^\infty e^{-sx} q_k(x)dx, \quad k = 1, 2, \ldots, n,$$

$$q_{ki}^*(s) = \int_0^\infty q_{ki}(x)dx, \quad k = -1, 0; \ i = 0, 1, \ldots, r-1; \quad \mathrm{Re}\, s \geq 0.$$

The unknown probabilities p_i^-, $i = 0, 1, \ldots, r - 1$, can be found by the condition of analyticalness of function $q_{00}^*(s)$ in domain $\mathrm{Re}\, s \geq 0$, where $|q_{00}^*(s)| < 1$. Consider the equation (see denominator in right-hand side of formula (11))

$$\left(1 - \frac{Us}{a} \right)^r = \varphi(s), \quad \mathrm{Re}\, s \geq 0, \tag{12}$$

where

$$\varphi(s) = \left[\left(\frac{\lambda + Ws}{\lambda} \right) \left(\frac{\lambda + Vs}{\lambda} \right)^{n-1} - \frac{Us}{\lambda + Vs} \left(\frac{a}{a + \lambda + Vs} \right)^r \right]^{-1}.$$

Put $z = 1 - \frac{Us}{a}$, $\mathrm{Re}\, z \leq 1$, and rewrite the Eq. 12

$$z^r - \varphi \left(\frac{a}{U} (1 - z) \right) = 0.$$

One can prove (with the help of Rouche's theorem) that the function in left-hand side of this equation has at least r roots on and inside a bounded domain (taking into account their multiplicity).

Let $z_0 = 1, z_1, \ldots, z_{r-1}$ be these roots. Then from Eq. 11, we obtain the system of equations for determination of probabilities p_i^-, $i = 0, 1, \ldots, r - 1$:

$$\sum_{j=1}^{r-1} \left(p_{j-1}^- - p_j^- \right) z_i^{-r+j} + p_{r-1}^- - p_0^- z_i^{-r} = 0, \ i = 1, 2, \ldots, r - 1.$$

This system and Eq. 8 must be solved jointly.

3 Optimum Lot Size of Spare Parts

One of the main goals in analysis of any inventory/production system is optimization of inventory policy. In our case it means the determination of optimum value of lot size n. The results obtained in the Section 2 allow us to formulate and solve the corresponding optimization problem.

First of all, note that we can calculate some indices of production process efficiency with the help of solution 7–11. Most important among them are:

(a) probability that production line is idle because of absence of spare parts at warehouse and emptiness of warehouse for final product storage, i.e. probability of market's losses (see (8))

$$p^- \equiv \sum_{i=0}^{r-1} p_i^-;$$ (13)

(b) rate of output flow of final product

$$U(1 - p^-);$$ (14)

(c) rate of input flow of spare parts batches

$$a(n) = a \sum_{i=0}^{r-1} \left(p_i^- + q_{-1,i}^*(0) + q_{0i}^*(0) \right) = r(a + \lambda)p_{00};$$ (15)

(d) rate of failures flow

$$\lambda(n) = \lambda \left\{ 1 - \left(\frac{a+\lambda}{\lambda} \right) \left[\frac{\lambda}{a} \left(r - \frac{a}{\lambda} \right) + \left(\frac{a}{a+\lambda} \right)^r \right] p_{00} \right\};$$ (16)

(e) mean inventory level of spare parts at warehouse

$$Ev = \lim_{t \to \infty} Ev(t) = \sum_{k=1}^{n} q_k^*(0)$$

$$= \frac{n}{\lambda} \left[\frac{(n-1)(a+\lambda)}{2} + a \left(\frac{a}{a+\lambda} \right)^{r-1} \right] p_{00};$$ (17)

(f) mean inventory level of final product at warehouse

$$E\xi = \lim_{t \to \infty} E\xi(t)$$

$$= -\frac{d}{ds} \left[\sum_{k=1}^{n} q_k^*(s) + \sum_{i=0}^{r-1} \left(q_{-1,i}^*(s) + q_{0i}^*(s) \right) \right] \Big|_{s=0}$$

$$= \left(1 + \frac{V}{U} \right) \frac{V}{\lambda} \left\{ \frac{n}{\lambda} \left[a - \frac{(n-1)(a+\lambda)}{2} \right] - \left(r + \frac{a}{\lambda} \right) \left(\frac{a}{a+\lambda} \right)^r \right\} p_{00}$$

$$+ \left(1 + \frac{V}{U} \right) \left(1 + \frac{a}{\lambda} \right) \left[n - \left(\frac{a}{a+\lambda} \right)^r \right] \left(-\frac{d}{ds} q_{00}^*(s) \Big|_{s=0} \right),$$ (18)

where

$$
-\frac{d}{ds} q_{00}^*(s)\Bigg|_{s=0} = \frac{\lambda}{U}\Bigg\{ \frac{U}{a+\lambda}\Bigg[\frac{V}{a+\lambda}\sum_{j=0}^{r-1} p_j^- - \frac{U}{a}\sum_{j=0}^{r-1}(r-j)p_j^- \Bigg]
$$

$$
-\Bigg[\frac{(n-1)(n-2)}{2}\left(\frac{V}{\lambda}\right)^2 - \left(\frac{W}{\lambda}\right)^2 + \frac{UW}{\lambda}\left(\frac{r}{a} + \frac{1}{\lambda}\left(\frac{a}{a+\lambda}\right)^r\right).
$$

$$
- UV\frac{a+\lambda(r+1)}{\lambda^2(a+\lambda)} - \left(\frac{U}{a}\right)^2\frac{r(r+1)}{2}\Bigg]\Bigg\}\Bigg[\left(\frac{a}{a+\lambda}\right)^r + \frac{r\lambda}{a} - 1 - \frac{nV}{U}\Bigg]^{-1}.
$$

Consider the following objective function

$$
\bar{C}(n) = \pi U p^- + (c_0 + en)a(n) + c_1\lambda(n) + c_2 Ev + c_3 E\xi, \qquad (19)
$$

where π is market price per unit of final product; e is purchase price per one spare part; c_0 is transportation costs for delivery of one batch of spare parts; c_1 is costs per one repair; c_2 is storage expenses per unit of time for storage of one spare part; c_3 is storage expenses per unit of time for storage of weight unit of final product.

Expression (19) is the total average production expenses per unit of time. Using the formulas (13–18), we can formulate the following optimization problem: it is required to find out the number n minimizing the expression (19) under condition (9). Note that for $r > 1$ the parameter n enters the function (19) through the expression $E\xi$ implicitly (see (18)).

4 Conclusion

The results obtained show that efficiency of logistical management of manufacturing systems may be increased by application of inventory control models including warehousing the nonrenewable spare parts for repair of unreliable units and optimization of their lot size.

The stochastic model considered in our paper describes relatively simple manufacturing system and may be generalized in many different ways. Some of them are pointed out below:

(a) manufacturing system with Markov modulated demand;
(b) taking into account the input flows of raw materials; e.g. model of such flow may be chosen as compound Poisson process (Prabhu 1998);
(c) several types of nonrenewable spare parts (practically it is possible to consider only two or three types because of high dimension of corresponding system of differential equations).

Though the models listed above are sufficiently complicated in mathematical aspect their analysis may be carried out by standard analytical technique including matrix calculus (Neuts 1994).

References

Ballou RH (1993) Business logistics management, 3rd Edn. Prentice-Hall Intl. Inc., New York

Gnedenko BV and Kovalenko IN (2005) An introduction to queueing theory. KomKniga, Moscow (in Russian)

Neuts MF (1994) Matrix-geometric solutions in stochastic models. An algorithmic approach, Dover Publications, Inc., New York

Orlisky J (1975) Materials Requirements Planning. McGraw-Hill, New York

Postan MYa (2006) Economic-mathematical models of multimodal transport. Astroprint, Odessa (in Russian)

Postan MYa (2008) Application of Markov drift processes to logistical systems modelling. Proc. of First Intl. Conf. "Dynamics in Logistics", LDIC 2007, Springer, Bremen: pp. 443–455

Prabhu NU (1998) Stochastic storage processes: queues, insurance risk, dams, and data communications, 2nd Edn. Springer, Berlin-Heidelberg-New York

Rolsky T, Schmidli H, Smidt V, and Teugels J (1998) Stochastic processes for insurance and finance. Wiley, New York

Part II
Routing, Collaboration and Control

Weighted Multiplicative Decision Function for Distributed Routing in Transport Logistics

Bernd-Ludwig Wenning, Henning Rekersbrink,
Andreas Timm-Giel and Carmelita Görg

1 Introduction

Routing in transport logistics is nowadays usually handled as a constrained optimization problem. This optimization problem is solved with the help of heuristic methods such as genetic algorithms, tabu search and others. If the optimization problem is dynamic in nature, e.g. because not all transport orders are known in advance, solutions to it are repeatedly calculated as time progresses, either in fixed time intervals (rolling horizon planning) or on demand. If the level of dynamics is high, these approaches are limited in their reactivity: Due to the time that is required to determine a new global solution, there is a limitation to the frequency of replanning. Here, a distributed approach that does local modifications to the original plan can have advantages.

This is where the paradigm of "Autonomous Cooperating Logistic Processes" (Scholz-Reiter et al. 2004) is targeted at. Within this paradigm, intelligence and decision-making capability is moved from the central dispatcher towards the individual actors in the logistic process, i.e. the vehicles and even the goods. That means they become autonomous in their decisions, and they have to cooperate in order to achieve their goals.

B.-L. Wenning (✉), A. Timm-Giel and C. Görg
Communication Networks, University of Bremen, Bremen, Germany
e-mail: wenn@comnets.uni-bremen.de

A. Timm-Giel
e-mail: atg@comnets.uni-bremen.de

C. Görg
e-mail: cg@comnets.uni-bremen.de

H. Rekersbrink
BIBA-Bremer Institut für Produktion und Logistik GmbH, Bremen, Germany
e-mail: rek@biba.uni-bremen.de

H.-J. Kreowski et al. (eds.), *Dynamics in Logistics*,
DOI: 10.1007/978-3-642-11996-5_11, © Springer-Verlag Berlin Heidelberg 2011

As a framework for the interaction of autonomous logistic entities, the "Distributed Logistic Routing Protocol" (DLRP) has been proposed.

The paper is structured as follows: Sect. 2 introduces the interaction of the entities using the DLRP. The multiplicative parameter aggregation function is presented in Sect. 3, and results of simulations using this function with DLRP are shown and discussed in Sect. 4. The paper ends with conclusions and an outlook to future work in Sect. 5.

2 The DLRP

The Distributed Logistic Routing Protocol (DLRP) (Scholz-Reiter et al. 2006; Wenning et al. 2007) is based on the assumption that the vehicles and the goods in a logistic network are equipped with devices capable of computing and communicating. Thereby, they are able to interact and decide autonomously.

In contrast to classical routing problems such as the Vehicle Routing Problem (VRP) or the Travelling Salesman Problem (TSP), the scenarios where the DLRP is applied are restricted in the connections that are existing between locations (vertices) in the logistic network, i.e. scenarios are not only defined by a set of vertices, but by a graph connecting those. In reality, the vertices may be logistic distribution centers, and the edges the main motorway connections between them.

Vehicles and goods determine their routes by using a route discovery messaging that is similar to source routing methods in ad-hoc communication networks: When a vehicle or a goods item needs a route, it sends out a route request to the nearest vertex, which forwards this request to the neighbor vertices, which in turn do the same. Before forwarding an incoming route request, the vertex adds current context information to the request, including knowledge about other vehicles and goods that have announced to travel on the same route. So by the time the route request reaches the destination vertex, it has collected information about the complete route that it has travelled. Based on the information collected in the route request, the destination vertex sends a reply to the vehicle or good, which then can make a decision. After having made a decision, the chosen route is announced to all vertices that are involved in this route.

Consequently, the vertices play an important role in the DLRP: They act as information brokers. Vehicles and goods announce their intended routes to the vertices, where other vehicles and goods can access them to retrieve information which is relevant for their future planning. This facilitates the mutual interdependence of vehicle and goods routes. For more details about the DLRP, refer to Scholz-Reiter et al. (2006) and Wenning et al. (2007).

The DLRP has shown to be able to achieve competitive results compared to Tabu Search in adapted vehicle routing problems (Scholz-Reiter et al. 2008).

3 The Decision Function

In Scholz-Reiter et al. (2008), the route decisions were based on the shortest path in case of the goods' decisions, and on vehicle utilization in case of the vehicles' decisions. No time constraints or other decision criteria are used there. This is a largely simplified decision strategy. In reality, there are usually multiple criteria that have to be considered to achieve decisions which lead to a good logistic performance.

If multiple criteria are of interest, these criteria have to be combined in a decision system that leads to a unique decision. Several ways of combining are possible, for example sequential use of criteria, fuzzy logic, additive or multiplicative combination. A sequential use of criteria has the disadvantage that the sequence leads to a fixed prioritization of those parameters that are first used in the sequence. Fuzzy logic usually results in a set of fuzzy levels, and to avoid indifferent cases (multiple alternatives on the same fuzzy level) as much as possible, a high granularity of fuzzy levels and a large set of corresponding fuzzy rules are required. An additive or multiplicative combination of criteria does not prioritize criteria, nor does it have a limited granularity of output values. This makes it favorable to use such a way of combining the criteria.

Assuming that each of the criteria should be able to make a route impossible if its value is inacceptable, a multiplicative aggregation is the more practical option. In an aggregation of different criteria, the value ranges of those criteria are usually different. Therefore, they have to be mapped to a common range to avoid that one criterion dominates the decision.

Based on the constraints and assuming k criteria, the function

$$U_j = \prod_{i=1}^{k} \left(f_{s,i}\left(c_{i,j} \right) \right)^{w_i} \tag{1}$$

is defined as the Multi-Criteria Context Decision (MCCD) function for the decision alternative j. In this function, $c_{i,j}$ represents the value of criterion i for alternative j, $f_{s,i}$ is a function that scales the criterion values to a common range and w_i is a weight to adjust the importance of the criterion for the decision. A scaling function instead of a simple scaling factor is used because the criteria may have significantly different characteristics. Especially if the value range of a criterion is unbounded at the lower end, the upper end or both, a scaling factor is not sufficient to map the criterion into a bounded range. The scaling function as well as the weight is specific for the respective criterion. The target value range that the criteria are mapped to by the scaling functions is the interval [0, 1], with 1 being the best and 0 being the worst value.

3.1 Decision Criteria

This generalized decision function is now applied to the specifics of the distributed routing in logistics. For vehicle routing, three criteria are taken into account:

- The revenue the vehicle is expecting, which is based on the goods' offers and the transport costs per km. Revenue per km is used here because they are a better representation of economical efficiency than absolute revenue values. The revenue values can be positive or negative (the latter is the case if the transport costs are higher than the price the goods offer). Negative revenues, however, mean that it is not useful for the vehicle to travel on this route. Therefore, the scaling function has to map negative values to 0. For positive revenues, the scaled value has to approach 1 for increasing revenue. A scaling based on the Error Function (erf) was applied for the positive revenue here.
- The ecological impact. Efficient utilization of a vehicle's cargo space reduces the pollution per tkm. As the ecological impact can consist of various effects, and not all of them are well measurable or even well understood, only the carbon dioxide output is considered here, as this can be easily calculated if the vehicles' fuel consumption is known. Low carbon dioxide output is preferred, while high output should be avoided. Here, a scaling function based on the Error Function Complement (erfc) is used. This function has the center of its slope at the targeted carbon dioxide maximum.
- The reliability. Based on historic data collected during previous transports on a route, it can be estimated whether the expected revenue can really be achieved. This reliability is a probability, and as such, its values are already in the target interval, so no further scaling is required.

For goods routing, there are three criteria as well:

- The route costs. These costs depend on the offers the goods make towards the vehicles, storage costs, transshipment costs and delay fines. The goods' offers are supposed to depend on the available budget and on the urgency. As the costs have a lower limit (0) and an upper limit (the budget), and this range can be scaled to the [0, 1] interval by using a linear scaling.
- The risk of damage. Each transshipment operation implies a risk that the goods may be damaged. Additionally, there is a damage risk related to the transport itself. It is assumed that there is a maximum acceptable risk that should under no circumstances be exceeded. Therefore, the scaling function is set to 0 for all risk values above the maximum acceptable risk. Between "no risk" and the maximum, a linear scaling is used that maps "no risk" to 1 and the maximum acceptable risk to 0.
- The risk of being delayed. This risk may be deduced from knowledge about how long it takes in average to travel on a specific route. This knowledge is based on feedback from previous transports. Based on the historic travel time statistics and the time that is still left for an in-time delivery, a probability of being delayed is calculated. The scaling function that is applied here has to map a low delay risk to 1 and a definite delay to a low, but nonzero value. It has to be nonzero because otherwise, goods that are already certain to be delayed on any route would not get a route any more.

4 Simulations

For simulative evaluation, a scenario was used that is based on a topology that has been first introduced in Wenning et al. (2005). This topology represents 18 cities in Germany and major highway connections between them (see Fig. 1).

25,000 goods are to be transported in this scenario. They are not all known from the beginning, but are generated during runtime at a rate of 23 goods per time unit. Each possible source–destination pair is present among the goods. 12 vehicles with a capacity of 12 goods each and a maximum speed of 100 km per time unit are present in the scenario. The delivery time window is 25 time units for each piece of good (the goods may be delivered anytime between 0 and 25 time units after entering the scenario).

The following results were obtained in simulations where all criteria are equally weighted with a weight of 1. Here, the average vehicle utilization is 0.7827. Figure 2 shows that after a transient phase in the beginning of the simulation, the utilization varies around this average value.

The performance with respect to the goods' deliveries is best represented by the delivery delays. The cumulative distribution of the goods' delivery delays can be seen in Fig. 3. In this figure, the delays are displayed with respect to the goods' due times, i.e. the delivery is late if the delay is greater than 0, otherwise it is on time.

The figure shows that around 70% of the deliveries are on time in this configuration. Note that no optimizations of criteria weights are done here yet. To improve the timeliness, weight variations were done for the criteria influencing the goods' route decisions. There are two criteria that are related to the timeliness: The delay risk and the costs.

Figure 4 shows the CDFs for different weights on the delay risk criterion, while all other criteria are weighted with 1.

As it can be seen from the figure, varying the weight for the delay risk does not show much influence on the logistic performance. While this discovery seems surprising, it can be explained because the risk only covers the question if, and not how much the delivery will be delayed. The cost, on the other hand, increases with

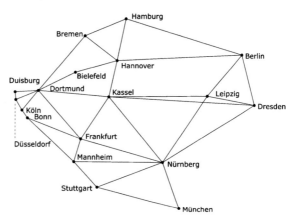

Fig. 1 Logistic scenario topology (Wenning et al. 2005)

Fig. 2 Vehicle utilization

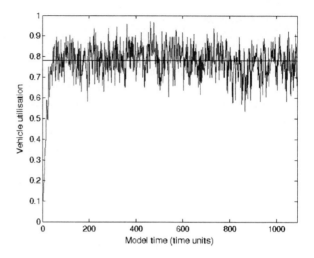

Fig. 3 CDF of the delivery delay

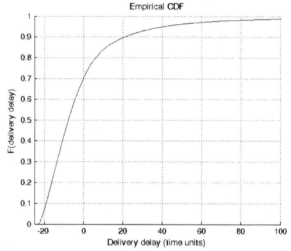

longer delays due to higher storage costs and delay fines. Therefore, changing the weight of the costs in the decision can be more suitable to improve the timeliness. Figure 5 proves this.

By tuning the cost weight, the percentage of timely deliveries can improved, as shown in the figure. With a very high weight, the percentage approaches 80%. However, the average vehicle capacity utilization is reduced to 0.6988. This means that the vehicles take the goods on more direct paths, and the load consolidation potential is lower. A side-effect is that to ensure the timely delivery of more goods in total, some other goods are not transported at all. These goods would require costly individual transports.

To position the proposed route decision function in comparison to other routing approaches, the results achieved here were placed into the comparison chart which was introduced in Rekersbrink et al. (2009). Figure 6 shows this.

Fig. 4 CDFs of the delivery delay with different weights for the delay risk

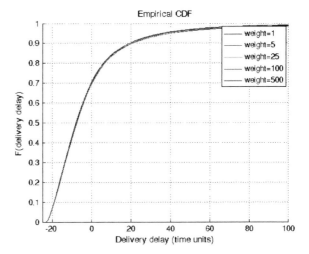

Fig. 5 CDFs of the delivery delay with different weights for the cost criterion

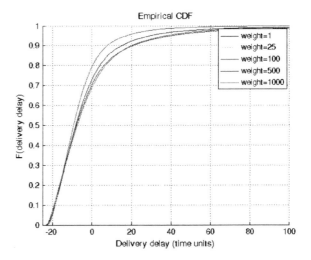

It can be seen from the figure that the proposed decision function leads to good results compared with the other methods investigated in Rekersbrink et al. (2009). The red star represents the decision function with equally weighted criteria; the black star represents it with a weight of 1,000 on the cost criterion.

5 Conclusions and Outlook

A multi-criteria decision function for autonomous routing with DLRP has been introduced in this paper. Simulation results have shown that the decision function performs well in comparison to other approaches and can be further improved by a fine-tuning of weights. Further research will include adaptive

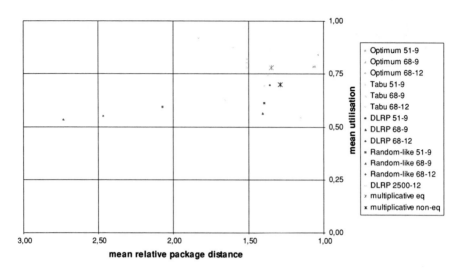

Fig. 6 Positioning of the MCCD decision approach (*red star* and *black star*) in the comparison chart introduced in Rekersbrink et al. (2009)

tuning of weights during runtime, and investigations on topologies of different scales.

Acknowledgments This research was supported by the German Research Foundation (DFG) as part of the Collaborative Research Centre 637 "Autonomous Cooperating Logistic Processes".

References

Rekersbrink H, Makuschewitz T, Scholz-Reiter B (2009) A distributed routing concept for vehicle routing problems. In: Logistics Research 1/1, Springer, pp. 45–52.

Scholz-Reiter B, Windt K, Freitag M (2004) Autonomous Logistic Processes: New demands and first approaches. In: Monostori L (ed) Proceedings 37th CIRP international seminar on manufacturing systems, pp. 357–362.

Scholz-Reiter B, Rekersbrink H, Freitag M (2006) Internet routing protocols as an autonomous control approach for transport networks. In: Proc. of the 5th CIRP international seminar on intelligent computation in manufacturing engineering, pp. 341–345.

Scholz-Reiter B, Rekersbrink H, Wenning B-L, Makuschewitz T (2008) A survey of autonomous control algorithms by means of adapted vehicle routing problems. In: Proceedings of the 9th Biennial ASME Conference on Engineering Systems Design and Analysis ESDA 08, published on CD-ROM.

Wenning B-L, Görg C, Peters K (2005) Ereignisdiskrete Modellierung von Selbststeuerung in Transportnetzen. In: Industrie Management 21/5, pp. 53–56.

Wenning B-L, Rekersbrink H, Timm-Giel A, Görg C, Scholz-Reiter B (2007) Autonomous control by means of distributed routing. In: Hülsmann M, Windt K (eds) Understanding Autonomous Cooperation and Control in Logistics—the Impact on Management, Information and Communication and Material Flow, pp. 335–345.

Distributed Decision Making in Combined Vehicle Routing and Break Scheduling

Christoph Manuel Meyer, Herbert Kopfer,
Adrianus Leendert Kok and Marco Schutten

1 Introduction

In practice, apart from the task of vehicle routing and scheduling, also the problem of scheduling breaks and rest periods has to be addressed by planners when creating vehicle schedules. According to the European legislation, when creating vehicle schedules planners have to make sure that drivers can adhere to the legislation on driving and working hours as laid down in Regulation (EC) No 561/2006 and in Directive 2002/15/EC. We call the arising planning problem the problem of combined vehicle routing and break scheduling. It comprises three subproblems, namely the clustering of customer requests, the routing of the vehicles, and the scheduling of breaks and rest periods (Meyer and Kopfer 2008). A main characteristic of the problem of combined vehicle routing and break scheduling is that these planning tasks are usually divided over several decision making units (DMUs), namely planners and drivers. Therefore, the problem is characterized by hierarchies in distributed decision making. To analyze this problem, we apply the framework for distributed decision making as presented by Schneeweiss (2003). The objective of this paper is to investigate the effects of

C. M. Meyer (✉) and H. Kopfer
University of Bremen, Wilhelm-Herbst-Str. 5, 28359 Bremen, Germany
e-mail: cmmeyer@uni-bremen.de

H. Kopfer
e-mail: kopfer@uni-bremen.de

A. L. Kok and M. Schutten
Operational Methods for Production and Logistics, University of Twente,
P.O. Box 217, 7500AE, Enschede, The Netherlands
e-mail: a.l.kok@utwente.nl

M. Schutten
e-mail: j.m.j.schutten@utwente.nl

H.-J. Kreowski et al. (eds.), *Dynamics in Logistics*,
DOI: 10.1007/978-3-642-11996-5_12, © Springer-Verlag Berlin Heidelberg 2011

different degrees of anticipation of the drivers' planning behaviour both on the planner's and on the drivers' objectives.

The paper is structured as follows. Section 2 presents the European legislation on driving and working hours in road transportation. Section 3 embeds the problem of combined vehicle routing and break scheduling into the framework for distributed decision making. In Sect. 4, computational experiments illustrate the effects of different planning approaches by the planner. Section 5 summarizes the main findings and gives some conclusions.

2 EC Legislation on Driving and Working Hours

The European social legislation for drivers in road transportation mainly comprises two legal acts. Regulation (EC) No 561/2006 lays down rules on drivers' driving hours and Directive 2002/15/EC restricts working hours of persons engaged in road transportation.

EC Regulation No 561/2006 concerns three different time horizons: single driving periods and daily and weekly driving times. Figure 1 depicts the relationship between these different time horizons.

The regulation restricts the driving time in each single driving period to 4.5 h. Drivers are obliged to take a break of at least 45 min after each driving period. Optionally, this break can be divided into two parts of at least 15 and 30 min, respectively. A driving period ends, when a break of sufficient length has been taken.

The daily driving time is restricted to 9 h. However, there is the optional rule that twice a week, i.e. twice between Monday 0:00 am and Sunday 12:00 pm, the daily driving time may be extended to 10 h. Daily driving times are defined as the accumulated driving time between two daily or between a daily and a weekly rest period respectively. A daily driving time ends when a daily rest period is taken or a weekly rest period starts. Within 24 h after the end of a daily or weekly rest period the next daily rest period must have been taken. A regular daily rest period is defined as a period of at least 11 h in which a driver may freely dispose of his time. A reduced daily rest period is a rest period of at least 9 h. The regulation provides

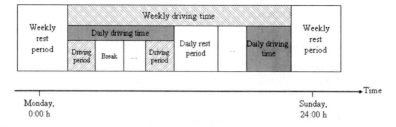

Fig. 1 Relation of the different time horizons (Kopfer et al. 2007)

the option to take up to three reduced daily rest periods between 2 weekly rest periods. Moreover, it allows splitting a regular daily rest period into two parts of at least 3 h and 9 h, respectively.

The weekly driving time is limited to a maximum of 56 h. Additionally, the maximum driving time of any two consecutive weeks must not exceed 90 h. The weekly driving time is defined as the accumulated driving time during a week, i.e. between Monday, 0:00 am and Sunday, 12:00 pm. A weekly rest period is a recreation period in which a driver may freely decide how to spend his time. The regular length of a weekly rest period is at least 45 h; the reduced duration is at least 24 h. A driver is allowed to use this optional reduction once in any two consecutive weeks. Reductions have to be compensated by equal extensions of other rest periods of at least 9 h before the end of the third week following the week considered. A weekly rest period has to be started within 144 h after the end of the previous weekly rest period.

EC Regulation No 561/2006 only comprises restrictions on driving times. As driving times are considered as working times, they are also affected by Directive 2002/15/EC, which contains restrictions on weekly working times and breaks. In the directive the working time is defined as the time devoted to all road transport activities, i.e. driving time, time for loading and unloading, for assisting passengers while boarding and disembarking from the vehicle, time spent for cleaning and technical maintenance, and the time a driver has to wait at the workstation when the end of the waiting time is not foreseeable. The directive postulates that after a working time of no more than 6 h workers have to take a break. The total duration of breaks during working periods of 6 to 9 h must equal at least 30 min. If the daily working time exceeds 9 h the total break time has to amount to at least 45 min. These break times can be divided into parts of at least 15 min. Consequently, a break which meets the requirements of EC Regulation No 561/2006 also satisfies Directive 2002/15/EC.

Furthermore, the directive restricts the weekly working time to a maximum of 60 h. Moreover, an average working time of 48 h per week over a period of 4 months must not be exceeded. When creating vehicle routes, planners have to make sure that both driving time restrictions and working time restrictions for drivers are satisfied.

3 Combined Vehicle Routing and Break Scheduling as a Problem of Distributed Decision Making

As mentioned before, in combined vehicle routing and break scheduling three interconnected planning problems have to be solved: the clustering of customer requests, the routing of vehicles, and the planning of breaks and rest periods for the drivers. These problems can be solved either simultaneously or in sequence. In the case of sequential planning, the possibility of solving two of the three planning problems simultaneously remains. However, not all sequences are reasonable in practice since the requirements for breaks and rest periods arise from the duration of the routes for the drivers. Therefore, the break scheduling should be performed last.

Apart from the three interconnected planning problems, there is another factor that adds to the complexity of combined vehicle routing and break scheduling: usually the planning process is divided over two DMUs, namely the planner and the driver. Therefore, the overall problem is characterized by hierarchical structures in distributed decision making. These hierarchies can be found both in the relationship between schedulers and drivers and in the structure of the planning problems to be solved. In the following the framework of Schneeweiss (2003) is used to analyze the decision problem.

In this framework for distributed decision making, two DMUs are considered. In the case of hierarchies in distributed decision making, these DMUs are situated on different levels. The top-level takes its decision first and instructs the base-level with the resulting plans. Subsequently, the base-level takes its decision based on the frame set by the top-level's decisions. However, when performing its planning, the top-level can try to anticipate the subsequent planning of the base-level in order to avoid infeasibilities on the base-level or if the base-level's decisions also influence the top-level's objectives. To consider the base-level's planning, the top-level can apply some sort of anticipation function. This anticipation function needs not be a precise representation of the base-level's planning model but can also be an approximation. In the following two different degrees of anticipation will be suggested for the problem of combined vehicle routing and break scheduling.

According to the classification in Schneeweiss (2003), the planning situation between planners and drivers can be described as a situation with several DMUs, in which a conflict-free team situation can be assumed. This results in a situation of organizational hierarchies in distributed decision making. The encountered information asymmetry mainly results from the fact that when taking their decisions, drivers have more accurate information about when it is possible to schedule breaks than the planner has.

In practice usually two different divisions of the subproblems over planners and drivers are encountered. The clustering of customer requests is typically performed by the planners. Moreover, the break scheduling is always carried out by the drivers for two reasons. First, drivers know best when they require a break or rest period. Therefore, leaving this autonomy to the driver seems reasonable. Second, a planner does not know exactly when it is possible for drivers to take a break. Drivers cannot stop their vehicles directly on the highway but require a service area. Consequently, in practice this task cannot be performed by the planners. The only task that can possibly be carried out by both DMUs is the routing. A rough conceptualization by whom the routing is performed for vehicle routing problems (including a central depot) can be made according to the characteristics of the transports. In the case of full truckload transports, only one possible route exists for each vehicle. Therefore, this task needs not be considered in the total planning process of the planner. In less than truckload transports, the routing is mainly

Fig. 2 Hierarchical planning
situation

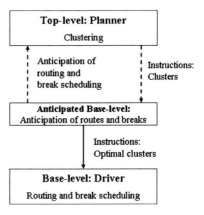

carried out by the planners. In parcel services and other services operating in a
restricted area, the routing is mainly carried out by the driver, especially if the
locations of the customers are very close to each other and if the set of customers is
not the same from day to day. For the remainder of this paper we concentrate on
this last situation. Figure 2 depicts this division of the tasks between the DMUs
using the framework of distributed decision making by Schneeweiss (2003).

In distributed decision making, different decision levels are considered. In our
case the planner constitutes the top-level. His objective is to create vehicle
schedules using as few vehicles as possible. The planner carries out the clustering
of the customer requests. He derives his optimal instructions and advises the
drivers which customers they have to service. When creating the customer clusters,
he has to make sure that the drivers can service all customer requests within their
delivery time windows and can also adhere to the European social legislation.
Therefore, the planner has to anticipate the planning behavior of the drivers, who
constitute the base-level. To accomplish this task, he uses an anticipation function
to take into account the routing and break scheduling that will be performed by the
drivers subsequently. He does this to avoid creating infeasible plans with respect to
the base-level's behaviour. The base-level receives the top-level's instructions and
carries out the routing and break scheduling within the clusters it is assigned using
some sort of planning model. We assume that each driver's objective is to minimize the travel distance.

The planner considers the base-level's planning model using anticipation
functions. These anticipation functions are approximations of the expected base-
level's planning model and need not be precise representations. Schneeweiss
(2003) distinguishes between four different degrees of anticipation: perfect reactive anticipation, approximately perfect reactive anticipation, implicit reactive
anticipation, and non-reactive anticipation. The first three take into account the
base-level's behaviour via some sort of anticipation function. Non-reactive
anticipation means that no anticipation function exists but that some general
features of the base-level may be taken into account in the top-level's objective
function.

For further analysis we consider only two different degrees of anticipation. First, in perfect reactive anticipation the mathematical structure of the base-level's planning model is completely considered (Schneeweiss 2003). In combined vehicle routing and break scheduling we model this situation such that the planner minimizes the number of vehicles used. Thereby, for each vehicle, he takes into account the drivers' task of finding a shortest route exploiting all optional rules of the legislation on driving and working hours as described in Sect. 2. So when creating the clusters the planner uses the drivers' planning model that tries to find the minimum travel distance under consideration of the EC social legislation including all optional rules. The drivers may still improve on these routes and break schedules, since they only focus on their specific route and break schedule, while the planner has to distribute his computational power over the clustering problem and several different routing and break scheduling problems.

Second, in the case of approximately perfect reactive anticipation the anticipation function uses some approximate solution procedure of the base-level's planning model (Schneeweiss 2003). In our case the driver's planning tasks of routing and break scheduling are approximated by the planner. Therefore, as an approximation of the driver's planning model we use a model that finds the shortest travel distance including only the basic rules of the EC social legislation. Omitting the complex set of optional rules simplifies the planner's task. However, when carrying out the routing and break scheduling, the drivers do use the full planning model including all optional rules. By anticipating the drivers' planning model including only the basic rules of the social legislation the planner makes sure that a feasible solution for the whole planning problem can be found by the drivers since the application of the optional rules by the drivers will cause an enlargement of the solution space compared to the solution space considered by the planner. We assume that the planner also communicates his routes and break schedules to the drivers, but the drivers do not have to follow these routes and break schedules, trying to reoptimize the routes according to their objectives. In a dynamic planning scenario the driver will also try to adapt the schedules to actual situations.

In Sect. 4, we analyze the described scenarios with some computational experiments. Our approach to addressing the planner's problem is to solve a vehicle routing problem with time windows (VRPTW) and EC social legislation. Both anticipation functions allow drivers to find feasible vehicle routes and break schedules. However, the effects of the different degrees of anticipation on the objective functions are investigated at both levels: at the top-level, i.e., the number of vehicles used, and at the base-level, i.e., the total travel distances.

4 Computational Experiments

To solve the customer clustering problem, we apply the dynamic programming algorithm presented by Kok et al. (2009). This algorithm is based on the restricted dynamic programming framework proposed by Gromicho et al. (2008) in which the idea of dynamic programming for the travelling salesman problem is applied and the number of states to be expanded in each stage is restricted. To use the dynamic programming approach to solve vehicle routing problems, the giant-tour representation of vehicle routing solutions proposed by Funke et al. (2005) is used. The algorithm includes the EC legislation on drivers' driving and working hours as described in Sect. 2. It applies a local perspective when scheduling breaks and rest periods which fits well into the concept of dynamic programming. The algorithmic parameters are set such that it first minimizes the number of vehicles used and second the total distance travelled.

After the customer clusters are generated using the above algorithm, they are given to the drivers and in these clusters the drivers carry out the routing and break scheduling also using the algorithm by Kok et al. (2009) where only one vehicle is allowed. Moreover, we assume that the planner communicates the routes and break schedules he establishes to the drivers. If a driver cannot improve upon the routes suggested to him in terms of his objective function, i.e. if a driver cannot reduce his travel distance, he follows the planner's advice. To test the scenarios, the Solomon (1987) test problems for the VRPTW are used in the adjusted form proposed by Goel (2009).

Table 1 presents the average numbers of vehicles used for the different problem types for the two anticipation functions. The Solomon instances consist of 6 problem types in which the C-instances have clustered customer nodes, the R-instances have randomly located customer nodes, and in the RC-instances the customer nodes are semi-clustered. The difference between the one- and two-instances is that the demands and distances in the two-instances are, on average, smaller than in the one-instances, allowing for longer (and, as a consequence, fewer) vehicle routes. The results indicate the change in the planner's objective, i.e., the number of vehicles used, by using the two different anticipation functions.

The results show a strong reduction in the number of vehicle routes (5% on average) if the perfect anticipation function is used by the planner. Therefore, this case is superior to the case of approximately perfect anticipation in terms of the planners' objective value.

Table 2 presents the resulting average total travel distances for the vehicle routes found by the drivers. Again, the perfect anticipation function results in the

Table 1 Planner's objective

Problem sets (# of instances)	Average # of vehicles: perfect reactive anticipation	Average # of vehicles: approximately perfect reactive anticipation
C1 (9)	10.00	10.22
C2 (8)	5.25	6.00
R1 (12)	9.25	9.83
R2 (11)	7.27	7.82
RC1 (8)	9.88	10.25
RC2 (8)	8.25	8.38
All (56)	8.36	8.80

Table 2 Drivers' objective

Problem sets (# of instances)	Average travel distance: perfect reactive anticipation	Average travel distance: approximately perfect reactive anticipation
C1 (9)	927.23	948.56
C2 (8)	780.59	836.32
R1 (12)	1130.52	1152.03
R2 (11)	1084.17	1091.19
RC1 (8)	1323.96	1291.30
RC2 (8)	1238.99	1257.80
All (56)	1081.89	1097.28

Table 3 Improvements found by the drivers (rerouting)

Problem sets (# of instances)	Perfect reactive anticipation		Approximately perfect reactive anticipation	
	% routes improved	Average improvement (%)	% routes improved	Average improvement (%)
C1 (9)	4.44	0.73	8.79	1.14
C2 (8)	5.00	0.60	3.33	4.93
R1 (12)	13.20	2.14	13.40	2.08
R2 (11)	18.26	0.66	17.79	1.78
RC1 (8)	10.58	2.79	16.21	2.28
RC2 (8)	5.12	2.02	27.27	1.68
All (56)	9.94	1.62	14.86	1.89

best vehicle routes, also in terms of the drivers' objective. The average total travel distance over all problem instances is reduced by 1.4%.

To analyze the impact of the rerouting performed by the drivers, we determine the percentage of vehicle routes for which drivers' found better vehicle routes in terms of travel distances by rerouting. We also determine the average improvement in travel distance for these routes. Table 3 presents these results.

The results show that the improvements found by the drivers are significant. In case of perfect reactive anticipation 9.94% of the routes are improved and the average improvement of these routes is 1.62%. The improvements are even larger in case of approximately perfect reactive anticipation. This is due to the fact that the planner does not exploit the optional rules of the EC social legislation in this case. Therefore, using also the optional rules of the social legislation, the drivers can improve the routes even further.

5 Conclusions

We analyzed the problem of combined vehicle routing and break scheduling from a distributed decision making perspective. The problem was embedded into the framework for distributed decision making proposed by Schneeweiss (2003). This framework is very suitable for the analysis of this problem from a practical point of view. We incorporated different degrees of anticipation of the drivers' planning model into the scheduler's planning procedure. Our computational experiments showed that a more accurate anticipation function results in better vehicle routes and break schedules. This holds both for the planner's and the drivers' objectives: the perfect reactive anticipation function clearly dominates the approximately perfect anticipation function.

Acknowledgments This work was financially supported by the German Research Foundation (DFG) as part of the Collaborative Research Centre 637 "Autonomous Cooperating Logistics Processes—A Paradigm Shift and its Limitations" (subproject B9) and by Stichting Transumo through the project ketensynchronisatie.

References

Directive 2002/15/EC of the European Parliament and of the Council of 11 March 2002 on the organisation of the working time of persons performing mobile road transport activities, Official Journal of the European Communities L 80/35, 23.3.2002.

Funke, B., Grünert, T., Irnich, S. (2005). Local Search for Vehicle Routing and Scheduling Problems: Review and Conceptual Integration. Journal of heuristics, 11(4), 267–306.

Goel, A. (2009). Vehicle Scheduling and Routing with Drivers' Working Hours. Transportation Science, 43(1), 17–26.

Gromicho, J., van Hoorn, J.J., Kok, A.L., Schutten, J.M.J. (2008). Restricted dnamic programming: a flexible framework for solving realistic VRPs. Beta Working Paper 266.

Kok, A.L., Meyer, C.M., Kopfer, H., Schutten, J.M.J. (2009). Dynamic Programming Algorithm for the Vehicle Routing Problem with Time Windows and EC Social Legislation. Beta Working Paper 270.

Kopfer, H., Meyer, C.M., Wagenknecht, A. (2007). Die EU-Sozialvorschriften und ihr Einfluss auf die Tourenplanung. Logistik Management, 9(2), 32–47.

Meyer, C.M., Kopfer, H. (2008). Drivers' autonomy for planning and rest periods in vehicle routing. In Ivanov, D., Jahns, C., Straube, F., Procenko, O., Sergeev, V. (Eds), Logistics and Supply Chain Management: Trends in Germany and Russia (pp. 343–352). Saint Petersburg: Publishing House of the Saint Petersburg State Polytechnical University.

Regulation (EC) No 561/2006 of the European Parliament and of the Council of 15 March 2006 on the harmonisation of certain social legislation relating to road transport and amending Council Regulations (EEC) No 3821/85 and (EC) No 2135/98 and repealing Council Regulation (EEC) No 3820/85, Official Journal of the European Union L 102/1, 11.4.2006.

Schneeweiss, C. (2003). Distributed decision making—a unified approach. European Journal of Operational Research, 150(2), 237–252.

Solomon, M.M. (1987). Algorithms for the vehicle routing and scheduling problems with time window constraints. Operation Research, 35(2), 254–265.

Dynamic Routing Applied to Mobile Field Service

Auro C. Raduan and Nicolau D. F. Gualda

1 Introduction

The vehicle routing problem has received great attention due its complexity as well as its critical role in services and merchandises distribution. The internet boom besides the increasing use of electronic commerce, gives rise, inside the companies, the adoption of real time logistics decisions. The routing activities, happening before the vehicles dispatch, now acquire a new dimension, provided that computer and communications technologies enable instantaneous interchange of information between clients, demands, dispatch processes and vehicles. The vehicles, even after leaving the base and during all the journey, can have its positioning constantly traced and messages interchanged with the base. This fact provides significant value-added benefits to the operations as a whole in a cost-effective way. Therefore, the need for dynamic routing becomes a breakthrough in the current practices of mobile service environment. This paper presents features of dynamic routing, its application to mobile field services and variables and strategies that impact the whole performance. A computer model becomes an important tool to identify the best operational strategy according to cost or service level and to dimension resources in a setting of diverse spatial and temporal demand configurations.

A. C. Raduan (✉)
Programa de Mestrado em Engenharia de Sistemas Logísticos, Escola Politécnica da,
Universidade de São Paulo, Av. Prof. Almeida Prado, Travessa 2, no 83,
CEP 05508-900 São Paulo, SP, Brazil
e-mail: auro.raduan@poli.usp.br

N. D. F. Gualda
Departamento de Engenharia de Transportes, Escola Politécnica da Universidade de,
São Paulo, Av. Prof. Almeida Prado, Travessa 2, no 83, CEP 05508-900
São Paulo, SP, Brazil
e-mail: ngualda@usp.br

H.-J. Kreowski et al. (eds.), *Dynamics in Logistics*,
DOI: 10.1007/978-3-642-11996-5_13, © Springer-Verlag Berlin Heidelberg 2011

2 Dynamic Vehicle Routing Problem (DVRP) and the Dynamic Repairman Travelling Problem (DRTP)

Psaraftis (1988) defines vehicle routing problem (VRP) as a process of dispatching vehicles in a way that multiple demands for service arrive in real time. Psaraftis (1995) explains the main differences between static routing and dynamic routing:

- Time dimension is essential. At least, the dispatcher should know the spatial position of all the vehicles and its availability, and the arriving of new ones.
- The problem is open-ended. In the dynamic setting there is no limit. Instead of routes there are paths for the vehicles to follow.
- Future information may be imprecise or unknown. In the dynamic routing, the future is seldom known with sureness. Probabilistic information may be available.
- Near-term events are more important, most recent information earns more importance comparing to those of the future.
- Information update mechanisms are essential with respect to real time characteristic of the dynamic routing.
- Re-sequencing and reassigning decisions may be necessary.
- Faster computation times are necessary. The dispatcher needs solutions in few minutes or few seconds.
- Indefinite deferment mechanisms are essential to avoid that the service due to unfavorable conditions be postponed indefinitely.
- Objective function may be different. In dynamic setting it might be reasonable to optimize only over known inputs. The information about the future is imprecise.
- Time constraints may be different and tend to be softer. In the dynamic cases it is more important to accomplish the service than deny it.
- Flexibiliy to vary vehicle fleet size is lower in the dynamic routing, due to the fact all fleet is sent to the field in order to serve or wait for service.
- Queuing considerations may be important. In dynamic is necessary to deal with queuing problems due to congestion.

Lund et al. (1996) and Larsen (2000) defines degree of dynamism as a relationship between the number of dynamic requests and the total number of requests. This relationship provides various types of dynamic applications and defines the objective function in a dynamic system, as shown in Fig. 1.

The dynamic travelling repairman problem (DTRP) was introduced by Bertsimas and Van Ryzin (1991). According to the authors the model has several characteristics: (1) the objective is to minimize waiting time, not travel cost, (2) future demands are stochastic, (3) demands vary over time (dynamic), (4) policies have to be implemented in real time, and (5) the problem involves queuing. The authors explore several strategies concerning dispatch, vehicle positioning and partitioning of the service area. In another paper (1993) they explore the multiple vehicles branch of the same problem. Papastavrou (1996) proposes a

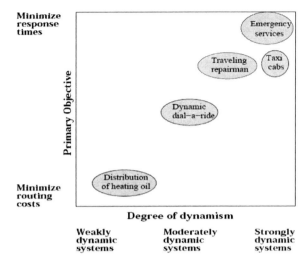

Fig. 1 The relationship between the parameters of the objective function and the degree of dynamism in several dynamic applications. *Source*: Larsen (2000)

DTRP policy called generation policy that demonstrated to be suitable for light and heavy traffic (queuing terms). Although the literature is scarce of specific applications about DTRP many of the employed techniques come from general dynamic routing problems and this is the reason to present the following references.

Regan et al. (1995) explore the benefit of real time rerouting of a vehicle previously assigned to another destination. This strategy of deviation is also studied by Ichoua et al. (2000).

Spivey and Powell (2004) solve simple dynamic routing problems through sequences of assignment problems and mention the value of advance information as in the case when the dispatcher holds a truck near an high demanding area. Branke et al. (2005) conclude that a good waiting strategy can increase the probability of serving an additional customer also reducing his attendance cost. Fleischmann et al. (2004) detail a model for a dynamic system for combined pick up and delivery of goods with time windows using dynamic travel times.

3 Methodology

3.1 The Conceptual and the Computational Model

Dynamic problems are common in practice and worths the adoption of various operational strategies, such as: (1) dispatching, when there are different ways of assigning a vehicle to a service request, (2) waiting, when a request wait some time in the hope of finding a nearest vehicle, (3) partitioning the area of service in quarters assigning dedicated vehicles for each of them, and (4) diverting, the vehicle to a closer or more urgent request instead of the planned one.

There are, a priori, no advantages of an operational strategy to the other, because the efficiency or effectiveness of each mainly depends on a set of attributes related to the demand profile and available resources. Therefore, it justifies the development of a computational model of discrete simulation capable of assessing this complex scenario of operational strategies, demands and resources. It means to model the problem and study the effect of variables and strategies over assignment and routing. To this end, the main components of the computer model are presented below:

- Demand: A file that represents the immediate requests of service.
- Servers: The entity that represents the vehicles and technicians.
- External constraints: Data that shape the model for processing.
- Dynamic model: This is the main component, and represents the logical and mathematical relationships between all variables.
- Indicators: numbers representing the performance of the scenario studied.

The dynamic model component in possession of requests to be serviced and available servers will perform the processing according to operational strategies and the informed variables like servers' speed, number of partitions and servers. The strategies considered in this model are limited to items a, b, c and d as follows:

a. FCFS—first come first served, which serves primarily the older requests, when availability of server.
b. NN—nearest neighbor, which serves the request that is closest to the server.
c. NN-W, NN combined with a waiting strategy delaying till certain limit of time looking for the availability of a server close to the request. This allows in some cases to increase the likelihood of getting a server closer to the request also reducing the distance travelled.
d. The strategy of partitioning (PART), that resembles FCFS for server/request divides the attendance area, the Euclidean plane, in a number of quarters assigning to each quarter a fixed number of dedicated vehicles. The main objective is to position the vehicles near the potential demand points inside each quarter. This is important when the model is applied for emergency services.

In PART, the initial positioning of the servers is at the centroid of each quarter.

In the other three strategies, the initial positioning of the vehicle will be coordinated in the Euclidean plane (0,0) and, after service, will remain in the coordinates of the last request served.

3.2 Modelling the Nearest Neighbor Strategy (NN)

In NN strategy, a process of obtaining the path of least time is triggered for each combination server/request as follows. It happens each time at least one server is available and there is a pending request. A heuristic calculate the Euclidean distance between each combination of server/request and divide the various distances found by the average speed reported for the processing, getting the travel time for each alternative. The values obtained are recorded in a vector or a matrix of charges, according to the number of available servers and requests in the queue. If only one server and several requests, the heuristic pick up the pair that gives the shortest time path as shown in Fig. 2. Similarly, when there are multiple servers and a single request. However, the occurrence of multiple requests and servers, in the matrix of charges, implies the taking in of the optimal assignment problem (Dantzig 1963) from integer linear programming utilizing binary variables as depicted below.

$$\text{Minimize:} \sum_{(i,j)\text{in}A} c_{ij} \cdot x_{ij}$$

$$\sum_{j:(i,j)\text{in}A} x_{ij} = 1, \quad \text{for any } i$$

$$\sum_{i:(i,j)\text{in}A} x_{ij} = 1, \quad \text{for any } j$$

$$x_{ij} = 0,1 \quad \text{for any } (i,j) \text{ in } A$$

$$x_{ij} = \begin{cases} 1 & \text{if } i \text{ is assigned to j that is, Server } i \text{ is assigned to Request j.} \\ 0 & \text{not assigned} \end{cases}$$

for A being a set of all the possible assignments from i (server) to j (request) and c_{ij} the cost of assignment (or charge) between i and j.

The assignment problems are solved in polinomial time (Ahuja et al. 1993), and its computational complexity is $O(N^3)$, where N is equal to the number of nodes (Knuth 1998), in this problem represented by servers or requests which is greater. This problem is solved using the algorithm of Kuhn, also called the Hungarian

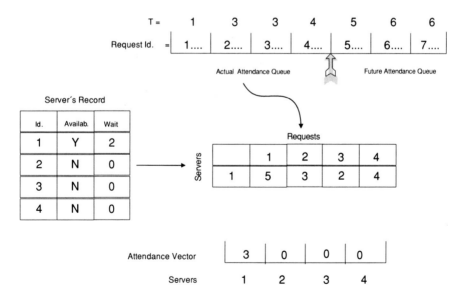

Fig. 2 The attendance when one server for multiple requests

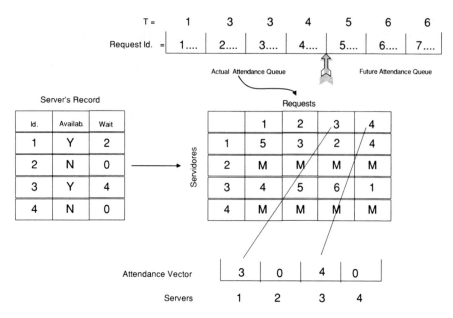

Fig. 3 The treatment of multiple servers and requests through the matrix of charge

method (Kuhn 1955) utilizing the balanced (square) matrix of charges. To create matrix cells for requests or servers not available at each decision time are used maximum (M) values. Figure 3 shows the case where M charges have been created for requests connected to servers 2 and 4 that were unavailable at time of mounting costs of the matrix. This algorithm achieves the lowest total time to service all requests by the available servers at that moment of decision. Beyond the simplicity of Khun's method, it also achieves a low computational run time. An experiment using the developed computational model for a square matrix of order 100 reached 12 s in a intel core duo microcomputer with 2 Gb of RAM.

When a never accepted request occurs due to unfavorable distance related to the others, resulting in indefinite deferment as Psaraftis (1995), a factor of reduction R was defined which can priorize a request in case its waiting time reaches an informed threshold. Factor R is the coefficient that results one when multiplied to the deferred request charge in a way it will be chosen by a server.

3.3 Modelling FCFS and PART Strategies

In FCFS strategy, the attendance priority is for the idle server waiting longer. The idle server will choose the oldest immediate request in actual service queue. Once established the pair server/request, the request id. will be recorded in attendance vector and the process continues seeking idle servers and old requests till no more idle server nor requests exist in queue. Figure 4 illustrates the allocation process

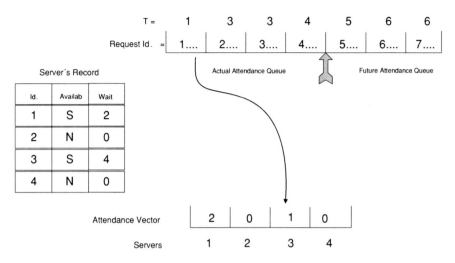

Fig. 4 The choosing process of the pair server/request in FCFS strategy

server/request. In this example, the server id. 3 is available and waiting for 4 TU (time units), the oldest and therefore entitled to choose the request that entered the queue for a long time, which in this case is the id. request 1 (arrived before the others). The record in the attendance vector is done by placing the number of request in the index corresponding to the relative entry position to the server id., in this case the number 3.

The PART strategy applies the same algorithm described for FCFS in each attendance area and its dedicated servers.

4 Application

4.1 Setting an Instance

The arrival intervals and service times for actual and projected requests for service presented in this work were generated randomly through a rand() function coded in a C++ program. The same happened with the server and request location coordinates in a Euclidean plan, set to size 100 by 100. Figure 5 shows that the points considered in the instance, are not evenly distributed over the available area. The present instance considered 50 requests with mean arrival rate = 0.575 (λ) and mean rate of attendance = 0.026 (μ).

4.2 Application

The computational model was used to simulate a processing system of the mentioned instance serviced by a variable number of servers (1 or 6–15 servers)

Fig. 5 Points of demand
(requests) used, distributed in
Euclidean plan with density
0.005 (number of points/area)

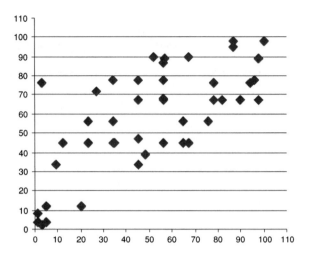

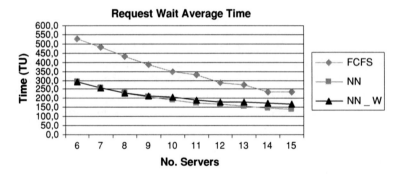

Fig. 6 The request wait average time of requests for each strategy

according to the operational strategies. The speed informed was 30 DU (Distance Units) by 1 TU (30 km/h) and the wait limit of time equal to 30 TU for NN-W strategy. Figure 6 demonstrates that the NN strategy results in less waiting time for requests. This is not always the case with the NN-W strategy, since the system "delays" waiting for a server near the request.

For cost of transportation, reflected in the graph of Fig. 7 as the distance, one observes that the NN strategies results in lower total distance travelled. In another simulation, illustrated in Table 1, where total time means the time spent to service all requests and Process. Time is the run time, it is observed that the strategy of partitioning with four quarters and with three servers in each quadrant has less total distance travelled. It is due to proximity between requests and servers. However, due to lack of servers in areas of high demand (see Fig. 5) the total time for answering requests overcome the time of the other strategies.

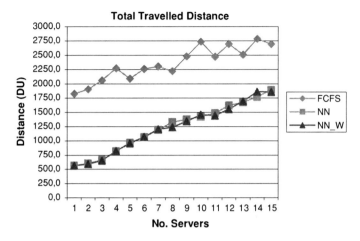

Fig. 7 The total distance travelled in each strategy

Table 1 Comparison of Strategies for 50 requests and 12 servers

Strategy	Total time (TU)	Waiting time (TU)	Total travelled distance (UD)	Process. time (s)
FCFS	537.0	283.7	2716.1	0.5
NN	361.0	165.9	1617.9	0.8
NN-W(30)	392.0	180.5	1553.2	8.9
PART(4)	601.0	200.2	1164.4	0.6

4.3 When to Apply: FCFS, NN, NN-e or PART

Applying other instances to the computer model is possible to evaluate cases in which one strategy is preferable than other. Strategy FCFS results in better performance when time intervals between requests are long or when the request locations are nearby. NN and NN-e is preferable when the time intervals are short and the distances between points of service are distant. PART is a way to position the vehicles close to the requests improving the response time. Its tradeoff is the low level of server utilization when scarce demand in the partition.

5 Conclusions

This article presented a modelling approach for analysis of the problem of real time critical timing requests for mobile field service. It described the core subjects of a computational model that evaluate the performance of several operational strategies (FCFS, NN, NN-W and PART) against various instance profiles, number of servers and diverse conditions. An algorithm based on the Hungarian method was implemented to optimize the matrix produced by the NN strategy, which resulted in the lowest combined response time for a set of servers and requests.

References

Ahuja RK, Magnanti TL, Orlin JB (1993) Network Flows: Theory, Algorithms and Requests Prentice Hall, Englewood Cliffs.

Bertsimas DJ, Van Ryzin G (1991) Stochastic and Dynamic Vehicle Routing Problem in the Euclidean Plane. Operations Research 39(4):601–615.

Bertsimas DJ, Van Ryzin G (1993) Stochastic and Dynamic Vehicle Routing Problem in the Euclidean Plane with Multiple Capacitated Vehicles. Operations Research 41(1):601–615.

Branke J, Middendorf M, Noedth G, Dessouky M (2005) Waiting Strategies for Dynamic Vehicle Routing. Transportation Science 39(3):298–311.

Dantzig GB (1963) Linear Programming and Extensions. Princeton University Press.

Fleischmann B, Gnutzmann S, Sandvoss E (2004) Dynamic Vehicle Routing Based on Online Traffic Information. Transportation Science 38(4):420–433.

Ichoua S, Gendreau M, Potvin JY (2000) Diversion Issues in Real Time Vehicle Dispatching. Transportation Science 34:426–438.

Knuth DE (1998) The Art of Computer Programming. Addison Wesley, Reading.

Kuhn HW (1955) The Hungarian Method for the Assignment Problem. Naval Research Logistic Quarterly 2:83–97.

Larsen A (2000) The Dynamic Routing Problem—Thesis for the Degree on Ph.D. Math. Model. Dpt. Denmark Tech. University.

Lund K, Madsen OBG, Rygaard JM (1996) Vehicle Routing Problems with Varying Degrees of Dynamism. Technical Report IMM, Technical University of Denmark.

Papastavrou JD (1996) A Stochastic and Dynamic Routing Policy Using Branching Processes with State Dependent Immigration. European Journal of Operational Research 95(1):167–177.

Psaraftis HN (1988) Dynamic Vehicle Routing Problems, Vehicle Routing: Methods and Studies. Elsevier Science Publishers, Amsterdam.

Psaraftis HN (1995) Dynamic Vehicle Routing: Status and Prospects. Annals of Operations Research 61(1):61–143.

Regan AC, Mahmassani HS, Jaillet P (1995) Improving Efficiency of Commercial Vehicle Operations Using Real-Time Information: Potential Uses and Assignment Strategies. Transportation Research Record 1493:188–198.

Spivey MZ, Powell WB (2004) The Dynamic Assignment Problem. Transportation Science, 38(4):399–419.

Intelligent Agent Control and Coordination with User-Configurable Key Performance Indicators

Florian Pantke

1 Introduction

In industrial and logistic applications, Multi-Agent Systems (MASs) are a means to implement distributed autonomous control of complex processes. The decomposition of a task into several subtasks and their assignment to multiple agents can largely reduce the absolute processing time due to exploitation of parallelism and distributed resources, or even enable the whole undertaking to be completed within the available time in the first place (Weiss 1999). However, a problem might not be decomposable into completely independent subtasks as certain interconnections concerning the overall goal satisfaction or quality of solution might persist between the problem parts. As a result of given resource constraints (e.g., limited computing power, memory or communication bandwidth), it might be necessary to simplify the task by dropping or relaxing some of these relationships and to accept trade-offs between solution quality and resource consumption instead of seeking an optimal, but practically unachievable outcome. For example, in (Xuan et al. 2001; Xuan and Lesser 2002) a scenario is discussed where the agents in a MAS periodically need to exchange information about the current state of their partially observable environment in order to construct a complete global view from their limited local views of the world. Due to constraints on the available communication bandwidth, they cannot do this as frequently as it may be necessary for producing the best possible outcome, so a deliberate trade-off between resource consumption and global solution quality has to be made.

In addition to possible losses in solution quality introduced by simplified decompositions of hard problems that, with the given resources, are not manageable otherwise, multiple competing subgoals may already exist in the original

F. Pantke (✉)
Center for Computing Technologies (TZI), University of Bremen,
Am Fallturm 1, 28359 Bremen, Germany
e-mail: fpantke@tzi.de

H.-J. Kreowski et al. (eds.), *Dynamics in Logistics*,
DOI: 10.1007/978-3-642-11996-5_14, © Springer-Verlag Berlin Heidelberg 2011

problem formulation, as it often is the case with applications in logistic settings. In general, the agents of a MAS need not be cooperative and share the same set of global objectives; they can also act in a self-interested and competitive way by following only their individual local goals. Therefore, the designer or user of a MAS might have certain expectations about two types of agent behavior: the local behavior of each individual agent on the one hand, which is determined only by the respective agent's sequence of executed actions, and the global behavior of a set of multiple agents on the other hand, corresponding to all actions performed by a certain agent group. With proper knowledge within the system's specific application domain, these demands can be expressed as a set of qualitative and quantitative performance criteria, which can then be applied to assess the local and global agent behavior and to also proactively direct it towards certain desired outcomes, when appropriately implemented into the MAS at design time or passed to it as explicit objectives at runtime. In general, the formulation of such goals results in multi-criteria decision problems that need to be solved by the agents, with objectives possibly conflicting at both the local and global agent level.

Due to the emergent, not entirely predictable behavior of many complex MASs, different users of the system may have different views and requirements regarding what constitutes an acceptable trade-off, or they may need to experiment with different goal settings before the system starts exhibiting the desired behavior. This calls for a flexible and highly configurable goal system that can be set up and fine tuned by the user of the MAS at runtime, instead of being hard-wired into the intelligent agents at design time.

While publications like (Timm 2004) deal with the identification and resolution of logical (i.e., qualitative) goal conflicts in MASs, the paper at hand focuses on agent control and coordination by the optimization of numerical objectives and the determination of suitable quantitative trade-offs. A software-framework for local and global agent control based on the concept of numerical key figures as quantitative performance indicators is presented. It allows the user of a MAS to define at runtime numerical performance indicators and quantitative objectives attached to them, which can then be incorporated into the agents' goal system in order to influence the agents' actions. The central parts of the framework have been implemented as a Java programming library, which can be used, e.g., in conjunction with the JADE agent platform (Bellifemine et al. 2007) to construct a MAS in the Java programming language.

This paper is organized as follows: First, a short introduction to key figure systems is given in Sect. 2. Section 3 then gives an overview of the different components of the key figure software framework. One of these components, the local key figure assessment model used by the individual agents of the MAS for local agent control, is described in detail in Sect. 4. Section 5 presents four different approaches to global agent control and coordination, which differ in the way the global key figure assessment model is distributed among the agents. Three of these approaches are then compared in a production logistics scenario in Sect. 6. The results were obtained by integrating the framework into the IntaPS MAS (Lorenzen et al. 2006). Finally, Sect. 7 concludes this paper with a summary and outlook.

2 Key Figure Systems

In business performance management, key figure systems constitute tools for quantitative modeling, aggregation, and consolidation of selected facts of interest. If key figures are associated with appropriate target values, they become instruments for communicating quantitative goals within the organization and for evaluating achieved outcomes with respect to these goals. As they permit direct comparison between prescribed target values and achieved values, they can be used as numerical performance indicators, i.e., for explicitly quantifying degrees of goal satisfaction (Franceschini et al. 2007). Since key figures are also very commonly used for the evaluation and control of industrial and logistic processes (consider, e.g., key figures like mean job processing time or machine utilization in production logistics), it seems quite natural to adopt the key figure approach also for control and coordination tasks within MASs.

3 The Framework Components

Figure 1 depicts the components of the key figure software framework incorporated into a generalized local agent control cycle. The framework is suited for use with different types of agent control and decision-making, e.g., its components can be integrated into a BDI control cycle (Rao and Georgeff 1995) as well as into Markov decision processes. The framework follows a cooperative approach in which the agents are organized in a hierarchy, report key figure values upwards within this structure, and pass generated target values and derivatives downwards. Self-interested agents are given the possibility to engage in appropriate trade-offs between the global set of objectives and their individual local goals. Although this still allows, in principle, for the task environment to be highly competitive, it requires the agents to have at least *some* interest in mutual cooperation.[1] The following components are shown in Fig. 1.

3.1 Key Figure Assessment Model

The global key figure assessment model plays a vital role within the framework. It contains user-defined key figures and numerical objectives monitoring and guiding the global agent behavior. The calculation rules of the (non-atomic) key figures can be specified by the user as textual formulas; internally they are represented as computational graphs. The global view, which consolidates key figures measured

---

[1] Otherwise the key figure based global coordination process specified by the framework would be ineffective and methods from the field of mechanism design need to be applied.

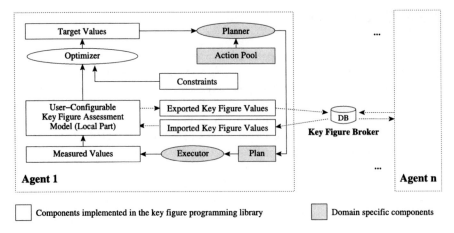

Fig. 1 The framework components

and published by numerous different agents, is constructed by assigning each of the individual agents a user-defined local key figure assessment model and then combining these into a distributed or centrally managed global model, taking into account the agents' current organizational hierarchy. Hence, the global assessment model is given by a mutable hierarchy of multiple local assessment models. The internal structure of the key figure assessment model is described in more detail in Sect. 4.

3.2 Measured Values

In the terminology of the framework, measurands are those key figures that provide the atomic input values, called the measured values, into the global key figure assessment model. These values are part of the agents' sensory input and are not calculated using values of other key figures. Instead, they are directly measured by the agents themselves and can be influenced by the agents' local actions. The set of available measurands is specific to a system's application domain and is usually predefined at design time.

3.3 Imported and Exported Key Figure Values

In order to construct a (semi-) global view on the system, agents can also acquire values of key figures measured or calculated by other agents. The measurement procedures and calculation rules of imported key figures are not known to the importing agents, so these operations have no counterpart in the computational graph representation of the importing agents' local key figure assessment models. Like the measured values, imported key figure values are atomic inputs into the local models.

3.4 Key Figure Broker

During runtime of the MAS, all agents periodically update the values of their key figures at application dependent synchronized time points. At each update the current key figure values and gradients are assigned a unique time stamp and are permanently recorded by the key figure broker. This component serves as a system-wide blackboard and facilitates the exchange of key figure values and derivatives between agents. It is backed-up by a relational database, which can also be accessed by external statistical tools for a posterior analysis of the system runs. The key figure programming library uses the open source database implementation MySQL (MySQL 2009).

3.5 Optimizer

The local key figure assessment model of each agent typically contains several local key figure objectives and thereby forms a multi-criteria optimization problem, which may possess several Pareto optimal solutions with respect to the agent's atomic measurands. In order to maximize (local) goal satisfaction, either a scalarization of the assessment model can directly be used as a plan metric in the agent's planning process, or a set of Pareto optimal target values can be determined for all local measurands and then used as atomic numerical goals in the planning stage. The optimizer component implements the mathematical optimization methods required for the latter approach. The programming library uses Differential Evolution (Price et al. 2005) as the standard optimization algorithm and can easily be extended with other (e.g., gradient-based) methods.

3.6 Target Values

An agent can determine target values for its own measurands as well as for (not necessarily atomic) key figures imported from other agents. The generation and negotiation of target values for imported key figures is necessary for the maximization of global goal achievement, as otherwise it cannot be guaranteed that all agents strive for the same optimum of the global assessment model. Thus, the target values produced by the optimizer not only represent a local medium-term goal commitment but also serve a global coordination purpose among the agents.

3.7 Constraints

The constraint set defines the feasible region for the optimization problem to be solved by the optimizer. In the simplest case, this set contains only boundary constraints, defining an upper and lower limit for each of the agent's local measurands and imported key figures. Boundaries for exported non-atomic key figures can be derived from the calculation rules using interval arithmetic (Hansen and Walster 2004).

3.8 Planner, Action Pool, Plan and Executor

These components are not contained in the key figure programming library as they may require highly domain specific implementations and representations. The task of the planner is to create, from a pool of available actions or plan templates, a (partially) ordered sequence of actions, called a plan, which maximizes (local) goal satisfaction when eventually carried out by the executor component of the agent. After complete or partial plan execution, all key figure values are updated according to the most current sensory input, and a new control cycle begins.

4 The Key Figure Assessment Model

To allow for user-configurable key figures and key figure objectives, the local parts of the key figure assessment model assigned to the individual agents are implemented as computational graphs. Figure 2 shows the basic structure of such a graph. For each key figure whose value the respective agent imports from a different agent, the graph contains a key figure proxy, which stands in for the remote key figure and automatically obtains its current value through the key figure broker. The current values of the agent's local measurands, on the other hand, are determined and then stored in the key figure broker component by the agent itself. The framework subsumes the measurands and key figure proxies under the term basic key figure. They constitute the leaves of the computational graph.

In the terminology of the framework, key figures whose values are computed from the values of other key figures according to some explicit calculation rule are called composed key figures. The calculation rules can consist of functions from a basic set of arithmetical and statistical operations such as calculation of sum, product, average, median, etc. This set of standard operations offered by the programming library can easily be extended by the programmers of the specific agent system.

In order to guide the values of selected basic or composed key figures into one of possibly several desired directions, key figure goals can be assigned to these key figures. For each goal the user of the MAS can define a piecewise linear function

Fig. 2 The structure of the local key figure assessment model

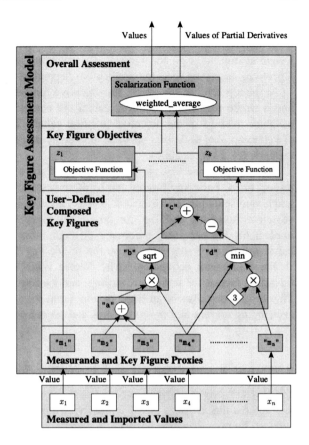

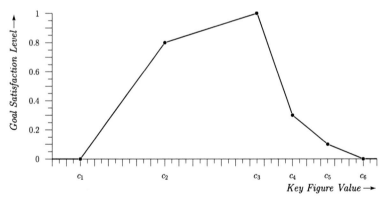

Fig. 3 Example for an objective function of a key figure goal

(PWLF) as an objective function, which maps the respective key figure's value to a goal satisfaction level between 0 and 1. An example of such a function is shown in Fig. 3. The user describes the PWLF by setting up its control points

$(c_1, v_1),...,(c_n, v_n)$. He can also assign a weight between 0 and 1 to the key figure goal, representing the goal's initial importance. The agent system may adjust the weight at runtime, e.g., for dynamically balancing local against global key figure objectives according to given application dependent policies.

At the top level of the graph, the objective functions of all goals are combined into a single real scalar function, called the overall assessment, by forming the weighted average of the objective function values. This allows the determination of Pareto optimal trade-offs between the key figure goals using scalar optimization methods (cf. Ehrgott 2005).

The values of the first partial derivatives of the scalar overall assessment with respect to the basic key figures can be directly calculated from the computational graph representation (cf. Griewank 1989). This also holds for the derivatives of the objective functions and the composed key figures. The programming library implements the necessary automatic differentiation techniques, so that gradient based optimization methods can be run on the key figure assessment model.

The framework defines an XML (World Wide Web Consortium 2008) configuration format for setting up, storing, and loading the local assessment models. This format can also be utilized in communication acts between the agents for exchanging current key figure settings and properties.

5 Global Agent Control and Coordination

Key figures that create a (semi-)global view on a group of several agents can be assessed by organizing the agents into a hierarchy as it is shown in Fig. 4. A group agent is introduced, which imports all necessary key figures from its subagents by creating collections of key figure proxies in its group assessment model. Group key figures depending on key figure values from different subagents can then be defined, with group key figure goals assigned to them where desired. This organization principle can be applied several times for establishing agent hierarchies of more than two levels.

Since agent hierarchies may change over the course of time, the framework offers mechanisms to automatically acquire and consolidate all key figures with a given name from all current members of a group. For instance, the following textual key figure formula, as it is syntactically accepted by the key figure programming library, defines a new group key figure GroupOrderCount, which takes the current values of all key figures named LocalOrderCount from all members of the agent group MACHINES and calculates their sum:

```
GroupOrderCount=sum(LocalOrderCount[$MACHINES])
```

While the local assessment models are sufficient for guiding the behavior of the agents into the directions given by their local key figure goals, the maximization of the global goal satisfaction requires that parts of the group key figure objectives,

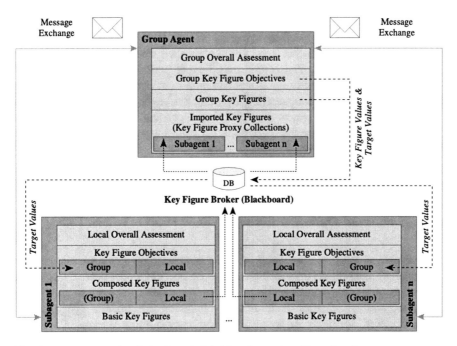

Fig. 4 Assessment and optimization of global key figures in a hierarchy of agents

managed by the group agent, are fed back into the local assessment models of the individual group members in some way. There exist several possibilities for doing this: The framework specifies four different approaches how the global key figure assessment model can be distributed among and managed by the agents in order to attain global key figure based agent control and coordination. They are described in the following subsections.

5.1 Centralized Optimization without Local Models (I)

One possibility for implementing global agent control and coordination is to let the group agent at the highest level in the organizational hierarchy manage the entire global key figure assessment model in a centralized way. For this purpose it imports all local assessment model configurations (e.g., via message exchange using the framework's XML configuration format) into its own assessment model and arranges them in a graph that replicates the agents' current organizational hierarchy. Whenever the latter is changed at runtime, the performed changes must also be reflected in the central assessment model.

In this approach the subagents of the central group agent need to publish the current values of only their local measurands through the key figure broker.

All composed key figures as well as key figure objectives defined in their local assessment model configuration are already present in the central assessment model, so these agents do not need to maintain their own representation of their local part of this model. The central group agent periodically determines target values for its imported measurands and communicates them to the other agents in the hierarchy. This can be done either by using the key figure broker as a system-wide blackboard, or by means of application dependent message exchange. The subagents then directly use the received target values as atomic numerical goals in their planning stage. This approach is only applicable when the planning component supports numerical goals.

5.2 Distributed Optimization (II)

A downside of the first approach is that the central group agent must import the values of all measurands on which the global assessment model depends, calculate the values of all composed key figures within this model, and react to all changes in the agents' organizational hierarchy by updating the model's computational graph structure. For large MASs this task may be prohibitive in terms of available communication bandwidth as well as computing time and memory. Furthermore, when the system's complexity grows, approach (I) might not scale well on the already installed hardware infrastructure.

The second approach uses a distributed global assessment model. Each individual agent in the organizational hierarchy maintains its own representation of only its local part of the global assessment model. If the composed key figures of this local part depend on key figures assessed by other agents, the remote key figures are imported using the key figure proxies described in Sect. 4. Consequently, the importing agent has no knowledge of the calculation rules and measurement processes of the imported key figures. Hence, the local models of the other agents appear as black boxes. Each agent periodically runs an optimization process on its local assessment model and thereby generates target values for its basic key figures. The target values for key figures imported from its subagents are then communicated to or negotiated with the respective subagents, which integrate these target values as derived group key figure goals into their local model (cf. Fig. 4). This is done by setting up a suitable PWLF (e.g., an inverse V-shape) as an objective function for each derived target value based goal.

While the black-box approach may lead to a notably higher distribution of the overall communication and computation load among the agents by possibly eliminating local bottlenecks, it may also lead to the generation of non-reachable sets of target values. Since the calculation rules of imported composed key figures have no representation in the local assessment model of the importing agent, it can happen that subsets of the target values determined by the optimizer component of this agent cannot be achieved all at the same time by the agents actually computing these composed key figures. This may have a negative effect on the system's global goal satisfaction.

5.3 Centralized Optimization with Local Models (III)

The third method is identical to approach (I), except that all agents in the organizational hierarchy do maintain a representation of their local parts of the centrally optimized global assessment model, and target values generated by the central group agent are integrated into the local models as key figure goals with appropriate objective functions as described for approach (II).

This third approach allows the local models to contain additional composed key figures and key figure goals, which do not appear in the central representation of the global model. Thus, the individual agents are able to deliberatively balance their exclusive local (i.e., self-interested) goals against the system's global key figure objectives. After the agents have received the current target values from the central group agent, they can either start a further local optimization process in which they determine target values for their local measurands for use in the planning stage, or they may directly use the local scalar overall assessment as a plan metric or as a utility function. Thus, approach (III) is also applicable in situations where no numerical planner is available to the agents, e.g., when the actions are instead selected in Markov decision processes. While it is more flexible than approach (I), it suffers from the same resource bottlenecks.

5.4 Redundant Distributed Optimization (IV)

Another downside of the first and third approach is the fact that the central group agent constitutes a single point of failure (SPoF). When this agent becomes defective or is temporarily cut from communication, the operability of the entire MAS is at risk. A solution to this would be to eliminate the SPoF by replicating the group agent several times and then synchronize the redundant group agents' decisions concerning the chosen target values (e.g., by domain specific negotiation or failure detection protocols). Taking this principle to its extreme, each agent within the MAS could mirror the entire global assessment model as it is managed by the central group agent in approach (I). This, however, results in an at least n-fold increase in required communication bandwidth for a system of n agents, since each agent has to acquire the current values of all its imported measurands from the other agents.

If no domain specific communication protocols for synchronizing the generated target values among the agents are used along with this approach, it may happen that different agents work with different sets of target values and thereby strive for different optima of the global scalar overall assessment. This may result in a significant deterioration in global agent performance with respect to the global key figure goals.

6 Application and Evaluation

The key figure programming library has been integrated into the IntaPS MAS for process planning and production control (Lorenzen et al. 2006), i.e., into a production logistics intelligent agent scenario. Although the IntaPS Resource Agents use a very simple planning method based on the selection of parameterized plan templates, it was shown that the key figure framework, in fact, is suitable for directing the local and global agent behavior into the desired directions specified by the user-configured key figure objectives. Repeated simulation runs with different sets of key figure based objectives, aimed at local and global job balancing according to different criteria, were performed with the above-described agent control approaches (II), (III), and (IV).[2] In all cases the differences in goal achievement between the three tested methods were statistically significant, and highly significant in most cases. The centralized optimization approach (III) accomplished the highest goal satisfaction in all evaluation scenarios, whereas the redundant approach (IV) implemented without any coordination or negotiation mechanisms between the agents consistently produced the worst results. This shows that the latter approach is not very useful without the specification and utilization of domain specific protocols for target value synchronization.

For the tested goal configurations, the results achieved with the distributed black box approach (II) were only slightly worse than those obtained with approach (III). Although, at times, an IntaPS Group Agent determined target values that were not practically achievable by its subagents, the IntaPS Resource Agents simulating the shop floor machines, due to the structure of the locally encapsulated computation rules of the composed key figures, the subagents were often able to reach key figure values sufficiently close to these target values, so that a good global goal satisfaction level could still be accomplished. While this observation may be attributed to synergistic effects between the particular agent control processes present in the IntaPS system and the tested key figure and goal configurations, it might not hold in general for other goal configurations and agents systems.

Figure 5 illustrates the results achieved by the three control approaches for one of the tested goal configurations, using the IntaPS Group Agent's global composed key figure _TotalTypeDiff_ as an example. The plots in the figure depict the number of occurrences of the different attained key figure values over 100 independent simulation runs of 20 discrete time slots each, i.e., they show a statistical

---

[2] Approach (I) is not applicable to the agent control mechanisms in IntaPS as no sophisticated numerical planning techniques are used by this MAS.

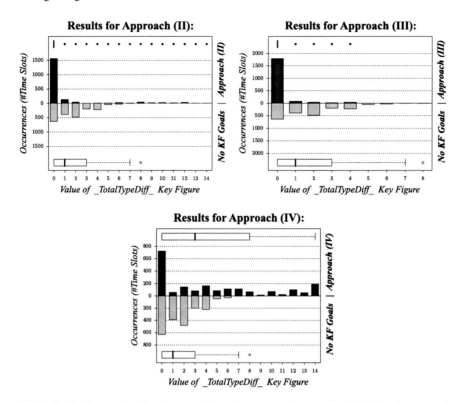

Fig. 5 Evaluation results: The three tested key figure control approaches (black bars) compared to the default IntaPS Resource Agent control strategy without key figure based goals (grey bars)

sample over a total of 2000 discrete time slots for each of the tested approaches. The _TotalTypeDiff_ key figure states the absolute difference between the current number of orders of type A and orders of type B currently assigned to all machines on the shop floor. In this particular goal configuration, the key figure value was to be minimized, i.e., the optimal value for _TotalTypeDiff_ is 0, expressing the intent that in every discrete time slot the amount of allocated orders is ideally the same for both order types. To assess the impact of the user-defined key figure goals on the behavior of the IntaPS agents, an additional 100 independent runs of 20 time slots were performed without any key figure based objectives, resulting in a (nearly-)random allocation of the two order types. In the figure the results of these runs are shown at the bottom half of each plot.

7 Conclusion

In this paper a software framework is presented that applies the concept of key figure systems to the control and coordination of intelligent agents in MASs.

An overview of the framework components is given and four approaches to global agent control and coordination are specified, which differ in the way the global key figure assessment model is distributed among the agents. The domain independent parts of the framework have been implemented in a Java programming library that allows the definition and optimization of user-defined key figures and key figure objectives at runtime of the agent system.

The framework has been integrated into the IntaPS agent system for process planning and production control, in which three of the specified control approaches were evaluated and compared using multiple representative test configurations. The representative results of one of these key figure assessment model configurations are presented in this paper, using the example of one selected global key figure. It was shown that the key figure framework is in fact suitable for directing the local and global agent behavior into the desired directions specified by the user-configured key figure objectives, and that the three tested control approaches (II), (III), and (IV) statistically significantly differ in their attained level of goal achievement.

As has been shown in the evaluation, control approach (IV) requires the implementation of suitable application dependent protocols for target value negotiation among the agents. The specification of such protocols is not part of the key figure framework and remains as future work.

References

Bellifemine F, Caire G, Greenwood D (2007) Developing Multi-Agent Systems with JADE. John Wiley & Sons, Chichester, UK

Ehrgott M (2005) Multicriteria Optimization, 2nd edn. Springer, Berlin, Germany

Franceschini F, Galetto M, Maisano D (2007) Management by Measurement—Designing Key Indicators and Performance Measurement Systems. Springer, Berlin, Germany

Griewank A (1989) On Automatic Differentiation. In Iri M, Tanabe K (eds) Mathematical Programming: Recent Developments and Applications. Kluwer, Amsterdam, Netherlands, pp 83–108

Hansen E, Walster GW (2004) Global Optimization Using Interval Analysis, 2nd edn. Marcel Dekker, New York, NY, USA

Lorenzen L, Scholz T, Timm IJ, Rudzio H, Woelk PO, Denkena B, Herzog O (2006) Integrated Process planning and production control. In Kirn S, Herzog O, Lockemann P, Spaniol O (eds) Multiagent Engineering—Theory and Application in Enterprises. Springer, Berlin, Germany, pp 91–114

MySQL (2009) MySQL: The world's most popular open source database. http://www.mysql.com/. Accessed 28 February 2009

Price KV, Storn RM, Lampinen JA (2005) Differential Evolution—A Practical Approach to Global Optimization. Springer, Berlin, Germany

Rao AS, Georgeff MP (1995) BDI Agents: From Theory to Practice. In Lesser V (ed) Proceedings of the First International Conference on Multiagent Systems (ICMAS '95). San Francisco, CA, USA, pp 312–319

Timm IJ (2004) Dynamisches Konfliktmanagement als Verhaltenssteuerung Intelligenter Agenten. Dissertationen in der Künstlichen Intelligenz (DISKI). infix—AKA Verlagsgruppe, Köln, Germany

Weiss G (ed) (1999) Multiagent Systems—A Modern Approach to Distributed Artificial Intelligence. MIT Press, Cambridge, MA, USA

World Wide Web Consortium (2008) Extensible Markup Language (XML) 1.0 (Fifth Edition) W3C Recommendation. http://www.w3.org/TR/2008/REC-xml-20081126/. Accessed 28 February 2009

Xuan P, Lesser V (2002) Multi-Agent Policies: From Centralized Ones to Decentralized Ones. In Proceedings of the First International Joint Conference on Autonomous Agents and Multiagent Systems (AAMAS '02), New York, NY, USA, pp 1098–1105

Xuan P, Lesser V, Zilberstein S (2001) Communication Decisions in Multi-agent Cooperation: Model and Experiments. In Müller JP, Andre E, Sen S, Frasson C (eds) Proceedings of the Fifth International Conference on Autonomous Agents (AGENTS '01), Montreal, Canada, pp 616–623

Stockout Costs in Logistics Unconsidered

Stockout Costs do Affect Service Level

Henner Gärtner, Rouven Nickel and Peter Nyhuis

1 Introduction

Inventory and procurement costs essentially determine the overall logistic costs. The follow-up costs of missing parts in production have not been considered adequately in the discussion about cost-drivers of production logistics in the past. Stockouts in manufacturing do disrupt scheduled production, leading to rescheduling, repeated exchange of dies, and downtimes (Reichmann 1997). Though stockout costs have neither been isolated from the production costs nor regarded as part of the logistic costs (Gudehus 2008). As a consequence, industries traditionally consider only inventory and procurement costs for computing the optimal inventory level of parts being stored in a warehouse.

The research question consists of quantifying the impact of stockout costs on the weighted service level (Lutz 2002) of the receiving store. The approach is to measure both sides of the scale in regard to the service level: the costs of inventory and procurement logistics on the one hand and the damage of stockouts in production on the other hand. Since inventory costs tend to be high at a high service level and stockout costs are high at a low service level, the cost minimized service level needs to be computed.

The developed logistic model allows to quantify the relation between the service level and the logistic costs for a given part, product or supplier and production characteristics. The quantification is built on the framework of Lutz's service level curve which describes the coherence between the inventory level and the service level towards production (Lutz 2002). The model seeks to point out the most relevant parameters (i.e. delivery date deviation, requirement rate deviation, material price). The example of typical A, B, and C parts demonstrates that the

H. Gärtner (✉), R. Nickel and P. Nyhuis
IPH-Institute of Integrated Production Hannover GmbH, Hollerithallee 6, 30419,
Hannover, Germany
e-mail: gaertner@iph-hannover.de

H.-J. Kreowski et al. (eds.), *Dynamics in Logistics*,
DOI: 10.1007/978-3-642-11996-5_15, © Springer-Verlag Berlin Heidelberg 2011

enhancement of stockout costs significantly shifts the point of the cost minimized service level. The model's practical relevance is to deduce decisions for the procurement, the warehousing, and the production process.

2 Supplier and Production Characteristics

The delivery characteristics of suppliers and the requirement characteristics of production have a direct impact on the costs of logistics and the achievable service level. The general stock model allows the description of deviations as independent characteristics (Nyhuis and Wiendahl 2008). The general stock model plots the inventory against time with large incoming quantities and ideally infinitesimal small outgoing quantities. Figure 1a illustrates delayed deliveries using the parameter maximum positive input due date deviation. Figure 1b shows under-deliveries by the parameter maximum negative input quantity deviation. Figure 1c demonstrates the effect of the maximum positive demand rate deviation from the production's mean demand rate deviation on the inventory progression starting with the given re-order inventory level. The higher the lead times are, the less deviations can be buffered by the use of the available stock. In the given example, the safety stock needs to be used to fulfill all parts requirements (Fig. 1c).

3 Service Level and Inventory Level

The term service level is being introduced to measure the logistic performance of the receiving store towards production. The service level indicates the probability by which the production's requirements can be met by the available stock in terms of time and quantity (Gudehus 2008; Lutz 2002). More than just considering the

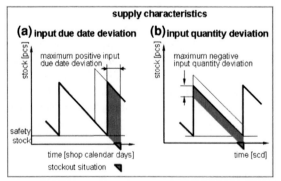
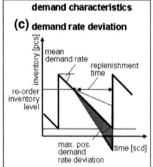

Fig. 1 Supply and requirements characteristics in the general stock model (Nyhuis and Wiendahl 2008)

number of stock release orders according to the inventory level, the weighted service level takes the parts quantities of each stock release order into account (Lutz 2002). The weighted service level is a common key figure in production.

4 The Service Level Operating Curve

The characteristic curve of the service level empirically shown by Lutz explains the quantitative relation between the mean inventory and the weighted service level (Lutz 2002). The service level operating curve belongs to the category of output rate curves which are used to assess operating conditions in production and logistics. The progression of the service level operating curve (service level with regard to the inventory) is similar to the progression of an engine's output rate curve (output rate with regard to number of revolutions). Though, the service level does not lapse beyond the turnoff point. Instead, a 100% logistic service level is obtained with a further increasing inventory (Nyhuis and Wiendahl 2008).

5 Components of Logistic Costs

In order to align which service level is aspirable for which enterprise, the logistic performance—measured by the service level—and the induced logistic costs need to be deliberated. Therefore, all logistic cost components influenced by the service level need to be analysed.

Indisputably, procurement and inventory costs are part of the logistic costs and need to be described as a function of service level. Enhancing logistic costs by the aspect of stockout costs is practically unsolved, even though the literature describes stockout costs since the end of the 1970s (Alscher and Schneider 1981). Up to now stockout costs remain anonymous within the production costs and are not being assigned to the logistic costs (Gudehus 2008). They arise when an agreed service level is not adhered to the production. Since stockout costs can accumulate to an extensive cost proportion (Goerke 2005; Esser 1996) they are subject to the logistic model proposed in this paper.

The procurement costs of an enterprise can be evaluated using a process cost approach. In the field of procurement, process costs originate from the parts disposition (i.e. identifying requirements), the purchasing (i.e. releasing purchase orders) and the receipt (i.e. verifying the goods and checking the invoices). For simplification, the number of purchase order items is identified as the uniform cost driver. For low service levels the procurement costs are high because a low inventory level requires an extensive amount of high priority purchase orders and a close monitoring of suppliers regarding the delivery date adherence.

In the warehouse, logistic costs can be measured as a percental markup between 19 and 30% on the stock value. They consist of inventory costs (in the interval of

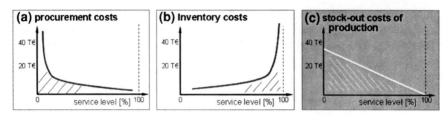

Fig. 2 Components of logistic costs in the service level-cost-diagram

6–9%), costs of deterioration (3–5%), costs of loss and transport (2–4% each) as well as costs of amortization, inventory control, tax, and insurance (1–2% each) (Delfmann 2003). The maximum values of the given intervals are often reached at a high service level being accompanied with a high inventory. Therefore, the inventory costs exponentially increase with a rising service level.

Stockout costs in production evolve from replanning, stopping, halting, and restarting production (Reichmann 1997) when parts requirements are not met in time and in quantity. The existence of stockout costs can be expressed by the discontent of production on missing parts. The reason for the discontent is that production stops or stalls once a month in 80% of Germany's production companies (Goerke 2005). The industry can neither describe the influence variables nor quantify the loss evolving from undelivered parts. Science has not yet revised its first approach of the 1970s by which stockout costs are considered as a linear declining function of the service level (Alscher and Schneider 1981).

The service level-cost-diagram allows a unique display format for all logistic costs by outlining the logistic costs on the ordinate and the service level on the abscissa. Hence, the procurement, inventory, and stockout costs are described as functions of the service level (Fig. 2).

6 Cost Minimized Service Level

The consequences of neglecting stockout costs in the logistic model become obvious, if at first the procurement and the inventory costs alone are added to the logistic cost function (Fig. 3a). The cost minimized service level can be found in the minimum of this function. Logistic efficient positioning between a high service level and low logistic costs seems feasible. Adding in the stockout costs (Fig. 3b) results in a new progression of the logistic cost function (Fig. 3c). The cost minimized service level can be found by leaping onto the new, higher situated function and shifting along this function down to the cost minimum. Due to the progression of the stockout cost function, the new cost minimized service level is always found at a higher service level. As a consequence, enterprises with low service levels are even more far away from the aspirable cost minimized service level as expected.

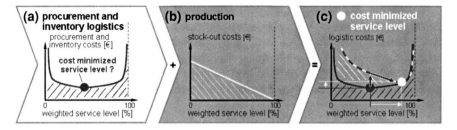

Fig. 3 Enhancement of logistic costs by stockout costs

Fig. 4 Exemplary cost minimized service levels for *A*, *B*,
and *C* parts

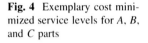

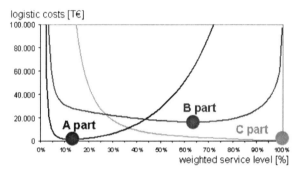

Figure 4 displays typical progressions of the logistic cost function for A, B, and C parts as well as the individual point of the cost minimized service level. The evaluation of the model reveals that process costs of procurement essentially determine the overall logistic costs for low value C parts with high quantities required. In order to avoid these process costs the service level needs to be set close to 100%.

For the B parts with a mean parts value and typically mean request rates bellied curve progression in a wide service level interval. An altered lead time considerably shifts the point of the cost minimized service level. According to the current research, other parameters (i.e. input due date deviation) show a comparably low influence. Enhancing the logistic model by stockout costs as proposed has a significant impact on B parts because their key component of the logistic cost function is made up of stockout costs.

7 Assembling Many Parts to One Product

The proposed logistic model does not only allow application on one article but furthermore the extension on entire lists of items. When machines and installations require the assembly of hundreds and thousands of parts to one product, the service level gains considerable relevance. 91% of production disruptions in the assembly can be traced back to parts deficiencies. Besides faulty or wrong items, 57% of the parts deficiencies are caused by missing parts (Esser 1996).

Production disruptions caused by missing parts can only be eliminated, if all items on the component or product list of items are available. The so called usage based service level describes the logistic performance of all items needed to assemble one product by multiplying the item based service levels. If an item based service level of i.e. 90% (70%) is given for a list of three items to be assembled, then the usage based service level amounts to:

$$SL_{usage} = 90\%_{A1} * 90\%_{A2} * 90\%_{A3}$$
$$= 73\% \left(SL_{usage} = 70\%_{A1} * 70\%_{A2} * 70\%_{A3} = 34\%\right).$$

8 Decision Support

By varying the logistic and cost parameters (i.e. input due date deviation, stockout cost rate), the proposed model allows decision support by setting up and confronting various scenarios. The scenarios demonstrate the contribution of an activity (i.e. lowering the input due date deviation) either to increasing service level or to lowering logistic costs. The additional logistic costs needed to reach an aspired higher service level shall be quantified so that an enterprise can take position between two target values logistic costs and logistic performance.

To achieve the aspired cost minimized service level, the purchasing department can readjust lot sizes. The receiving store can reset the inventory level. Structural changes in production can contribute to reducing stockout costs in the long run because the model supports decision making for increased flexibility of the organization (i.e. shop agreements which support flexible working hours) and new investments. Taking stockout costs into account may increase the profitability of an investment (i.e. for an additional equipment). Stockout costs can be included into make-or-buy decisions. The purchasing department may address additional stockout costs in negotiations with suppliers which are caused by insufficient logistic performance.

Once the impact between the logistic costs enhanced by the stockout costs on the service level is revealed, producing enterprises can take position at the cost minimized service level. Thereby inventory stock will be reduced, stockout costs in production will be avoided, and a cost optimal high service level can be achieved.

References

Alscher J, Schneider H (1981) Zur Diskussion von Fehlmengenkosten und Servicegrad. Fachbereich Operations Research, vol. 3, University of Berlin.
Delfmann W (2003) Controlling von Logistikprozessen: Analyse und Bewertung logistischer Kosten und Leistungen. Schäffer-Poeschel, Stuttgart.

Esser H (1996) Integration von Produktionslogistik und Montageplanung und -steuerung. Shaker, Aachen.

Goerke M (2005) Trumpfkarte ausspielen. MM Logistik, vol. 4, no. 7: 26–28.

Gudehus T (2008) Logistics—Principles, Strategies, Operations. Springer, Berlin.

Lutz S (2002) Kennliniengestütztes Lagermanagement. In: VDI-Verlag (eds) Fortschritt-Berichte VDI, vol. 13 Fördertechnik/Logistik, no. 53, Düsseldorf.

Nyhuis P, Wiendahl H-P (2008) Fundamentals of Production Logistics—Theory, Tools and Applications. Springer, Berlin.

Reichmann T (1997) Controlling—Concepts of Management Control, Controllership, and Ratios. Vahlen, München.

Performance Measurement
for Interorganisational Collaborations
of SMEs

Yilmaz Uygun and Andreas Schmidt

1 Introduction

In today's globalised and dynamic economy, small and medium-sized enterprises (SME) face enormous difficulties to compete on the market. Competitors from all over the world are able to provide products and services of similar quality and functionality. Simulataneously, companies are facing a significant development towards shorter product life-cycles, increasing customer focus, higher flexibility and faster innovations. Companies have seen that the willingness and ability to collaborate with other organisations is a fundamental requirement for the "health" and success of a company. Alliances with other organisations help especially small and medium-sized companies to maintain their collaborative advantage. Furthermore, collaborations result in creating additional value for all partners (Dreyer and Busi 2005).

Interorganisational relationship supports companies in concentrating on their core competencies while forming long-term partnerships with other businesses that poses complimentary competencies leads to sustainable opportunities and advantages. Consequently, the formation rate of inter-firm collaborations has dramatically increased in recent years (Ireland et al. 2002; Draulans et al. 2003). It is even acknowledged that the 21st century will be an age of competition between value chains and supply chains rather that between single companies. Despite the growing importance of inter-firm relationships, the success rate remains low. Although the failure rate is denoted differently in different literature, it is clear that more than 50% of all collaborations fail (Draulans et al. 2003). Organisations find that there are many unpredictable barriers to their integration strategy. Own empirical research on small and medium-sized companies on this topic have shown that companies do not employ systems or frameworks to measure or

Y. Uygun (✉) and A. Schmidt
Chair of Factory Organisation, Technische Universität Dortmund, Dortmund, Germany
e-mail: uygun@lfo.tu-dortmund.de

H.-J. Kreowski et al. (eds.), *Dynamics in Logistics*,
DOI: 10.1007/978-3-642-11996-5_16, © Springer-Verlag Berlin Heidelberg 2011

manage collaborations holistically, mainly due to the inadequacy of the existing performance measurement systems.

According to this, the paper explores the nature of collaboration and its complexity. Drivers as well as obstacles of forming collaborations are pointed out. It goes onto review performance measurement in order to develop performance measures and a performance measurement framework, both to be used by small and medium sized companies to support their collaborative advantage.

2 Interorganisational Collaboration

2.1 Theoretical Backdrop of Interorganisational Collaboration

A theoretical and practical examination of *interorganisational collaboration* reveals a high degree of heterogeneity within this topic. Heterogeneity is generated primarily by two factors: the high number of seeming synonyms and different, sometimes conflictive, associations that are linked with collaboration. A literature review provides many words that are used for the same idea. Lower (2008) writes thereto: "teamwork, partnership, group effort, alliance and cooperation are all synonyms for collaboration". In most cases, more than one term is used within one paper, indicating the use as synonyms. In some cases, a term can be a synonym to 'collaboration', in other occasions there exist key differences.

The term collaboration is widely used in many fields of research. The multi-disciplinary nature of the term leads to either a very broad or to a very narrow discussion, when the term is defined for the specific purpose. The description of collaboration in an industrial environment may differ significantly from the use in a public or psychological surrounding. Hagel et al. (2002) consequently write: "often, collaboration refers to any situation where companies interact with each other to support broad business objectives", just to add "by this definition, nearly every company is a collaborative enterprise." The above stated description seems to be one of the least common denominators that researchers and practitioners can agree upon. However, an in-depth study of collaboration requires a more specific and adapted definition.

In social sciences and philosophy collaboration is often regarded as an opposition to conflict (e.g. Curseu et al. 2005) or the definitions are focused on problem solving (e.g. Gray 1989). Literature concerning the public sector tends to neglect financial or business-related aspects (e.g. Luther 2005). Other, business-related papers devote the definitions on supply chain aspects, which cover only certain aspects of the topic (e.g. Mentzer et al. 2000) or they merely put collaboration in contrast to competition (e.g. Camarinha-Matos and Afsarmanesh 2008). The problem that results from this heterogeneity is that a too broad definition also applies to nearly all synonyms, whereas the more narrow definitions focus on certain aspects and are therefore unsatisfying. Cohen and Roussel (2004) pose the same question: "Why is it so hard to define collaboration?" just to answer it straight away: "Because it can be many things and involve many types of partners."

Before a valid definition can be provided, the surrounding parameters need to be identified. The underlying assumption includes the participation of many types of companies and organisations, as for example small, medium-sized and large companies, producing companies, service providers or retailers or public institutions such as universities. The number of involved partners is two or more, with a focus on bilateral relationships within this paper. Furthermore, as this paper stresses the needs of SME, at least one partner in the collaboration is a small or medium-sized company. A definition that matches these guidelines and moreover mentions basic characteristics of collaboration is provided by Parkinson (2006):

Collaboration is a mutually beneficial and well-defined relationship entered into by two or more organizations to achieve common goals. The relationship includes a commitment to mutual relationships and goals; a jointly developed structure and shared responsibility; mutual authority and accountability for success; and sharing of resources and rewards.

It is, however, to be pointed out that many of the results within this paper also apply for more loose or more intense relationships. This may include any form of relationship between organisations as well as within companies.

2.2 Classification and Characteristics of Collaboration

Collaboration requires the participation of at least two otherwise independent organisations. When exactly two entities (e.g. companies) are considered, it is about a bilateral collaboration. More than two companies are involved the situation is that of a supply chain or a network of companies, which is considerably more complex than the bilateral collaboration. This paper focuses on the bilateral situation.

Especially a particular triple of terms is viewed with care: most frequently, scientific papers distinguish between the three terms coordination, cooperation, and collaboration; e.g. Dreyer and Busi (2005), Fong et al. (2007), Luther (2005), Schuh et al. (2007). Often, communication is added as a pre-stage. This paper seizes this concept and provides descriptions for the above mentioned "four C's", as presented in Fig. 1 The figure shows a continuum with growing intensity, integrity, and complexity from coordination towards collaboration.

Compared to *communication* and *coordination*, *collaboration* has more strict and formal requirements and has a long-term devotion. It is the highest level of collective human or organisational interaction. Collaboration aims at achieving a common goal with shared resources and risks and comprehensive, thorough planning. Due to the long-term horizon the integrity between the partners is very high, the relationship is very intense and all partners face a new structuring of internal and external processes. The high intensity leads to a significantly higher complexity compared to the other levels.

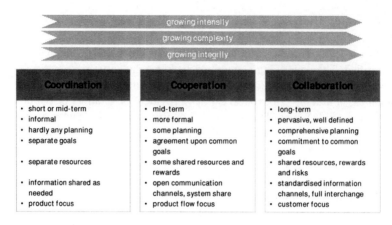

Fig. 1 Distinct characteristics between coordination, cooperation and collaboration

As important as to clearly define the term is to expose the *characteristics* of an interorganisational collaboration. Describing characteristics can lead to a better and more comprehensive understanding about the nature of collaboration and its prerequisites. These characteristics or main themes are in general involvement of more than one organisation, more than the sum of the parts, definition of a common goal, long term relationship, mutual commitment, mutual benefit, and interdependencies.[1]

In general, collaborations are a median between market and hierarchy (Dreyer and Busi 2005; Sydow 1992; Wildemann 1996; Schönsleben 1998; Corsten and Gössinger 2001) *Market* structure is characterised by arms-length relationships between the participants and is regulated by price discussions, confrontation, and coordination. *Hierarchy* means a strong vertical integration of two companies, characterised by authority and control. Somewhere between these extremities collaboration is settled, merging elements from both ends of the line.

2.3 Motivation, Obstacles, and Success Factors

For a better understanding of the reasons for companies to establish partnerships or networks, it is advised to differentiate between two basic *motivation* factors—internal and external ones (Fig. 2). External factors are parameters that force a company to collaborate. In the case the company chooses not to collaborate, it will face perceivable difficulties or penalties. In this situation the company has no other choice than to establish a close relationship with another organisation. These are external influences, coming in from the market, the customer, from competitors or

---

[1] For the definition of specific characteristics refer to e.g. Stadtler and Kilger (2002), Dussauge and Garrette (1999), Ireland et al. (2002).

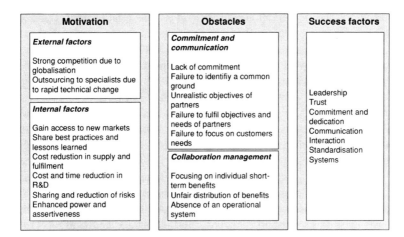

Fig. 2 Motivation, obstacles and success factors of collaboration

from the government. The basic idea is that the company will be worse off if no collaboration is undertaken. If a company however anticipates potential benefit from collaborating, then these reasons are internal factors. Collaboration will add bonuses, but abandonment will not affect the organisation negatively. It is the companies own choice and it is not compelled to do so. Both internal and external factors may apply at the same time and depending on the type of company, on the industry sector or the country, one side can outweigh the other.

Having pointed out some reasons for companies to collaborate in the first place, there have to be analysed the reasons, why collaborations are highly risky and most organisations fail to take the full advantage. Most people and companies are not aware of one very important fact: It is not natural to collaborate. Consequently, risks and weaknesses or typical reasons for failures of collaborations are pointed out that every collaborating party should be aware of. A collection of key failure reasons (*obstacles*) is presented by Bititci et al. (2005) as a result of a literature review. They identified eight most commonly quoted failures (Fig. 2). Roughly, the eight reasons can be divided into two areas. The first five reasons deal with commitment and communication. If these two elements are missing or poorly developed the collaboration has a chance to fail due to derived problems. Therefore, objectives and value propositions need to be communicated to build up and maintain trust, respect, and dedication. The last three reasons relate to collaboration management. A strong management set up to guide and control the relationship is necessary to keep focus on strategic goals and to organise benefits. A management system operating for the collaboration rather than for single organisation can contribute to the management.

As stated before, a large amount of interorganisational collaborations fail and cannot fulfil the expectations of the involved parties. There are obstacles that impede the success of collaborative ventures. Collaboration management can reduce these risks by promoting critical *success factors* being of strategic importance for the perpetuation and support of the relationship. The literature review provides many sources that deal with these success factors. For example, Shelbourn et al. (2007) name six factors, Kanter (1994) proposed eight criteria for successful relationships, Power et al. (2001) defined factors critical for successful agile organisations in managing their supply chains. A summary of the literature leads to collaboration drivers as they are presented in Fig. 2.

One of the most frequent success drivers for collaboration is leadership (e.g. Parkinson 2006; Parung and Bititci 2008; Mentzer et al. 2000). Without a clearly defined leadership that governs the collaboration, activities are likely to be ineffective and inefficient. Harvey and Koubek (2000) explain: "There are many steps that lead to any collaboration between multiple organizations, such as negotiating and developing the collaboration agreement, administering the collaboration project, and actually forming the group to complete the collaborative task." All these tasks need to be coordinated. Dussauge and Garrette (1999) by citing Killing (1982) state that the balance of power within a partnership has a great impact on its performance. Observing the management of joint ventures (JV) as a special form of collaboration, they point out three types of leadership: (1) dominant parent JV (2) shared management JV, and (3) independent JV through autonomous manager. A similar classification is proposed by Kuhn and Uygun (2008). They distinguish within a portfolio between central/decentral and hierarchical/heterarchical constellation; namely (1) committee (central and heterarchical), (2) leader (central and hierarchical), (3) dependent teams (decentral and hierarchical) and (4) independent teams (decentral and heterarchical). A relevant research on international joint ventures (IJV) reveals that the most successful collaborations are those governed by an independent manager and that those collaborations with a shared management show the worst performance. Furthermore, 50% of the jointly managed collaborations fail. Here, the conclusion can be drawn that intense and long-term collaborations need an independent management by an external manager. If this cannot be established, the dominant partner must take over the governance to secure high performance and stability. (Dussauge and Garrette 1999; Killing 1983)

2.4 Complexity in Collaborations

In general, complexity in collaboration consists of eight elements; uncertainty, dynamics, multiplicity, variety, interactions, interdependencies, instability and unpredictability.[2] Collaboration management is constantly faced with these elements of complexity. According to Schuh et al. (2008) a 'complex' system must

---

[2] For the specific aspects refer to Gulati (1998), Bititci et al. (2003), Schuh et al. (2008) Middleton-Kelly (2003); Andriani and Passiante (2004); Roetzheim (2007), Klein and Adelman (2005), Pfeffer and Salancik (1978), Banff (2003), Thompson (1967), Das and Teng (2001), Nkhata et al. (2008), McCutchen et al. (2008)

contain at least the first two elements, uncertainty and dynamics and any of the other elements. You can say that complexity rises with the number of involved elements. There are basically two strategies to encounter complexity, that is, reduce the effects or handle complexity. In this context, viewing performance is one way to manage complexity within an organisation. Kaplan and Norton (1992) has lead this awareness to the Balanced Scorecard (BSC), their structured approach to view performance.

A performance measurement system (PMS), like the BSC, is a tool that provides this structured insight into the many aspects of an organisation. In their review about performance measurement in SMEs, Garengo et al. (2005) state, the PMS "can play a key role in supporting a rational approach to growing complexity and qualitative improvement in SMEs". They propose the application of a measurement system to encounter the growing complexity. In the following performance measurement will be analysed so as to derive a specific PMS for collaboration of SMEs.

3 Performance Measurement

The change in the perception of performance and its measurement during the early 1990s has led to a number of new frameworks that adapted to the new challenges of business enterprises (Fig. 3). Along with the general view of corporate performance also the underlying performance measures have changed. The use of accounting measurement systems and the resulting focus on financial metrics impeded managers to notice the decline in non-financial areas (Eccles 1991; Fisher 1992). Customer satisfaction, quality, in-time-delivery, and

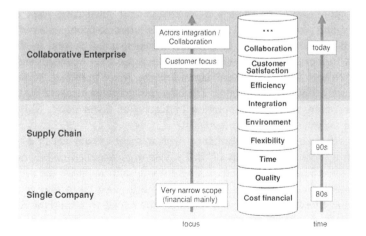

Fig. 3 Evolution of performance measurement (according to Busi and Bititci 2006)

innovation could not be monitored with the mere application of accounting systems. Furthermore, the shift towards quality and flexibility required a different way to measure performance and corporate success.

3.1 Basic Requirements of a Performance Measurement Systems in Collaborations of SMEs

Many authors proposed basic guidelines for how the PMS should look like and for the requirements for the used measures. In the following, some of these guidelines are briefly listed, followed by a summarised proposal for the criteria. Basic requirements for the development of a performance measurement system are presented in Oakland (1993), Tangen (2004), and Kennerly and Neely (2004). Summarised, the following basic specifications can be made: A performance measurement system must support strategic objectives, have a balanced set of measures, be simple and easy accessible, contain clear and unambiguous measures, have a limited and manageable number of measures, provide feedback for driving improvement, consider measures beyond organisational boundaries, contain stakeholder perspectives, and regard processes.

In addition to this, a performance measurement system for collaborations of SMEs has to be developed according to the needs of SMEs. Personal interviews with some SMEs and consultants specialised in collaboration in UK and Germany revealed interesting information. The interview partner stress that none of the companies are collaborating according to the understanding within this paper. However, every company had external business partners, and they had great impact on the well-being and success of the company. Furthermore, the interviews have shown three main issues: Firstly, collaboration management takes place on peer-to-peer basis including a high amount of inter-personal communication. Secondly, companies use models to systemise and standardise their processes, workflows, and projects. Thirdly, small and medium-sized companies often do not have the power and authority to establish a management system used by every partner. A further detection is that the companies see interorganisational relations rather as projects than as long-term commitments. Apart from that, performance is not measured within collaborations. The interview partners measure the success by taking the output of the collaboration into account. Even more, SME measure issues and collect data rather reluctantly.

The development of a performance measurement system needs a substantial basis with a focus on the needs of SMEs. Interorganisational collaborations turn out to be specifically different than single organisations. Thus, the requirements and surrounding conditions are different and need to be taken into account. The international literature review and the interviews lead to important findings. In detail, a PMS for SMEs and collaboration must feature eight essential requirements like presented in Table 1.

Table 1 Basic requirements for PMS in collaborations of SMEs

Simplicity	Framework must be
	• easily accessible and easy to be operated
	• seamlessly integrable into existing systems, standard procedures and methodologies
	• avoid additional efforts in establishing and measuring data and metrics
Communication	• communication as most important fundament of collaboration in SMEs must be a component of a measurement and management framework
	• communication can be represented in various ways, including communication channels, frequencies, media etc.
Common goals	• definition of and dedication on common goals and objectives is a key principle of collaboration to ensure that activities remain effective and goal-oriented
	• deriving milestones and intermediate goals
Organisation structure	• organisation structure combines all data indicating who is involved in the collaboration (participants and actors)
	• leadership, responsibilities, roles, competences, and abilities are assigned to the actors giving a holistic view over the structure
	• participants and actors can be displayed on individual as well as on business unit level
Rewards	• rewards must be in-line with the previously set goals
	• definition of both financially and non-financially rewards
	• most commonly used indicators for success needs to be included into the PMS
Contributions	• balance of contribution between the partners to the collaboration is a great indicator for the share of rewards
	• contribution of physical goods like material or finance, but also more intangible goods like reputation, trust and respect and know-how
Interactions	• definition of interactions (information, material and services)
	• standardising of processes, workflows and the supporting systems
Cross-company	• explicit focus on cross-company interactions

3.2 Evaluating Different Performance Measurement Systems

In order to meet the abovementioned requirements, it is appropriate to analyse the existing performance measurement systems. Keller and Hellingrath (2007) identified an extensive list of performance measurement frameworks, like the *Balanced Scorecard (BSC)* with its four perspectives (Kaplan and Norton 1992, 1996a, 1996b), the *Performance Prism* with its five perspectives (Neely et al. 2001, 2002), the *Integrated Performance Measurement System (IPMS)* with its integration of four hierarchically aligned business levels (Bititci and Carrie 1997), the *SMART Pyramid* with its internally and externally focused measures of performance (Cross and Lynch 1988), the *Performance Measurement Matrix* with its external and internal dimensions (Keegan et al. 1989), the *Results and Determinants Framework (RDF)* with the distinction of two types of performance measures (Fitzgerald et al. 1991) or the *Inputs, Process, Outputs, and Outcome Framework (IPOO)* with four stages of performance measures (Brown 1996) and similar ones (Parung and Bititci 2008; Saiz et al. 2007; Simatupang and Sridharan 2004; and Wood and Gray 1991). Apart from these performance measurement framework there also exist relevant collaborative frameworks, like derivates and adoptions of the BSC to supply chains and networks (e.g. Weber et al. 2002; Jehle et al. 2002), the *Three Elements Framework* with three kinds of measurement (Parung and Bititci 2008), the *Performance Measurement System linked to Enterprise Networks (PMS-EN)* with its three dimensions (Saiz et al. 2007), the *Collaborative Index* with three interrelated dimensions (Simatupang and Sridharan 2004) or the *Antecedent-Process-Outcome Model (APOM)* (Wood and Gray 1991).

Each framework has distinct differences compared to the others. It can be concluded that different performance measurement systems have different approaches. Which framework is best suited depends on the situation, the organisation, the target. Figure 4 compares the deduced features with the capabilities of the existing frameworks. Traditional PMS for single organisations and those for collaborative environments are separated. (● for full match, ◗ for half match, ○ for no match) The eight features (Table 1) a displayed in the rows, the measurement frameworks are placed in the columns. Figure 4 provides several results. Firstly, every considered framework has its weakness and none is ideally suited according to the required features. Expectedly, the more traditional PMS have focus on single companies and thus do not consider systems beyond the organisational borders. Secondly, nearly all frameworks do not reflect organisation structures satisfyingly. The frameworks with the best results are the framework with the three elements by Parung and Bititci (2008) and the Antecedent-Process-Outcome Model by Wood and Gray (1991). However, these two models were originally not designed as performance measurement systems but more as supporting and systemising frameworks. Good results were also achieved by the

Features	Performance Measurement Systems							Collaborative Frameworks			
	BSC	IPMS	SMART	Prism	Matrix	RDF	IPOO	Three Elements	PMS-EN	C-Index	APOM
Simplicity	◗	◗	●	●	●	●	●	●	○	●	●
Communication	◗	○	○	◗	○	○	◗	●	◗	●	●
Common Goals	◗	○	●	◗	○	◗	◗	◗	●	○	○
Organisation structure	◗	●	●	◗	○	○	○	○	◗	○	○
Rewards	●	◗	◗	●	◗	◗	●	●	●	◗	●
Contribution	●	◗	◗	●	●	◗	●	●	●	◗	●
Interactions	○	●	○	●	○	◗	●	●	◗	◗	●
Cross-company	○	◗	○	○	◗	○	○	●	●	●	●

Fig. 4 Evaluation of given performance measurement frameworks

Performance Prism (Neely et al. 2001) and the inputs, processes, outputs, and outcome model by Brown (1996).

As none of the frameworks described seems ideally suited to support collaboration measurement and management for small and medium-sized companies, a new framework is created within this paper. This framework will combine strengths of the existing systems taking all other required eight features (Table 1) into consideration.

4 Development of a Performance Measurement Framework for Collaborations of SMEs

Based on the analysis above, the performance measurement framework developed within this paper consists of four main elements and which integrates holistically the features of the analysed PMS (Fig. 5): *contribution, process, benefit,* and *structure*—according to the quartet input, process, output, and outcome respectively. A focus on four main areas pays contribution to the required simplicity. All four elements combined make the performance of the collaboration. The performance stands as a fifth element in the centre but is not attached with performance indicators. The four elements are logically arranged around the performance. Not only are the closely connected with the centred performance, they are also interlinked.

4.1 Structure

The element of structure is the starting and end point of the performance measurement framework. Every collaboration is defined by common goals and

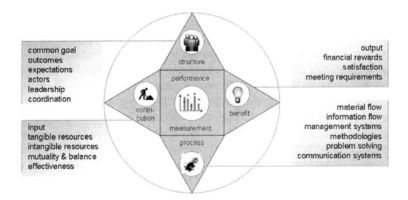

Fig. 5 The proposed performance measurement system with key elements

objectives, that represent the purpose and raison d'être of the partnership. These strategic goals and derived objectives form the basis of a vision or mission that ties the partners together. Here, participants name their expectations and desired outcomes. As this element contains a strong strategic connotation, it is displayed at the top of the framework.

All actions and activities within the collaborative environment need to be aligned with the goals. The goals are the basis for measuring effectiveness, therefore it marks the starting point. During the execution of the collaboration, all benefits and processes must meet the defined expectations, that is, be effective. Writing down and formulating the common goal of a partnership gives the relationship a deeper meaning and a strong basis. The agreement upon goals, objectives, milestones, and expectations also helps to reduce "hidden agendas". In addition to the goals and objectives the participants and actors are listed here. 'Actors' means 'individuals'. Every actor in the collaboration can influence the system. Also, every actor takes on a certain role within the social system of the collaboration. Each role conveys responsibilities, duties, and competencies and requires capabilities and qualification. Incorporating participants and actors into the framework helps to get insight into the organisation structure of both each organisation and the venture.

The element structure also embodies the superordinate management of the collaboration. Leadership is an important component of any collaboration. The type of leadership is defined here as well as its responsibilities are depicted. The management coordinates activities and processes and controls the "health" and state of the relationship. It compares achieved objectives with previously defined objectives and ensures that all activities strive towards the common goals. That means the outcome of activities is measured.

4.2 Contribution

The element contribution is the equivalent to the input into the collaborative system. These input parameters describe everything that is inserted into the system. The input can be divided into tangible and intangible parameters. Material and technical capabilities as well as information, data, and finance are more tangible resources. Organisations also contribute rather intangible resources like commitment, capabilities, knowledge, experience, and reputation. The framework can provide a list of the major contribution of every participant.

The listing, though, is not sufficient to display performance. It is necessary for partners in a collaboration to build up and maintain mutuality and balance. A successful collaboration means that all involved partners contribute equally and every partner benefits from the contribution of the respective partners. If sizable and long-term imbalance can be observed, it may pose a negative impact on the overall success of the partnership. In addition to the balance of input parameter, effectiveness is a second major precondition. Ineffective contribution and inadequate use of resources have a negative effect on the performance. Therefore, any contribution must be in an adequate proportion to the goals and objectives of the collaboration.

4.3 Benefits

Benefits from the collaborative venture are an important indicator for success. For companies, output is the most obvious and accessible parameter to measure performance and often might even be the only one. Benefits from the collaboration can be as multifaceted as the contribution and depend heavily on the type and objective of the partnership. Generally, it is the added value for the participants. Usually, organisations pay attention to financial profit, but collaborative benefit goes beyond mere financial measures. Often, companies gain access to new markets and customers as the collaboration enables them to produce other goods than without it. Here, partners can exploit synergy effects leading to generally better benefits that each partner is able to gain alone.

After displaying the collaborative benefits within the framework, these benefits must be compared with the strategic goals and objectives. Here, the differentiation between output and outcome becomes clear. Output means immediate and intuitive results. Outcome means interpreting the results and comparing them with the degree to which they satisfy the requirements and objective of the collaboration. So, the element benefit is directly linked back to the structure, where goals and objectives are defined.

4.4 Process

Input and output, in the case of this framework also called 'contribution and benefits', are linked with each other. Input parameters, however, do not automatically transform into benefits. There are activities within the system. Companies may pay attention to outputs and inputs, but pay less attention to the processes in-between. This element is concerned with the question, how things are done within the system.

Processes can be flows of material and information. The process element displays the way material and information is flowing within the partnership. Often, process management and reengineering is a fundament of enhancing performance within a system. Brannick et al. (1997) say about team performance: "Team performance measures must tap teamwork processes as well as outcomes since processes provide better diagnostic information." Organisations might be satisfied with the output of the collaboration, saying that therefore performance is high. However, regardless of changing outputs, changing processes that lead to outputs can boost performance.

The interviews have shown that companies work with standardised methodologies and workflows. These procedures find their places in the process element, showing how and when activities are carried out. Another important backbone of any collaboration is communication. Communication channels and systems are incorporated in the process element. This includes the agreement upon regular meetings as well as displaying, who communicates with whom. Communication is also about problem solving and conveying appropriate information. That is the reason, why communication is an important indicator for performance. A high degree of communication can indicate many problems that must be solved, but can also indicate sympathy and friendship. A low degree, though, can indicate interpersonal problems as well as a very efficient partnership with a high performance. This example shows that performance indicators need interpretation according to the situation.

Managing processes, more precisely managing communication, problem solving, information flow, and employing the right methodologies is a key factor for enhancing performance in collaboration. Particularly, potential problems are less likely to happen, and deviations from the set objectives can be avoided. Therefore, managing processes is an important step towards a future-oriented, predictive management, away from a reactive management based merely upon output.

4.5 Characteristics of the Measurement Framework

The above developed and described performance measurement framework is designed as an uncomplicated model. Sophisticated measurement systems are not acknowledged by SMEs. Instead of going one step further away from the needs of smaller companies, this framework pays contribution to the basic requirements worked out in this paper. Companies collaborate and cooperate but do not use performance measures to manage partnerships. With the help of an easily accessible yet not too simple framework, SMEs can be lead towards measuring and managing performance in collaborations in the first place.

The framework has adopted the input-process-output-outcome structure of other frameworks. The four elements are designed in a circular model, with the element structure as the starting and end point as shown in Fig. 6. The alignment and meaning of the elements is intuitive, with input and output in a straight line from left to right, processes below, supporting the framework and structure on top, coordinating the activities. The circular model implies a continuous measurement and management of the partnership. It resembles Deming's PDCA circle of continuous improvement. Likewise, performance within the collaboration can be improved steadily.

This framework must be seen as a guideline for SMEs to organise, structure, and control interorganisational collaborations. The enhancement and maintenance of performance and long-term success are seen as related improvements.

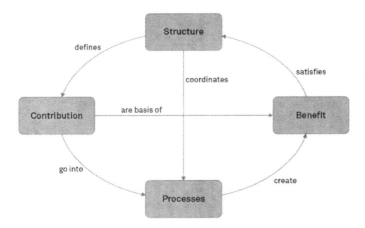

Fig. 6 The interdependencies between the parameters

4.6 Applying the Framework

After having described the performance measurement framework in the previous part, in the following, guidelines for implementing it and setting it up are provided. Establishing a framework is useless, when implementation fails and application starts with wrong preconditions. Essential directions are

- mutuality and commonness with the conjointly application and common goals and indicators,
- maintaining and continuously improving the framework according to the changes via regular meetings,
- self-reflection of goals and contribution,
- defining and assigning performance measures.

The major problem of measuring performance in collaboration is the absence of generally acknowledged measures. Every collaborative venture is different, and the broad approach within this paper makes it difficult to provide generally valid metrics. In literature there exist many measures and indicators which are grouped and listed in Table 2.

In this context, the indicators cannot be made too detailed. Different collaboration may require additional indicators and may not use one or more of the above mentioned at all. Table 2 is therefore a proposal for a set of indicators. Companies and organisations willing to install a performance measurement framework are advised to create their specifically own set of measures.

Table 2 Identified performance measures for collaborations

Performance measures	Literature	Leading questions and examples
Systems	Agitavi (2007), Bititci et al. (2003), Busi and Bititci (2006), Dreyer and Busi (2005), Hagel et al. (2002), Kale et al. (2001), Schuh et al. (2007), Vanpouke and Vereeke (2007)	Common accounting systems Collaborative information systems Integrated management systems Mutual measurement and evaluation systems Do systems operate across organisational boundaries? Can systems build up and maintain trust?
Decision making	Busi and Bititci (2006), Dussauge and Garrette (1999), Forker and Stannack (2000), Ireland et al. (2002), Parkinson (2006), Parung and Bititci (2008), Schuh et al. (2008)	Joint, dominant or external decision making Time to make and implement decisions Strategic and operational decisions Do decision makers have the appropriate competences? Are decision makers linked?
Goal orientation	Agitavi (2007), Busi and Bititci (2006), Dreyer and Busi (2005), Parkinson (2006), Segil et al. (2003), Schuh et al. (2008)	Definition of common goals and objectives Commitment to goals Actions and interactions strive to achieve the set goals
Participants and actors	Dreyer and Busi (2005), Khanna (1998), Hagel et al. (2002), Jap (1998), Kingsley and Klein (1997), Parung and Bititci (2008), Shelbourn (2007), Thomson and Perry (2006)	Who participates in and influences the collaboration? Does everybody have appropriate competences and abilities? Who is responsible for what? How are actors linked with each other?
Contribution	Dussauge and Garrette (1999), Jap (1998), Parung and Bititci (2008), Segil et al. (2003), Schuh et al. (2008)	Input into the collaborative system Mutuality and balance of contribution Sharing of resources, information and capacities
Communication	Agitavi (2007), Busi and Bititci (2006), Dreyer and Busi (2005), Dussauge and Garrette (1999), Hagel et al. (2002), Ireland et al. (2002), Luther (2005), Parkinson (2006), Parung and Bititci (2008), Schuh et al. (2008), Vanpouke and Vereeke (2007)	Communication channels Communication systems Shared information Timeliness of transmission of communication Regular meetings Standardised communication and terms

(continued)

Table 2 (continued)

Performance measures	Literature	Leading questions and examples
Interdependence	Gulati (1998), Ireland et al. (2002), Jap (1998), Segil et al. (2003), Vanpouke and Vereeke (2007)	Alignment of processes towards a common goal Degree to which processes intertwine Improvement of benefits and profits due to collaboration Ease to stop collaboration without losses
Leadership	Hansen and Nohria (2004), Parkinson (2006), Segil et al. (2003), Vanpouke and Vereeke (2007)	Who leads the collaboration? Has the leader the right competences and authority? Dominant parent, shared management or independence Planning and controlling tasks
Processes	Bititci et al. (2003), Busi and Bititci (2006), Draulans et al. (2003), Dreyer and Busi (2005), Gulati (1998), Hagel et al. (2002), Jap (1998), Schuh et al. (2008), Segil et al. (2003), Shelbourn (2007), Thomson and Perry (2006), Vanpouke and Vereeke (2007)	Standardised methodologies and workflows Milestones Interactions between partners Alignment of processes Flow of material and resources Knowledge and information transfer
Benefits	Agitavi (2007), Bititci et al. (2005), Dussauge and Garrette (1999), Gulati (1998), Hagel et al. (2002), Ireland et al. (2002), Jap (1998), Kanter (1994), Luther (2005), Schuh et al. (2008), Segil et al. (2003), Shelbourn (2007), Vanpouke and Vereeke (2007)	Revenue enhancement Cost reduction Exploitation of synergies Access to new markets, knowledge, patents etc Creation of additional value

5 Conclusion

The research carried out within this paper has lead to important findings for the topic of performance measurement for interorganisational collaborations of SMEs. The literature research on collaboration management has shown that collaborations on the one hand are important and vital for the life and success of enterprises in today's globalised economy. On the other hand collaborations are complex and difficult to manage. This leads to the conclusion that management systems need to be developed and installed to minimize risks and support the management. There are certain approaches to structure and systemise the management of collaborations, and literature revealed that performance measurement systems are generally capable of supporting management and reducing complexity in organisations.

The literature review within the topic of performance measurement has revealed the approaches of performance measurement systems within collaborative environments and networks. The interviews with small and medium-sized companies and consultants on this topic have shown that companies do not employ systems or frameworks to measure or manage collaborations holistically. Small and medium-sized companies have limited resources and capabilities and are not willing to deploy complicated systems. Existing measurement frameworks have turned out to be complicated or inappropriate for being used in measuring performance in collaborative environments and are mostly targeted on larger companies and networks. Here, it can be seen that scientific research is remarkably too far away from the requirements and surrounding conditions of smaller companies.

Due to these findings, a new, easily accessible, and appropriate performance measurement framework has been established. It consists of the four elements *structure, contribution, process,* and *benefit.* The framework can be used as a guideline to manage performance, provided all involved parties agree upon using it. Even more than a tool to measure performance, the framework is a guide to manage collaborations. It helps to systemise and structure relationships, providing a framework where all necessary issues concerning the alliance can be integrated into. So, the described framework is also applicable, when performance is not measured by numeric, tangible metrics.

References

Agitavi (2007). Collaboration for Global Success: A Tool Kit for Effective Industry Collaboration. ICT Cluster and Collaboration.

Andriani, P; Passiante, G. (2004). Complexity Theory and the Management of Networks: Proceedings of the Workshop on Organisational Networks as Distributed Systems of Knowledge. Imperial College Press.

Banff Executive Leadership Inc. (2003). Leading Collaboration—Living Interdependence. Leadership Acumen 15–December, 2003.

Bititci, U.S., Carrie, A.S., McDevitt, L. (1997). Integrated performance measurement systems: a development guide. International Journal of Operations & Production Management, Vol. 17 No. 5, pp 522–534.

Bititci U.S., Martinez, V., Albores, P., Mendibil, K. (2003). Creating and sustaining competitive advantage in collaborative systems: the what and the how. Production Planning & Control, Vol. 14, No 5, pp 410–421.

Bititci U.S., Parung J., Lopez U., Walters D., Kearney D. (2005). Managing Synergy in Collaborative Enterprises: Proceedings of the European Operations Management Association. EurOMA 2005, 19–21 June 2005, Budapest.

Brannick, M.T., Salas, E., Prince, C. (Eds.) (1997). Team Performance Assessment and Measurement, Lawrence Erlbaum, Mahwah, NJ.

Brown, M. (1996). Keeping Score: Using the Right Metrics to Drive World Class Performance, Quality Resources, New York, NY.

Busi, M., Bititci, U.S. (2006). Collaborative performance management: present gaps and future research. International Journal of Productivity and Performance Management Vol. 55 No. 1, pp 7–25.

Camarinha-Matos, L., Afsarmanesh, H. (2008). Collaborative Networks: Reference Modeling. Springer, p 56.

Cohen, S., Roussel, J. (2004). Strategic Supply Chain Management: The Five Disciplines for Top Performance. McGraw-Hill Professional.

Corsten, H.; Gössinger, R. (2001). Unternehmungsnetzwerke—Grundlagen-Ausgestaltungsformen-Instrumente. Nr. 38. Schriften zum Produktionsmanagement des Lehrstuhls für Produktionswirtschaft der Universität Kaiserslautern.

Cross, K.F., Lynch, R.L. (1988). The SMART way to define and sustain success. National Productivity Review, Vol. 9 No. 1, pp 23–33.

Curseu, P.L., Schrujter, S., Boros, S. (2005). The Meaning of Collaboration: A Study Using a Conceptual Mapping Technique. 18th conference of the International Association for Conflict Management, Seville, Spain.

Das, T. K., Teng, B.S. (2001). A risk perception model of alliance structuring. Journal of International Management, 7, pp 1–29.

Draulans, J., deMan, A., Volberda, H.W. (2003). Building Alliance Capability: Management Techniques for Superior Alliance Performance. Long Range Planning Journal Vol. 36 (2003), pp 151–166.

Dreyer, H.C., Busi, M. (2005). Supply Chain Collaboration: And what does it mean for Logistics? Productivity 2005, Norwegian University of Science and Technology (NTNU).

Dussauge, P., Garrette, B. (1999). Cooperative Strategy Competing Successfully Through Strategic Alliances. John Wiley & Sons, ltd, Chichester, New York.

Eccles, R.G. (1991). The performance measurement manifesto. Harvard Business Review, Vol. 69, pp 131–137.

Fisher, J. (1992). Use of non-financial performance measures. Journal of Cost Management, Vol. 6, Spring, pp 31–38.

Fitzgerald, L., Johnston, R., Brignall, S., Silvestro, R., Voss, C. (1991). Performance Measurement in Service Business, CIMA, London.

Fong, A., Valerdi, R., Srinivasan, J. (2007). Using a Boundary Object Framework to Analyze Interorganizational Collaboration. Paper, Massachusetts Institute of Technology, Lean Aerospace Initiative.

Forker, L.B., Stannack, P., (2000). Cooperation versus competition: do buyers and suppliers really see eye-to-eye? European Journal of Purchasing & Supply Management Vol. 6 (2000), pp 31–40.

Garengo, P., Biazzo, S., Bititci U.S. (2005). Performance Measurement Systems in SMEs: A Review for a Research Agenda. International Journal of Management Reviews, Vol. 7, No. 1, pp 25–47.

Gray, B. (1989). Collaborating: Finding Common Ground for Multiparty Problems. Jossey-Bass.

Gulati R. (1998). Alliances and Networks. Strategic Management Journal, Vol. 19, pp 293–317.

Hagel, J., Brown, J.S., Durchslag, S. (2002). Orchestrating Loosely Coupled Business Processes; The Secret to Successful Collaboration. Working Paper.

Hansen, M.T., Nohria, N. (2004). How to build collaborative advantage. MIT Sloan Management Review, Fall 2004, Vol. 49 No 1, pp 21–31.

Harvey, C.M., Koubek, R.J. (2000). Cognitive, social, and environmental attributes of distributed engineering collaboration: A review and proposed model of collaboration. Human Factors and Ergonomics in Manufacturing, Vol. 10(4), pp 369–393.

Ireland, R.D., Hitt, M.A., Vaidyanath, D. (2002). Alliance Management as a Source of Competitive Advantage. Journal of Management 28(3), pp 413–446.

Jap, S.D. (1998). 'Pie-Division' in Interorganizational Collaboration. Sloan School of Management, Massachusetts Institute of Technology

Jehle, E., Stüllenberg, F., Schulze im Hove, A. (2002). Netzwerk-Balanced Scorecard als Instrument des Supply Chain Controlling. In: Supply Chain Management IV, 2002, S. 19–25.

Kale, P., Dyer, J., Singh, H. (2001). Value Creation and Success in Strategic Alliances: Alliancing Skills and the Role of Alliance Structure and Systems. European Management Journal Vol. 19, No. 5, pp 463–471.

Kanter, R.M. (1994). Collaborative Advantages: Successful partnerships manage the relationship, not just the deal. Harvard Business Review, July–Aug 1994, pp 96–108.

Kaplan, R.S., Norton, D.P. (1992). The Balanced Scorecard: Measures That Drive Performance. Harvard Business Review, Jan–Feb 1992, pp 71–79.

Kaplan, R.S., Norton, D.P. (1996a). Linking The Balanced Scorecard To Strategy. California Management Review. Vol 39, No 1. Fall 1996, pp 53–79.

Kaplan, R.S., Norton, D.P. (1996b). The Balanced Scorecard. Translating Strategy into Action. Harvard Business Press.

Keegan, D.P., Eiler, R.G., Jones, C.R. (1989). Are your performance measures obsolete? Management Accounting (US), Vol. 70, No. 12, June, pp 45–50.

Keller, M., Hellingrath, B. (2008). Kennzahlenbasierte Wirtschaftlichkeitsbewertung in Produktions- und Logistiknetzwerken. In: A. Otto; R. Obermaier (Eds.): Logistikmanagement—Analyse, Bewertung und Gestaltung logistischer Systeme. Wiesbaden: Deutscher Universitäts-Verlag. S. 51–75.

Kennerly, M., Neely, A. (2004). Performance measurement frameworks: A review. In: A. Neely(Ed.): Business performance measurement: Theory and practice. Cambridge University Press, pp 145–155.

Khanna, T. (1998). The scope of alliances. Organization Science, Vol. 9, pp 340–355.

Killing, J.P. (1982). How to Make a Global Joint Venture Work. Harvard Business Review, Vol. 60, No. 3, pp 120–127, May–June.

Killing, J.P. (1983). Strategies for Joint Venture Success. Croom Helm, London.

Kingsley, G., Klein, H.K. (1997). Inter-firm Collaboration as a Modernization Strategy: A Survey of Case Studies. Journal of Technology Transfer Vol. 23 (1), pp 65–74.

Klein, D.L., Adelman, L. (2005). A Collaboration Evaluation Framework. International Conference on Intelligence Analysis.

Kuhn, A., Uygun, Y. (2008). Robuste Netzwerke durch kollaborative Anwendung Ganzheitlicher Produktionssysteme. In: H.-Chr. Pfohl und Th. Wimmes (Eds.): Robuste und sichere Logistiksysteme—Wissenschaft und Praxis im Dialog. Bundesvereinigung Logistik. Schriftenreihe Wirtschaft und Logistik. Deutscher Verkehrs-Verlag: Hamburg. pp 472–488.

Lower, R. (2008). The collaboration advantage. Teradata Magazin Online, Vol. 7, No. 1.

Luther, V. (2005). Managing Collaboration at the Community Level: Action Steps for Community Builders. North Central Regional Center for Rural Development.

McCutchen, W.W., Swamidass, P.M., Teng, B.S. (2008). Strategic alliance termination and performance: The role of task complexity, nationality, and experience. Journal of High Technology Management Research 18 (2008), pp 191–202.

Mentzer, J.T., Foggin, J., Golicic, S. (2000). Collaboration: the enablers, impediments and benefits. Supply chain Management Review, Vol. 4, No. 4, pp 52–58.

Mitleton-Kelly, E (2003). Complex Systems and Evolutionary Perspectives on Organisations: The Application of Complexity Theory to Organisations. Emerald Group Publishing.

Neely, A., Adams, C., Crowe, P. (2001). The performance prism in practice: Measuring Business Excellence, Vol. 5 No. 2, pp 6–12.

Neely, A., Adams, C., Kennerley, M. (2002). The Performance Prism: The Scorecard for Measuring and Managing Business Success. Pearson Education.

Nkhata, A.B., Breen, C.M., Freimund, W.A. (2008). Resilient Social Relationships and Collaboration in the Management of Social–Ecological Systems. Ecology and Society 13(1): 2.

Oakland, J.S. (1993). Total Quality Management: The Route to Improving Performance. 2nd ed., Butterworth-Heinemann, New York, NY.

Parkinson, C. (2006). Building Successful Collaborations: A Guide to Collaboration among Non-Profit Agencies and Between Non-Profit Agencies and Businesses. Cambridge & North Dumfries Community Foundation.

Parung, J., Bititci, U.S. (2008). A metric for collaborative networks. Business Process Management Journal Vol. 14 No. 5, pp 654–674.

Pfeffer, J., Salancik, G.R. (1978). The external control of organisations: A resource dependence perspective. Harper & Row, New York.

Power, D.J., Sohal, A.S., Rahman, S. (2001). Critical success factors in agile supply chain management. International Journal of Physical Distribution & Logistics Management, Vol. 31 No. 4, pp 247–265.

Roetzheim, W. (2007). Why Things Are: How Complexity Theory Answers Life's Toughest Questions. Level4Press Inc.

Saiz, J.J.A., Bas, A.O., Rodríguez, R.R. (2007). Performance measurement system for enterprise networks. International Journal of Productivity and Performance Management Vol. 56 No. 4, pp 305–334.

Schönsleben, P. (1998). Integrales Logistikmanagement. Berlin: Springer.

Schuh, G. et al. (2008). Complexity-Based Modeling of Reconfigurable Collaborations in Production Industry. CIRP Annals—Manufacturing Technology 57, pp 445–450.

Schuh, G., Lu, S.C., Elmaraghy, W., Wilhelm, R. (2007). A Scientific Foundation of Collaborative Engineering. Annals of the CIRP Vol. 56/2/2007.

Segil, L., Goldsmith, M., Belasco, J. (2003). Partnering: The New Face of Leadership. American Management Association, New York.

Shelbourn, M., Bouchlaghem, N.M., Anumba, C., Carrillo, C. (2007). Planning and implementation of effective collaboration in construction projects. Construction Innovation Vol. 7 No. 4, pp 357–377.

Simatupang, T.M., Sridharan, R. (2004). The collaboration index: a measure for supply chain collaboration. International Journal of Physical Distribution & Logistics Management Vol. 35 No. 1, pp 44–62.

Stadtler, H., Kilger, C. (2002). Supply Chain Management and Advanced Planning: Concepts, Models, Software, and Case Studies, Springer, New York.

Sydow, J. (1992). Strategische Netzwerke—Evolution und Organisation. Wiesbaden: Gabler.

Tangen, S. (2004). Performance measurement: from philosophy to practice. International Journal of Productivity and Performance Management, Vol. 53 No. 8, pp 726–737.

Thompson, J.D. (1967). Organizations in action. McGraw-Hill, New York.

Thomson, A.M., Perry, J.L. (2006). Collaboration Processes: Inside the Black Box. Public Administration Review, December 2006.

Vanpouke, E., Vereeke, A. (2007). Creating Successful collaborative relationships. Working Paper, Universiteit Gent.

Weber, J., Bacher, A., Groll, M. (2002). Konzeption einer Balanced Scorecard für das Controlling von unternehmensübergreifenden Supply Chains. In: krp-Kostenrechnungspraxis, 46. Jg., 2002, H. 3, S. 133–141.

Wildemann, H. (1996). Entwicklungsstrategien für Zulieferunternehmen. 3. Auflage. München: TCW-Verlag.

Wood, D., Gray, B. (1991). Toward a Comprehensive Theory of Collaboration. Journal of Applied Behavioral Science Vol. 27 No 2, pp 139–62.

On the Formation of Operational Transport Collaboration Systems

Melanie Bloos and Herbert Kopfer

1 Introduction

With idle transports making up for 20–30% of transport distances in road haulage in Europe and average utilizations of loading spaces of around 60% planning improvements may lead to higher rentability and sustainability for enterprises competing in the road haulage sector (Berger and Bierwirth 2007). Two observed business trends aiming at a reduction of the aforementioned problems for small and medium sized enterprises are electronic freight exchange systems, such as Teleroute (www.teleroute.de), and business cooperation, such as TimoCom or IDS Scheer in Germany, respectively (www.timocom.de and www.ids-logistik.de). Hauliers' motivations for entering either system are improved planning for less-than-truckload (LTL) freight as well as competitive advantages in the dynamic and competitive transport market. With LTL freight, idle trips occur when the truck is completely unloaded at one destination and the next pick up location or the depot has to be reached. Statistical data for Germany shows that idle trips accounted up for approximately 20% of all road transports—or 5.7 billion kilometres—in 2007 (Berger and Bierwirth 2007). For all trips with loaded trucks the same statistic shows average utilization rates of 57% for transport performance (in ton kilometres) and 62% in terms of load capacity. The opportunity to exchange individual requests with other hauliers in order to achieve reductions in idle trip distances and increases in load capacity utilization can create significant cost savings for the hauliers.

M. Bloos (✉) and H. Kopfer
Lehrstuhl für Logistik, Universität Bremen, Wilhelm-Herbst Straße 5, 28359 Bremen, Germany
e-mail: bloos@uni-bremen.de

H. Kopfer
e-mail: kopfer@uni-bremen.de

H.-J. Kreowski et al. (eds.), *Dynamics in Logistics*,
DOI: 10.1007/978-3-642-11996-5_17, © Springer-Verlag Berlin Heidelberg 2011

The option of using electronic market places has been pointed out as solution to such problems for the transport sector (Bierwirth et al. 2002). A popular option already used in practice are the above mentioned freight exchange systems. Those systems act like market places offering individual hauliers, customers and freight forwarders the option to acquire or sell cargo. The other option is cooperation. Here, a special type of cooperation, so called operational transport collaboration, will be discussed, for which enterprises exchange short term planning information and customers' transport requests electronically. The term collaboration refers to this joint planning of the participants. An overview of differences and commonalities of the two options is provided in Table 1. Operational transport collaboration is also an electronic marketplace. The main difference to the freight exchange systems lies in the system setup and objectives. Freight exchange systems aim at enabling transactions between individual participants whereas in operational transport collaboration the intention is to seek the overall optimal solution considering the planning situation of all participants.

Operational transport collaboration is organisationally embedded into a cooperative framework that establishes rules for the exchange of information, requests and payments. This framework creates security and trust between the participants. The aim of this contribution is to discuss and describe the formation process of a framework for the system of operational transport collaboration.

We focus on the operational transport collaboration as cooperative system that includes the exchange of transportation requests on an operational level and aims at better planning solutions with reduced idle trips and increased utilization of trucks. Research on operational transport collaboration is introduced in Sect. 2. Looking at the formation process of such a system, we split the strategic decisions into two separate but related decision making problems: the decision making of individual hauliers on whether to enter such a system and the decision of a system provider on which form of operational transport collaboration system to offer in the market. We describe both decision making problems and a suitable heuristic decision making process in Sect. 3.

Table 1 Comparison of operational transport collaboration and electronic freight exchange

	Electronic freight exchange systems	Operational transport collaboration
Objectives	Platform for freight exchange or exchange of individual transport orders	(Close to) Optimal planning solution for all participants, market-based exchange of individual or bundled transport orders
Participants	Open system, anonymous	Closed system, well known
Transactions	One to one (negotiation or fixed price)	Many to many (re-allocation and pricing mechanism)
Organizational embedding	Between different organisations, vertically or horizontally embedded into transportation market	Between different organisations, horizontal cooperation in transportation market

2 Operational Transport Collaboration

Operational transport collaboration is established between several hauliers in the market cooperating in their operative planning. The hauliers receive transport requests from their customers on short notice and then plan the execution by solving the vehicle routing problems. The hauliers' route planning is subject to restrictions such as load capacity of the truck. As such the planner might be able to identify requests that if planned conforming to those restrictions increase the operative costs of the haulier significantly. The idea of operational transport collaboration is that some or all of the requests that have been identified as unsuitable are submitted to a central pool of which other hauliers can acquire requests and fulfil them in the name of the submitting haulier. At the same time, the submitting haulier has the chance to acquire herself requests from that central pool in order to increase the profitability of her existing tours by improving utilization and increasing revenues. The purpose of this request re-allocation in operational transport collaboration can then be described as a levelling of capacity amongst the participating hauliers with the overall objective of creating a cost minimal allocation of the customers' transport requests to hauliers. The general idea of operational transport collaboration can be found in Krajewska and Kopfer (2006), Schönberger (2005), Berger and Bierwirth (2007, 2008), and Gujo et al. (2007).

In Krajewska and Kopfer (2006), the authors describe operational transport collaboration as a three phase process. In the first phase, the pre-processing, all hauliers determine the execution cost of all their acquired requests by determining the monetary difference of the planning solution with and without the respective request. Based on those calculations they then select those requests they want to offer to collaborating partners plus those requests to be entered into a third fulfilment mode, namely that of sub-contracting. In the second phase, the coalition profit optimisation phase, the actual exchange takes place by accepting bids from all hauliers for all requests but their own available in the central pool. The re-allocation problem in Krajewska and Kopfer (2006) is formulated as combinatorial auction problem, similar to those in the models of Schönberger (2005) and Gujo et al. (2007). An alternative for which only individual requests are sold in a Vickrey auction is introduced in Berger and Bierwirth (2008). In the third phase of profit optimisation and profit sharing the resulting payments from the request exchange are determined and the monetary benefit of collaboration is divided between the participants.

The system of operational transport collaboration, as described above, needs to be embedded into an organisational framework that regulates legal and monetary matters between the participants. We refer to this framework as cooperation. According to Kopfer and Pankratz (1999) cooperation is established by legally and economically independent partners for the purpose of commonly achieving better results than the partners could achieve individually. As such, cooperation can be seen as a framework under which independent partners operate parts of their businesses jointly in order to fulfil a joint purpose for their mutual benefit. Because of the resulting interdependency of the participants' businesses, the cooperation has a mid- to long-term orientation.

3 Strategic Decision Making Problems

A decision making problem arises whenever a discrepancy between goals and actual performance is discovered or whenever a new possibility for better performance in existing goals is discovered. The first situation can be described as the risk of failure whereas the second situation refers to the discovery of a chance for improvement (Grünig et al. 2005).

The *formation process of a system of operational transport collaboration* in the transportation market can be described as two separate decision making problems. The decision of each haulier is on whether to enter a system of operational transport collaboration or not and if so which system for which conditions. The decision on the system's design and set-up is made by potential providers of such systems in the market. Both decisions are based on a long time horizon and as such they are strategic. To our knowledge, the problem of the formation of operational transport collaboration has not been studied so far. However, a similar problem in the airline industry for strategic alliances exists. Strategic alliance formation has been studied in general and for special industry sectors and the results of those studies provide valuable insights into the formation of operational transport collaboration.

Technical procedures exist that support the decision making process by identifying and evaluating alternatives. As discussed in Grünig et al. (2005) two main categories of decision procedures exist: analytical and heuristic. Common to both types of procedure is that the decision process aims at solving an existing problem by selecting an alternative that improves or at least maintains the current situation. The main difference between analytical and heuristic procedures is that analytical procedures have to comply with restrictive prerequisites, follow restrictive formal criteria in order to find optimal solutions and only support decisions with regards to quantitative goals. Examples of analytical procedures are cost-benefit analysis or solving allocation problems in production planning. Heuristic procedures, in contrast, can incorporate qualitative goals in the decision making and are adapted to each problem individually. As such, heuristic procedures are better suited for supporting strategic decisions.

A *decision on participation* in operational transport collaboration and the decision on the collaborative system's design can be based to a large degree on quantitative data such as expected revenues or efficiency of the re-allocation mechanism. However, in case of entering the cooperation, the decision maker will also make qualitative considerations regarding trust in partners and the security of customer data before agreeing to a cooperative system as discussed in Townsend (2003). Then, heuristic procedures have to be chosen.

Following the suggested heuristic procedure of Grünig et al. (2005) we characterize the decision making of hauliers that might lead to participation in operational transport collaboration. Therefore, we assume a situation in which independent providers in the market exist that offer platforms for the operational exchange of transportation requests. The situation is similar to that of strategic alliances that operate flights jointly. A customer may book a flight with one of the participating airlines but the flight might be operated by one of the partner airlines in the name of the original provider. In the case of operational transport collaboration, the platform providers then offer the technical solution for a re-allocation of requests, such as an online combinatorial auction (Gujo et al. 2007). Because of the similarity to airline alliances we use existing studies on strategic alliances in order to describe the decision making problem and process.

The generic decision making process is depicted in Fig. 1. The process starts with the discovery of the decision making problem, e.g. a challenge for future business operation such as the long-term successful operation in the transportation market by entering cooperation. Research on operational transport collaboration and strategic alliances identifies various *drivers to establishing cooperation*. Coming from a marketing perspective, Townsend (2003) identifies firm internal motives or drivers to forming alliances. The firm internal motives are further divided into the categories market, product, resource, knowledge, and transaction risk. *Market related* drivers include chances such as entering new markets or protecting the company's position in markets already served. Increased competition seems to be the main driver in the transport sector (Berger and Bierwirth 2007; Dahl and Derigs 2007; Kopfer and Pankratz 1999). This increase is due to globalization (the EU enlargement for Europe) and the cabotage right extension. With higher competition hauliers are forced to reduce their costs in order to offer competitive prices. The motive of entering new markets is also thinkable and found in transport cooperation. Hauliers might strive for a geographical extension by setting up cooperation with hauliers of different geographical markets in order to offer transports into those regions at competitive prices. This motive finds empirical support in the study of Wagener et al. (2002) on future trends in the transport sector conducted by the Delphi method where the experts rated the two options for small and medium sized enterprises of extending the geographical area

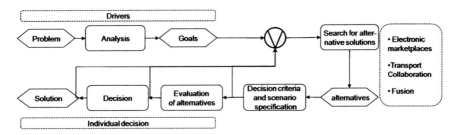

Fig. 1 Individual haulier's decision making depicted as Event-driven process chain

and extending the customer base by cooperation on average as "successful" strategy.

As Townsend (2003) states, the *product related* motives include "filling in gaps and broadening current product lines". The transfer to the transport industry is supported by the study of Wagener et al. (2002) which identifies demand trends to make the conditions of delivery the most important factor in future buying decisions. The most crucial factor identified in terms of the conditions of delivery is the ability to deliver within strict time windows. This trend might lead to even more idle transports, if the time windows do not match well into the planning of the haulier. The option of operational transport collaboration however extends the decision space since it also includes the incorporation of the order into other hauliers' plans. With an extension of the decision space more efficient solutions can be found while fulfilling the customer's service requirements.

Since we assume the cooperation to be based solely on terms of operational planning, an obvious influence on *resources* is not assumed here. Regarding *knowledge*, organisational learning is identified as one of the motives for cooperation in contrast to one-off arrangements in Townsend (2003). Organisational learning offers the possibility to create trust among the participants and to understand the mechanism and the resulting request re-allocations. Trust and knowledge of the system can also lead to a reduction in transaction risk. *Transaction risk* in terms of operational transport collaboration lies in the non-fulfilment of transferred requests or the failure to comply with the specified requirements (such as time windows) as well as in hauliers enticing customers away from their cooperation partners. Another driver to be added to the motive of knowledge is the protection of organisational knowledge and information. Entering operational transport collaboration implies the exchange of customer data which may be highly sensitive. For the decision process this means that one of the goals may be related to the highest possible degree of protection of the hauliers' information and knowledge.

For the decision making process it is crucial that the haulier identifies the reason for a strategic change first. These reasons will also determine the goals pursued and as stated, those goals are individual and differ from haulier to haulier. If the goals are known a search for alternative solutions starts, leading to a list of existing and available solutions. After listing the alternatives, a specification has to be made of how goal achievement can be measured. This leads to a statement of decision criteria and related potential parameter values (e.g. continuous scales for monetary outcome). In a dynamic business environment, chances and risks can turn up frequently. The realization of new chances and risks within the decision process may require an adjustment of the goals set and as such, the procedure leaves the formulation of quantitative goals and the specification of qualitative goals to the latest possible moment. Therefore, the decision procedure proceeds with the search and analysis of alternative solutions before deriving the decision criteria of the goals and possible future scenarios based on the latest developments.

At the same time, scenarios for alternative future states have to be made. Based on the decision criteria and the alternative scenarios the alternative solutions can be evaluated which finally leads to a decision and the realization of the preferred solution. The last three steps of decision criteria and scenario specification, evaluation and the final decision are omitted here since these steps are individual to each haulier and cannot be discussed in general. These three steps can also result in a new or additional search for solutions if criteria or possibilities previously not considered are found. This can be seen as strength of the heuristic procedure in dealing with strategic decisions since the underlying problems are complex and dynamic and alternatives might only be discovered in later stages of the process. Further, on the level of deciding on participation the possibility to negotiate conditions with the platform providers and other potential participants offers the chance to influence and adjust the alternative solutions.

Although the process of designing an operational transport collaboration system is like a black box to the hauliers, the criteria considered and to be fulfilled for deciding on a system have to be the same for hauliers and providers in order to establish a cooperative system jointly. In consequence, we describe the process of setting up a cooperative system of operational transport collaboration next.

For considerations regarding the *design of cooperative systems*, again research on strategic alliances is considered. Strategic alliances can be defined as "voluntary interfirm cooperative agreements, often characterized by inherent instability arising from uncertainty regarding a partner's future behavior and the absence of a higher authority to ensure compliance" (Parkhe 1993). This definition incorporates the most important aspects, namely those of opportunistic behaviour and the ability (or inability) to enforce the agreement. Opportunistic behaviour is the attempt of individual participants to manipulate the alliance or cooperation outcome towards an improvement of their own situation. The analysis of opportunistic behaviour may be conducted by means of Game Theory which is capable of analysing monetary incentives to cooperation as well as incentives to strategic behaviour that result in compliance or deviation with the agreement. In Parkhe (1993), three dimensions of explaining the establishment of cooperation from a game theoretic viewpoint are derived: the pattern of payoffs, the perception of future outcome and the number of players. The dimensions are illustrated by the Prisoner's Dilemma, in which two criminals are questioned separately by the police and can decide on whether to cooperate, that is not to confess the crime, or to deviate and thus confess. This scenario can be transferred and extended to operational transport collaboration. Then, two or more players decide individually on whether they act in conformity to the intended mechanism or whether they try to manipulate the results with their behaviour. The pattern of payoffs is then the monetary outcome the haulier expects from collaborative planning. The shadow of the future is related to the repetition of the collaborative planning and the effects a haulier's action has on future rounds. This leads to the aspects of trust in the exchange mechanism and its robustness against opportunistic behaviour and to trust in the partnering hauliers. The third dimension, the number of players, is not further discussed in Parkhe (1993), since the author only refers to a two person game. However, for operational transport collaboration, more than two participants are thinkable and realistic since improved planning through an extension of the decision space is sought. The discussion omits any self-interest the provider might have in setting up operational transport collaboration. However, for the system provider monetary goals are relevant and may well influence the selections made.

Using the existing models of operational transport collaboration and the three structural dimensions of Parkhe (1993), we suggest a split of the decision making problem of system design into three parallel subproblems. The decision making problem of system design is depicted in Fig. 2.

The first subproblem relates to the *cooperative framework* which specifies the participating partners, the duration, and the possibility of accepting new partners later on. In a first step possible frameworks are created. A potential framework could for example be a maximum of five participants, an exclusive cooperation,

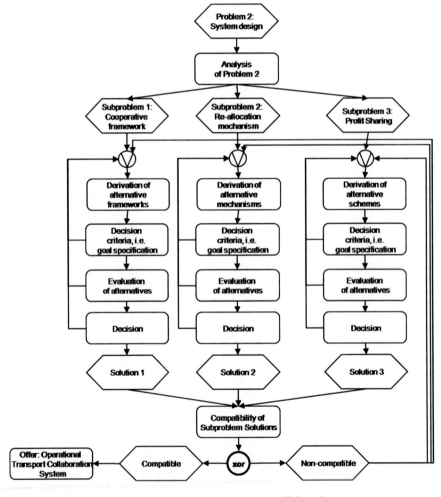

Fig. 2 Decision on the design of the operational transport collaboration system

and a contract based on the duration of 10 years. Then the decision criteria are determined, such as stability, robustness to strategic behaviour, and their evaluation is specified (for example score-based rankings of alternatives by senior management for qualitative criteria). The decision then leads to the preference of one of the frameworks as solution to the subproblem.

The second and third subproblems are solved in a similar manner. The second problem aims at finding a good *mechanism for re-allocating* customer requests. Alternative mechanisms include, e.g. Vickrey auctions, Generalized Vickrey Auctions or reverse auctions (Cramton et al. 2006). Decision criteria can be derived from economic considerations (Bloos and Kopfer 2009). The third subproblem compares different schemes for *profit sharing* such as the Shapley value or subsidies to cooperation participants as discussed in Krajewska et al. (2008).

Solutions to the individual subproblems influence each other and have to be combined carefully for the overall system design. The three subproblems have been suggested to simplify the decision making; however, the overall problem of system design requires one solution and as such compatible solutions to the subproblems. For example, if the solution to the first subproblem is an exclusive cooperation of two hauliers, then using a combinatorial auction is not likely to be a good solution to the second subproblem anymore since solving the combinatorial auction problem adds additional and unnecessary computational steps to the practical collaboration. Also, the solution to the third subproblem (profit sharing) will influence participants' behaviour and these changes might lead to changed performance regarding the efficiency of certain mechanisms. The resulting solution to the platform provider's decision problem is then an offer of a system for operational transport collaboration in the market. This offer is one of the alternative solutions considered by each haulier. The offer can be evaluated by the haulier regarding mainly qualitative decision criteria related to the cooperative framework but also regarding improvements of the planning and cost situation for different scenarios that include different degrees of participation of the cooperation partners. The participation can vary from one planning period to the next since it always involves the autonomously made operational decision on how many transportation requests to offer to cooperation partners and on how many to acquire from them.

4 Conclusions

The formation process of operational transport collaboration systems has been described as two distinct decision making problems—one for the haulier considering participation and one for the party offering the system. Although the decision making problems are distinct they are also interlinked, since the goal formulation and derived decision criteria of the haulier determine the acceptance and as such the success of the offered system. For both decision making problems not only quantitative goals such as expected cost reduction or generated revenue are relevant but also qualitative goals like mutual trust. Therefore, a framework for the heuristic decision making procedure has been suggested and aspects related to the procedure have been discussed for both problems. The procedure is capable of considering multiple quantitative as well as qualitative goals. Additionally, the procedure defers the specification of the decision criteria to the latest possible moment which improves its performance in dynamic environments. The discussion here is limited to very general goals with no decision criteria specified since those depend on the situation of the decision maker and her degree of risk aversion.

Acknowledgments This research was supported by the German Research Foundation (DFG) as part of the Collaborative Research Centre 637 "Autonomous Cooperating Logistics Processes— A Paradigm Shift and its Limitations" (Subproject B9).

References

S. Berger and C. Bierwirth. The collaborative carrier vehicle touring problem for capacitated traveling salesman tours. In R. Koschke, O. Herzog, K.-H. Rödiger, and M. Ronthaler, editors, *Informatik 2007, Bremen, 24–27 September*, volume 1 of *GI-Edition. Lecture Notes in Informatics*, pages 75–78, 2007.

S. Berger and C. Bierwirth. Ein Framework für die Koordination unabhängiger Transportdienstleister. In K. Inderfurth, G. Neumann, M. Schenk, G. Wäscher, and D. Ziems, editors, *Netzwerklogistik: Logistik aus technischer und ökonomischer Sicht. (13.Magdeburger Logistik-Tagung)*, pages 137–151, 2008.

C. Bierwirth, S. Schneider and H. Kopfer. Elektronische Transportmärkte: Aufgaben, Entwicklungsstand und Gestaltungsoptionen. *Wirtschaftsinformatik*, 4:335–344, 2002.

M. Bloos and H. Kopfer. Efficiency of transport collaboration mechanisms. *Communications of SIWN*, 6:23–28, 2009.

P. Cramton, Y. Shoham, and R. Steinberg, editors. *Combinatorial Auctions*. MIT Press, Cambridge, 2006.

S. Dahl and U. Derigs. Ein Decision Support System zur kooperativen Tourenplanung in Verbünden unabhängiger Transportdienstleister. In R. Koschke, O. Herzog, K.-H. Rödiger, and M. Ronthaler, editors, *Informatik 2007, Bremen, 24–27 September*, volume 1 of *GI-Edition. Lecture Notes in Informatics*, pages 57–61, 2007.

R. Grünig and R. Kühn. *Entscheidungsverfahren für komplexe Probleme. Ein heuristischer Ansatz.* Springer, Berlin, Heidelberg, New York, 2nd edition, 2005.

O. Gujo, M. Schwind, and J. Vykoukal. The design of incentives in a combinatorial exchange for intra-enterprise logistics services. In *IEEE Joint Conference on E-Commerce Technology (CEC'07) and Enterprise Computing, E-Commerce and E-Services (EEE '07); Tokyo, Japan, July*, pages 443–446, 2007.

H. Kopfer and G. Pankratz. Das Groupage-Problem kooperierender Verkehrsträger. In P. Kall and H.-J. Lüthi, editors, *Proceedings of OR'98*, pages 453–462, Berlin, Heidelberg, New York, 1999. Springer.

Kraftfahrt-Bundesamt. Verkehr deutscher Lastkraftfahrzeuge. Inlandsverkehr, Eigenschaft der Fahrt, Jahr 2007. Statistische Mitteilungen des KBA und des BAG, 11 2008.

M. Krajewska, H. Kopfer, G. Laporte, S. Ropke and G. Zaccour. Horizontal cooperation of freight carriers: request allocation and profit sharing. *Journal of the Operational Research Society*, 59:1483–1491, 2008.

M. A. Krajewska and H. Kopfer. Collaborating freight forwarding enterprises—request allocation and profit sharing. *OR Spectrum*, 28(3):301–317, 2006.

A. Parkhe. Strategic alliance structuring: A game theoretic and transaction cost examination of interfirm cooperation. *Academy of Management Journal*, 36(4):794–829, 1993.

J. Schönberger. *Operational Freight Carrier Planning. Basic Concepts, Optimization Models and Advanced Memetic Algorithms*. GOR Publications. Springer, Berlin, Heidelberg, New York, 2005.

J. D. Townsend. Understanding alliances: a review of international aspects in strategic marketing. *Marketing Intelligence & Planning*, 31(3):143–155, 2003.

N. Wagener, R. Wagner, D. Jahn, R. Lasch and A. Lemke. Endbericht zur Delphi-Studie "Der Transportmarkt im Wandel". February 2002.

Adaptive RBAC in Complex Event-Driven BPM Systems

Bernardo N. Yahya and Hyerim Bae

1 Introduction

The ability to complete a role-based task assignment in complex-event-driven systems will become essential in the future of business process management systems (BPMS). The next generation of BPMS, based on the real-time environment, requires an adaptive mechanism of access control regarding security, privacy, accuracy and conformity. Role-based access control (RBAC) has been introduced into many domains as a cost-effective information access system.

Generalized temporal RBAC (GTRBAC) (Joshi et al. 2003) offers temporary role assignment based on operations such as enabling–disabling, assignment–deassignment, and activation–deactivation by adding duration constraints, which are time-based semantics for role hierarchy and separation of duty relations. This approach can be considered to be an event-based access control approach. However, event occurrence in GTRBAC assumes a domain-specific set of simple events.

Some problems can occur when the process invocation is based on single events. Single purchase order events sometimes trigger an alert, sent from the retailer to the distribution center. However, the alert is meaningless without any confirmatory response event. Thus, the concurrencies of request and response events are considered as a single composite event. When the events occur sequentially, the activity state is changed to "complete" and the system renders the next activity state to "ready". In the case of product scarcity, it is better for the requester to publish the event instead of making a request to a certain distributor.

B. N. Yahya (✉) and H. Bae
Business & Service Computing Lab., Industrial Engineering, Pusan National University, 30-san Jangjeon-dong Geumjong-gu, Busan 609-735, South Korea
e-mail: bernardo@pusan.ac.kr

H. Bae
e-mail: hrbae@pusan.ac.kr

H.-J. Kreowski et al. (eds.), *Dynamics in Logistics*,
DOI: 10.1007/978-3-642-11996-5_18, © Springer-Verlag Berlin Heidelberg 2011

The first incoming subscribe event that satisfies the condition can automatically trigger the pertinent activity. Another problem relates to connection. A non-response to a request incurs waiting time, which can result in disconnection and customer low-quality service. Repetition of the same single event can minimize this disconnection problem and give alternative action when no any response occurs. In order to overcome all such problems and maintain an appropriate quality of service to customers, user-role assignment is necessary.

Figure 1 shows a typical retailer-distribution center logistics-process example. In order to invoke activity automatically in the inter-workflow process, event monitoring can host each process using rule-based language. A retail shop and a retail warehouse have a specific interaction regarding product placement on the shelf. The retail warehouse and purchasing interact with external entities such as a distribution center or manufacturer in order to manage inventory stock. Each event generates a notification, an alert, or even initiates a new workflow process.

Previous research proposed self-assignment with rule-based RBAC using single event. The main objective of this paper is to develop an adaptive user assignment using RBAC in complex-event-driven systems. This adaptive user assignment, which means self-assignment in workflow system, tries to minimize customer waiting time since this mechanism allow the expected occurrence event in collaborative environment to notify directly to specific user under certain condition defined in event-condition-action (ECA) rules. We expect that our approach can improve the quality-of-service (QoS) in collaborative environment workflow management system.

The remainder of this paper is as follows. Section 2 reviews previous research related to RBAC, event-driven systems and BPMS. Section 3 presents some

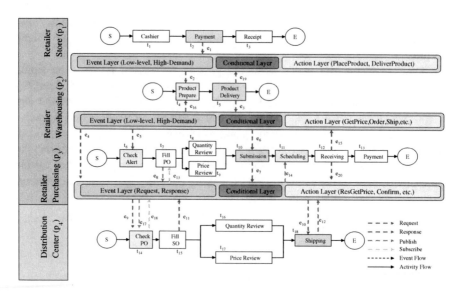

Fig. 1 Example of event-based inter-workflow process

definitions regarding complex-event-driven systems and user-role assignment methodology. Section 4 proposes a simple case study. Finally, Sect. 5 offers conclusions and outlines further research directions.

2 Related Work

Most of the related work deals with RBAC, workflow and event-driven systems. We incorporated previous RBAC research, and we here propose an adaptive RBAC based on an complex-event-driven system mechanism. Ferraiolo et al. (2003) explained traditional RBAC in the recent system environment. Al-Kahtani and Sandhu (2002) proposed an RBAC model for user-attribute-based user-role assignment.

Cruz et al. (2008) presented an idea regarding to collaborative environments, by which the dynamic assignment of a role is particularly suited to collaborative applications where the users' geospatial locations are of interest. Dynamic role assignment had been addressed in Joshi et al. (2003) as GTRBAC. Static and dynamic separation-of-duty are achieved using specific time-based constraint semantics. Shafiq et al. (2005) addressed RBAC for real-time systems, combines the event-based approach with Petri-net modeling.

Some supply chain process-related issues entailing RBAC mechanisms also have been proposed. Yahya et al. (2007) incorporated RBAC into the supply chain process, including document version handler access control approach. The research was extended to include event-based workflow systems that have been quickly emerging recently.

In order to understand event terminology in the workflow environment, Leune (2004) implemented event-based distributed workflow execution. Luckham (2002) defined an event as an occurrence of something that might incur another event. All occurring events are managed in order to invoke workflow as a real-time system. Kong et al. (2008) applied ECA rules to automatically invoke workflow using generated event, with regard to the generation of a rule-based approach. And, Bae et al. (2004) discussed the automatic control of ECA rules application in workflow processes.

3 Event and User-Role Assignment Definitions

Users are allowed to carry out an operation if the access control policy condition is satisfied. Complex events are associated with rules to enforce access control policy checking. In the literature, a number of event operators have been proposed based on the requirements of several application domains. Event detection semantics are based solely on time of occurrence and do not include attributes, predicates or their combinations. This paper will present some definitions to describe the mechanism of event-driven systems in workflow management using RBAC.

Definition 1 A process P denotes as a directed graph $p = (T, L, A)$ in the workflows are defined as follows:

- A set of tasks: $T = \{t_i \mid i = 1, ..., I\}$ where t_i represents the ith task and I is the total number of tasks in p
- A set of links: $L = \{l_k = (t_i, t_j) \mid t_i, t_j \in T, i \neq j\}$ where l_k represents a relation between two tasks, task i (t_i) preceding task j (t_j)
- A set of attributes: A_i is a set of task attributes
- $A_i = \{t_i, a_m \mid m = 1, ..., M\}$ is a set of task attributes, where t_i, a_m represents the mth attribute of task t_i, and M is the total number of t_i's attributes

Definition 2 (*Role-based access control (RBAC) definition*) The definition of RBAC obtained from Ferraiolo et al. (2003) is defined as follow:

- U, R, Pm, Pv, where U is a set of participants, R is a set of roles, Pm is a set of permissions and Pv is a set of privileges
- A set of user to role assignment $UA \subseteq U \times R$: $U \rightarrow 2^R$
- A set of permission to role assignment $PA \subseteq Pm \times R$: $R \rightarrow 2^{Pm}$
- A set of permission $Pm \subseteq Pv \times T$
- A partial ordering $RH \subseteq R_1 \times R_2$, where $R_1, R_2 \in R$, represented by the symbol: $\geq$, which defines role hierarchy. $R_1 \geq R_2$ implies that R_1 inherits permissions from R_2

Definition 3 (*Primitive event types*) An event is defined as any occurrence of interest. A primitive event is a single event that can be composed into a complex event. Request(Attr), Response(Attr), Publish(Attr) and Subscribe(Attr) are the typical primitive events that usually occur in the logistics environment.

All mentioned events in this paper are considered as level 2 (Luckham 2002), according to the network protocol activity. Luckham (2002) described the level 1 activities necessary in order for an event to be delivered, as the existence of an observable event in the network and of a server that is sending, transmitting, and delivering data. A requester can perform the activities *send data(send())*, *wait for acknowledgement (wait())*, *time out(TimeOut())*, and *resend (ReSend())* the same data. Readers who want to understand more about level 1 activity can refer to Luckham (2002).

Definition 4 (*Complex-event types*) A complex event is considered as an abstraction from other events called its *members*. A complex event *denotes* or *signifies* the set of its member events. Tombros et al. (1997) classified the event into six types. Our research simplifies that scheme into five types of primitive event combinations, as follows:

- OR(($e_1, e_2, ..., e_n$),k) is an exclusive-or event type, occurring when either k-of-the-n defined events arrive

- $SEQ(e_1,e_2, ...,e_n)$ is a *sequential* event type, occurring when n component events come in a specific time order
- $AND(e_1,e_2, ...,e_n)$ is a *concurrent* event type, occurring when n component events come at the same time
- $NEG(e1, (e2,e3))$ is a *negative* event type, occurring when e1 does not come in the interval defined by e2 and e3
- $REP(e1,n)$ is a *repetitive* event type, occurring when e1 comes a predefined (n) number of times

Events written in the Table 1 are obtained from Luckham (2002) specific on supply chain events. One consideration not to include the CCR event in Tombros et al. (1997) is somewhat rare to have the same time occurrence in the logistic area. Thus, instead of having two components in the same time, AND event type accommodate this kind of condition due to concurrent event.

Definition 5 (*ECA (event-condition-action) rules*) The listed ECA rules below regulate a complex-event detection system. Once an event occurs, the system checks complex-event detection (CED) and corresponds to the ECA rules. If the predefined condition is satisfied, the system can execute an action set in the ECA rule definitions (Bae et al. 2004)

ON (*event*)
IF conditionSet $= \{(condition\_expression)\}$
DO actionSet $= \{(user\text{-}rule\ assignment)\}$

Definition 6 (*Adaptive-assignment*) Adaptive-assignment is a tuple <CED, *func()*> where CED, again, is complex-event detection; and *func()* is an additional predicate function to either assign, execute or monitor the session status.

A finite set of *act*, standing for action, is denoted as $<e_i,ti>$ where e_i is a finite set of ith single or primitive events at time *ti*. Event e_i may hold specific attribute to be mapped into specific task or process. A finite set of *act* is included in the predicate function *func()* as shown in Table 2. The predicate function is categorized into three types: assignment, function and status information. Predicate should satisfy the condition hold in order to execute the implication hold. The details of condition and implication hold are not shown due to the page limitation.

Table 1 Example of complex events in logistics area

Complex event	Semantic
SEQ (*GetPrice(),ResGetPrice()*)	An activity complete when a *GetPrice()* request is followed by *ResGetPrice()*
AND (*ResGetPrice(),SubGetPrice()*)	The occurrence of *ResGetPrice* and *SubGetPrice* events at the same time
OR (*ResGetPrice(),SubGetPrice(),1*)	A predefined action is executed when either *ResGetPrice* or *SubGetPrice* event occurs
REP (*PlaceProd(),5*)	*PlaceProd()* event occurs 5 times with no response

Table 2 Example of predicate function

Predicate function	Semantic
Assigned (u,r,p,act)	u can acquire privilege p through role r with action act
Delegate (u_1,u_2)	Delegate the privilege of user u_1 to user u_2 who has the same role
Revoke (u_1,u_2)	Revoke the privilege of user u_1, which had been delegated to user u_2
Activate $(r_1(u_1),r_2(u_2))$	User u_1 who has role r_1 in the active state can activate u_2 who is in role r_2
Assign (u_1,u_2)	User u_1 is inactive, and the system elevates the access control to user u_2
Enable (r,act)	Role r is enabled at action act
U_assigned (u,r,act)	User u is assigned to role r at action act
P_assigned (p,r,act)	Privilege p is assigned to role r at action act
Can_activate (u,r,act)	User u can activate role r at action act
Can_acquire (u,p,act)	User u can acquire privilege p at action act
Can_be_acquired (p,r,act)	Permission p can be acquired through role r at action act
Active (u,r,s,act)	Role r is active in user u's session s at action act
Acquires (u,p,s,act)	User u acquires privilege p in session s at action act

Figure 2 illustrates the architecture of the system. It includes three components, complex event processing engine layer, access control layer and BPM layer. An activity could send a specification to parser for filtering the incoming detection event. Complex event process layer is responsible for receiving the primitive event and detect the complex event occurrence. The specification sent by parser in BPM layer is used to send event a report back to BPM layer for task assignment process. The access control layer decides the user who carries out privilege based upon the role and adaptive assignment access control policy. Thus, it assigns the user to execute the activity.

The complex-event detection (CED) algorithm is a simple search algorithm presented in Fig. 3. CED has its own rule to find the matched complex event. The recent time, denotes as *time(now)*, is stamped into the event attribute before generating the classes. Some classes, such as *Add, isBefore, isIncluded, is-NotInBetween*, and *occurrence* determine the complex-event classification. Function *Add(EventList[])* retrieves event from repository and appends the

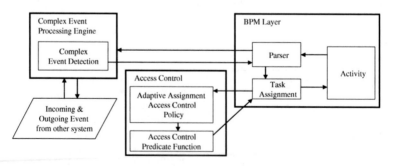

Fig. 2 System Architecture

```
- Input e_n
- Output CED (e_k, e_n, attribute) // CED = Complex Event Detection

    For any e_i in repository {
            IF (time(e_i) isBefore time(now)) AND (e_i hasNot invoked)) THEN {
                    Add(EventList[e_i]);}} // to store the event list temporarily

        For any e_i in EventList[]{
        For any CED that includes e_n {
        IF (e_i isBefore e_n) THEN { search SEQ(e_i,...,e_n, ...,e_{n+j});
                return SEQ(e_i,...,e_n);} //sequential complex event
        ELSE IF (time(e_i) == time (e_n)) THEN {search AND (e_i,...,e_n, ...,e_{n+j});
                return AND(e_i,...,e_n);}//consecutive complex event
        ELSE IF (e_i isIncluded in OR(e_i,...,e_n)) THEN {
                return OR(e_i,...,e_n, k);}//exclusive-or complex event
        ELSE IF (time(e_i) isNotInBetween time (e_n) AND time (e_{n+1})) THEN
                return NEG((e_n),(e_i,...,e_{n+j} );}//negative complex event
        ELSE IF (occurrence(e_i) > 1) THEN {
                return REP(e_i,...,e_n, k);}//repetitive complex event
        }
        }
        Store event e_n to repository;
```

Fig. 3 Complex-event detection (CED) algorithm

un-processed event into the *EventList* to be checked with other event. Class *is-Before* refers to a function to identify whether a precedence event occur before the next event. Class *isIncluded* tries to seek the event among the complex event detection OR logic. To find the negative complex event, class *isNotInBetween* checks whether an event is not in between the time range of precondition defined event. And, class *occurrence* enumerates the occurrence number of exactly the same event to detect the repetition complex event. If not every condition is satisfied, the event will be stored and the system will check for the next event occurrence.

4 Application

The logistics process in Fig. 1 describes the inter-workflow process that can contain adaptive RBAC. We considered the interaction between the retailer–shop and retailer–warehouse processes. Once a payment activity is completed, the system will update each product quantity to a new number of stocks. This quantity will be compared with the safety stock. And, if the recent quantity is lower than the safety stock, an order event is generated to place new product items on the shelf. Next, the system will check the remaining stock in the warehouse. If it is lower than the warehouse defined safety stock, then another event occurs to invoke the purchasing process. If a contract with a distribution centre has been established, the event could directly trigger the pertinent process of the other party. All requests should have a response in order to complete the activity mechanism.

Figure 4 simply represents the user, role, role hierarchy and privilege definitions both retailer and distribution center. And, it shows the user-role and role-permission assignments. Those detail adaptive assignments are exemplified in Fig. 5.

Figure 5 shows the different assignment problems based on the amount of event repetition to satisfy the service level agreement. The first example says that a delegation assignment should take place when a *PlaceProd* event occurs more than five times without any response since the first event occurrence.

Figure 5 also represent different type of complex-event adaptive assignment system. Once a user either publishes a request or requires a product price, one of the approach of complex-event-detection is to seize the first occurrence event either responses or subscription a product price. The response event is going to invoke role *financial* and the subscription event is going to invoke role *purchasing_manager* for further confirmation. Other possibility is event occurrence in the same time. If it happens, AND complex-event-detection checks the condition which event carry out the lower price attribute. When systems are waiting for response or subscription event, other single event can occur. During this period

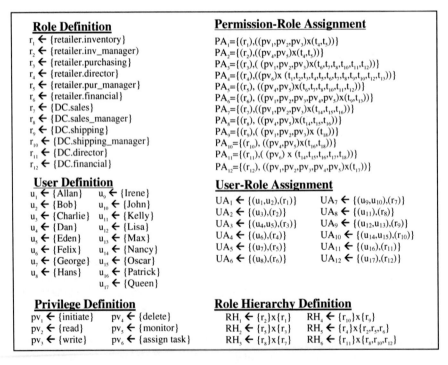

Fig. 4 Assignment of user, role, privilege in an entity

ON activity.completed('payment') //activity t_2
IF (sales.product(Pd).quantity < product(Pd).ShelfSafetyStock)
DO *PlaceProd*(Pd, S, '2009-01-22:08.15.23')

Adaptive-Assignment : <REP(PlaceProd,5), *delegate*(u_1, u_2) >
(Note: the other user in the same role replaces the user u1 task)

ON *Wait*(*PlaceProd*(Pd, S, '2009-01-22:08.15.23'), time(now)) // CED
IF ((*ReSend*(*PlaceProd*(Pd, S, '2009-01-22:08.15.23')) > 5) OR (time(now)) > event.firstTime(60 min)))
DO *Can_activate*(u_2,r,act) WHEN (Active (u_2,r,s,act)) isTrue AND (Active (u_1,r,s,act)) isFalse)

ON activity.completed('Fill PO') // activity t_7
IF (product(P).price == null)
DO *PubGetPrice*(R,Pd, Pr, Q, '2009-01-22:08.15.23') AND *GetPrice*(R,Pd,Q,Pr, DC, '2009-01-22:08.15.23')

Adaptive-Assignment : <OR (*ResGetPrice*(),*SubGetPrice*(),1), *Can_acquire*(u,p,act) >
<NEG(*Purchase*(),(*ResGetPrice*(), *SubGetPrice*()), *Can_activate*(George,pur_manager, *act*)>

ON *ResGetPrice*(Re,P, Pr, Q, time(now)) // CED
IF (price_review.state="ready")
DO *Can_acquire* (Hans,price_review.write, act) AND alert(); // (t_9.pv_3) (u_8.r_6)
ELSE
ON *SubGetPrice*(Re,Pd, Pr, Q, time(now))
IF *SubGetPrice*(Re,Pd, Pr, Q, time(now)) where StdPrice > Price(*ResGetPrice*)
DO *Can_acquire*(George.pur_manager, act) AND alert();// (t_9.pv_3) (u_7.r_5)

Fig. 5 Application of adaptive assignment policy

waiting time, system should not allow the occurrence of purchase event of the same product. If it occurs not in the range of both mentioned events, system will activate the *purchase_manager* role for special process treatment.

5 Conclusions

This paper presents adaptive RBAC in a complex-event-driven system capable of supporting BPM for logistic. Event detection to activate the complex-event rule processing engine is the most important component of this system. The detection of a complex event triggers the rule processing engine and executes an action if it satisfies the condition. The action deals with the adaptive RBAC system in order to assign tasks automatically and thereby improve process efficiency.

The main contribution of this paper is to develop an approach to automate the task assignment in order to improve the business process efficiency regarding to requester waiting time. Either non-response or specific important event is managed in adaptive assignment rule for self-assignment using RBAC. Human activity in logistic process such as purchasing, product order, product delivery, etc. may be executed in a better approach for satisfying customer. The simple application gives indirectly evidence of a better logistic process system performance since complex event occurrence handled by ECA rule can establish an automatic assignment mechanism.

Although task assignment performance was not considered in this paper, we could say that this approach brings a better QoS for customer in terms of time. The assignment approach, related to time efficiency and user workload, will be the focus of future research. Given the huge number of incoming events, such a workflow pattern demands consideration in order that the interoperability of complex-event detection and business process management can be simplified and improved.

Acknowledgment This work was supported by the Grant of the Korean Ministry of Education, Science and Technology (The Regional Core Research Program/Institute of Logistics Information Technology).

References

Al-Kahtani, M. A., Sandhu, R., 2002, A Model for Attribute-Based User-Role Assignment, Proceedings of the 18th ACSAC '02, pp. 353–362

Bae, J., Bae, H., Kang, S., Kim, Y., 2004, Automatic Control of Workflow Process using ECA rules, IEEE Trans. On Knowledge and Data Engineering, Vol. 16, No. 8, pp. 1010–1023

Cruz, I. F., Gjomemo, R., Lin, B., Orsini, M., 2008, A Constraint and Attribute Based Security Framework for Dynamic Role Assignment in Collaborative Environments, CollaborateCom, pp. 1–18

Ferraiolo, D. F., Kuhn, D. R., Chandramouli, R., 2003, Role-based Access Control, Artech House

Joshi, J. B. D., Bertino, E., Shafiq, B., Ghafoor, A., 2003, Dependencies and Separation of Duty Constraints in GTRBAC, SACMAT '03, pp. 51–64

Kong, J., Jung, J.Y., Park, J., 2008, Event-Driven Service Coordination for Business Process Integration in Ubiquitous Enterprises, Computers & Industrial Engineering, 57, pp. 14–26

Leune, K., 2004, An Event-based Framework for Service Oriented Computing, Infolab Technical Report Series, No. 14

Luckham, D., 2002, The Power of Events, Addison Wesley, Boston

Shafiq, B., Masood, A., Joshi, J., Ghafoor, A., 2005, A Role-Based Access Control Policy Verification Framework for Real-Time Systems, Proceedings of the 10th IEEE International Workshop on Object-Oriented Real-Time Dependable Systems, pp. 13–20

Tombros, D., Geppert, A., Dittrich, K. R., 1997, Semantics of Reactive Components in Event-Driven Workflow Execution, Proc. 9th Int'l Conference on Advanced Information Systems Engineering, pp. 409–422

Yahya, B. N., Kwon, M., Bae, H., 2007, RBAC for Supply Chain Process Monitoring, International Conference on Convergence Information Technology, Nov. 2007

A Preliminary Investigation
on a Bottleneck Concept
in Manufacturing Control

Bernd Scholz-Reiter, Katja Windt and Huaxin Liu

1 Introduction

Conventional manufacturing systems are characterised by central or hierarchical control methods, which show a wide range of weaknesses especially regarding the flexibility and adaptability to the unexpected events including unpredictable internal disturbance (e.g. equipment failures, rework etc.) and changing external environmental influence (e.g. variations in demand patterns, unsatisfied raw material delivery etc.). In order to cope with these challenges, the ongoing paradigm shift from a central control of "non-intelligent" system entities (e.g. orders, machines, cells etc.) in central structures towards a decentralized control of "intelligent" system entities in heterarchical structures (Scholz-Reiter et al. 2004). Along the paradigm shifting spectrum as shown in Fig. 1, the control systems in hybrid structures and heterarchical structures tend to be lower complex while improving flexibility, adaptability and fault tolerance more or less. Generally speaking, the more heterarchical the system structures are, the more described advantages the systems possess. Although it is now widely agreed that decentralized control systems make the best use of the described advantages to the utmost extent, they also suffer some drawbacks, in which the most significant one is the difficulty in optimizing system performance. Therefore, nowadays it becomes a popular issue to capture the positive aspects of both central and decentralized control systems. The majority studies attempt to either modify control system structures by introducing one level of hierarchy or to offer individual entities a more global view by gaining access to necessary global information. In this paper, we call both of them as hybrid control systems.

B. Scholz-Reiter (✉), K. Windt and H. Liu
International Graduate School (IGS) at Bremen University, Bremen, Germany
e-mail: bsr@biba.uni-bremen.de

H.-J. Kreowski et al. (eds.), *Dynamics in Logistics*,
DOI: 10.1007/978-3-642-11996-5_19, © Springer-Verlag Berlin Heidelberg 2011

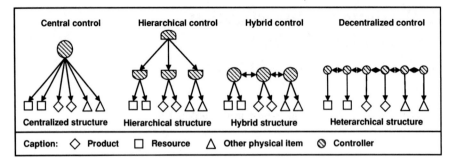

Fig. 1 Paradigm shifting spectrum

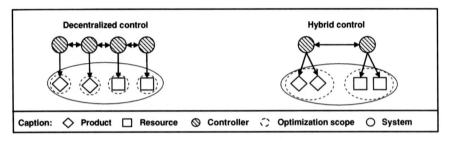

Fig. 2 Optimization scope

2 Literature Overview

2.1 Modification Control System Structures

In this category, several researches attempt to modify control system structures by backward shifting from heterarchical structures to hierarchical structures along the paradigm shifting spectrum. The main idea is to congregate fully distributed decision power of decentralized control systems over again and to delegate the congregated decision power to high-level controllers. Consequentially, the optimization scopes (i.e. the amount of entities, which cooperate with each other to maximize a certain objective value) extend from individual entities to the entities clusters controlled by high-level controllers, so as to enhance system performance as shown in Fig. 2.

For instance, Ottaway et al. (2000) describe an adaptive production control system (APCS), in which the control system structure transits dynamically between heterarchical structure and hierarchical structure. In their model there are job, resource and supervisory agents. When a resource agent determines that the resource is not being properly utilized, it requests a supervisory agent which has jurisdiction over the resources, so one level of hierarchy is introduced into system. They present the comparison results between a heterarchical control system and APCS and argue that the APCS performs better.

Maturana et al. (1999) propose a multi-agent architecture (MetaMorph) for distributed manufacturing system. They use two types of agents: resource agents for representing physical resources, and mediator agents for coordination. The individual resource agents register themselves with mediator agents and find other agents through mediator agents, which use brokering and recruiting mechanisms for coordination and play the role of system coordinator by encouraging cooperation among intelligent agents. In this way, virtual clusters or organizations of intelligent agents can be created and disbanded dynamically.

Brussel et al. (1998) presents the reference architecture, product-resource-order-staff architecture (PROSA). There are three types of basic holons: order, product, and resource holons. Each of the basic holons is responsible for logistics, technological planning, and the determination of resource capabilities respectively. The staff holons can provide centralized algorithm such as, scheduling algorithm and assist the basic holons. By including staff holons, the architecture of holonic manufacturing system (HMS) lies between hierarchical and heterarchical structure to compensate one of the major drawbacks of fully heterarchical structures: the lack of guaranteeing a certain global behaviour and performance (Scholz-Reiter and Freitag 2007).

Although these control systems achieve better optimization in contrast to decentralized control systems, the structure, as an abstractness of system (Brussel et al. 1998), is not enough to make a system, a structure with certain properties (Alexander 1983). Due to the sensitivity of cooperation protocols to system performance and the nonlinear dynamics, the development of effective protocols is quite difficult. Thus, by the sole means of modifying control system structures and without finding explicit and effective cooperation protocols, not only optimization task is underachieved, but also the complicity of communication and cooperation increases simultaneously. Most importantly, a subsystem, an entities cluster, in hybrid control systems is a metaphor for an entity in decentralized control systems. Consequentially, individual subsystems also attempt to achieve local objective without considering global performance.

2.2 Reconfiguration of Decision-making Based on Information

Since global optimization always relies on global information (Dilts et al. 1991; Lin and Solberg 1991), several researches develop the hybrid control systems in which individual entities make decisions based on not only local information but also global information by the means of scheduling. Such kind of information endows system entities a global view. As a result, the globally coherent local decisions (Duffie and Prabhu 1994) are made by individual entities so as to enhance system performance.

For example, Duffie and Prabhu (1996) present a look-ahead cooperative scheduling algorithm to enhance the global system performance of the heterarchical manufacturing systems. Ramaswamy and Joshi (1995) combine a centralised off-line scheduling algorithm based upon Lagrange relaxation together with a distributed on-line control based on market mechanisms. Bongaerts et al. (1998) propose a centralised reactive scheduler which sends the generated schedule to a set of order and resource agents in HMS environment. The centralised reactive scheduler tries to capture the effect of local decisions on a global performance measure by using partial derivatives of the global performance to the local decision parameter.

In these hybrid control systems, the properties of global information play a determinative role. If autonomous entities consider redundant global information in decision-making process, the computational burden and complexity will increase and compromise the benefits of heterarchical architectures. On the other hand, if autonomous entities make decisions based on insufficient or unfit global information, the made-decisions might be unwise or even wrong. Therefore, in contrast to the means of modifying control system structures, obtaining minimal global information becomes the premise of achieving optimization task in decentralized control systems.

3 Bottleneck Concept

In the above discussed hybrid control systems, individual entities are delegated indiscriminative rights and decision power to cooperate with each other due to unawareness of the difference of their positions. For instance, in the second category the hybrid control systems are organized as a cooperative heterarchy in which individual entities have equal rights of access to each other (Duffie and Prabhu 1994). Similarly, in the first category the same level controllers (e.g. high-level controllers or controllers controlled by a high-level controller) have also equal decision power.

However, nothing in this world develops absolutely evenly, and the theory of equilibrium should be opposed regarding to the dialectical relationship between whole and part. Alexander (1983) takes the organic whole (i.e. organism, biological species, society etc.) as an example to illustrate the various parts making up a whole may occupy by no means equal positions. He argues that every whole is composed of key part and non-key parts, and the key part plays a leading role in a whole and determines performance of a whole. The positive performance of key part results in positive performance of a whole; On the contrary, negative performance of key part results in negative performance of a whole. As a result, aiming at achieving positive performance of a whole, the improvement of negative performance of key part must be given first priority.

The term bottleneck as a metaphor for the key part which has the most negative influence upon global performance is therefore introduced. In order to cope with the common drawback of existing manufacturing control systems, individual entities should be aware of their bottleneck situations, and the bottleneck entities should also obtain the highest priority to achieve local high performance regarding to Spirkin. As noted already, if bottleneck information as the minimal global information is taken into account in decision-making process, global performance of decentralized control systems might be enhanced. That means that the optimization task might be accomplished by finding bottleneck entities and delegating the highest decision power to bottleneck entities.

The denotations of bottlenecks are quite diverse in terms of different goals, subjects and application fields. In order to verify the bottleneck concept, in this paper we mainly focus on the domain of production logistics by taking the throughput bottleneck as the research object. From a logistic perspective, manufacturing enterprises are trying to distinguish themselves from their competitors not only by manufacturing high quality products at low costs, but also increasingly by a superior logistic performance (Kim and Duffie 2005). The latter is primarily demonstrated by high delivery reliability and short delivery time, both of which necessitate short throughput time (TTP) (Wiendahl and Lutz 2002). Therefore, the goal to reduce the order TTP stands in the foreground of our investigation, and the TTP bottleneck work system is defined based on the relative proportion of operation throughput time (TTPRP). In the production system consisted by several work systems (WSs), the WSs with the maximum or minimum TTPRP are identified as the bottleneck or non-bottleneck WSs, respectively. The TTPRP of work system (WS) is derived from the sum of TTPRP of work centers (WCs), and the TTPRP of work center (WC) is calculated as follows:

$$\text{TTPRP}_j = \frac{(\text{TTP}_i \cdot n)_j}{\sum_{j=1}^{\text{NWC}} (\text{TTP}_i \cdot n)_j} \cdot 100$$

Based on Nyhuis and Wiendahl (2003) and Windt (2000) where TTPRP_j is the relative proportion of operation throughput time for work center j (%), TTP_i the unweighted throughput time of operation i (h), n the number of accomplished operations at work center (–), i, j the general variable, NWC the number of work centers (–).

4 Experiment

To test our ideas, a discrete event simulation model is developed using the practical data exported from Production Planning and Control (PPC) software (FAST/Pro). The simulation model describes a partial manufacturing system of a German hanger manufacturer and consists of four WSs which represent different production stages including manual turning, CNC turning, CNC drilling and CNC center as shown in Fig. 3. Each WS works in the given shift calendars and consists of different number of parallel WCs (i.e. machines). For simplicity, the following simulative experiments are carried out for 128 shop calendar days (SCD).

Fig. 3 A partial manufacturing system with complex
material flow

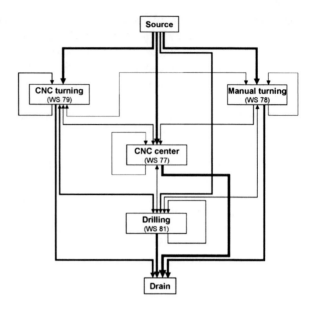

4.1 Scenarios Introduction

There is no doubt that the total amount of work in process (WIP) determines system throughput time. If bottleneck WSs do not exist, on the premise that total WIP is constant, the throughput time performance would not be different under the conditions: (1) every WS possesses equal decision power to cooperate with each other; (2) the bottleneck WSs possess the highest decision power; (3) the non-bottleneck WSs possess the highest decision power. On the contrary, the control system under condition 2 should perform better than under condition 3, and the best system performance should be achieved under condition 2.

In order to achieve short throughput time, in our model intelligent orders always prefer to be processed at the WC with the minimum WIP on each production stage; WSs and release station cooperate with each other try to keep the total WIP at a constant and relatively low level. In the cooperation process, only when orders arrive at buffer areas, the work content of current operation is taken into account into the total WIP, which may sometimes decrease to under the constant WIP level due to deficiency of incoming orders as well. Based on a long-term simulation, the constant WIP level is set at 200 working hours by reference to the sum of the average ideal minimum WIP ($WIPI_{min}$) levels of four WSs.

On the basis of the above defined control approaches and the constant WIP level, the following three scenarios are developed: In scenario 1, WSs are not aware of bottleneck situations and have equal decision power to cooperate with each other and release station; in scenario 2, the identified bottleneck WSs are distributed the highest decision power which endows bottleneck WSs with the priority to reduce workload by informing release station to change order release sequence. I.e. the orders, which are not processed at the bottleneck WSs according to the process plan, have the priority to release at first. In the case that all waiting orders must be accomplished at the bottleneck WS, release station releases the orders to keep total WIP at the constant level; in scenario 3, the identified non-bottleneck WSs are distributed the highest decision power and attempt to reduce workload.

4.2 First Evaluation

Table 1 provides the performance measures of mean (unweighted) order and operation throughput time. By comparing the performance measures of scenario 1, 2 and 3, the interesting facts are that the performance of different scenarios are significantly different, and scenario 2 performs much better than scenario 3. These indicate that on one hand, the bottleneck WSs do exist in manufacturing system, i.e. individual WSs do occupy unequal positions, on the other hand, the delegation of the highest decision power to bottleneck rather than non-bottleneck WSs leads to the improved throughput time.

Moreover, the results from scenario 2 exhibit the complex dynamics and nonlinear behaviour of bottlenecks as well as the faults of bottleneck configuration. To analyse these characteristics, the relation among throughput time, actual WIP and WIPI_{min} level will be introduced firstly. In accordance with the logistic operating curve (LOC), the WIPI_{min} level represents the WIP level necessary to achieve the minimal attainable throughput time, and the throughput time proportionally increases with increasing WIP if the actual WIP level exceeds the WIPI_{min} level (Nyhuis and Wiendahl 2003). The WIPI_{min} level of each WC is calculated as:

$$\text{WIPI}_{min} = \frac{\sum_{i=1}^{n} \left(\text{WC}_i^2 \right)}{\sum_{i=1}^{n} \text{WC}_i}$$

where WIPI_{min} is the ideal minimum WIP level of work center (h), WC_i the work content of operation i (h), n the number of accomplished operations at work center (–), i general variable.

Regarding to the manufacturing system operating curves (MSOC) (Schneider 2004), the WIPI_{min} level of a manufacturing system can also be derived from WIPI_{min} levels of its subsystems (e.g. work systems and work centers). In our case, since each WS is composed of not series-wound but parallel WCs, the WIPI_{min} level of WS is derived from the sum of WIPI_{min} levels of its WCs.

Table 1 Comparison performance measures

Scenario no.	Mean order TTP (h)	Mean operation TTP (h)
Scenario 1	9.47	7.77
Scenario 2	11.94	9.77
Scenario 3	12.47	10.18

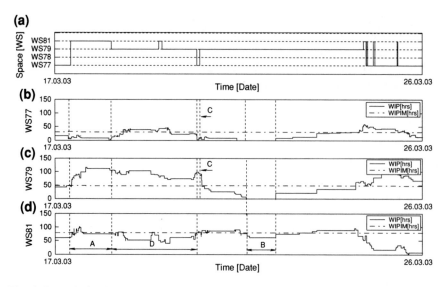

Fig. 4 Dynamic bottleneck in manufacturing system

Figure 4 presents the correlation between bottleneck situations, actual WIP and WIPI$_{min}$ levels in a randomly selected observation period. The complex dynamic characteristics of bottlenecks are demonstrated in Fig. 4a, where the length of horizontal line presents the duration of a certain WS being bottleneck WS. In the observation period, bottleneck shifts in time (X axis) and space (Y axis) and jumps among WSs due to the unexpected events.

Moreover, bottlenecks' nonlinear behaviour as well as the faults of bottleneck configuration is presented in Fig. 4. In the observation period A, WS 79 has the highest WIP and the maximum difference between WIP and WIPI$_{min}$ level and seems as if to be the bottleneck WS, which has the most negative influence on throughput time performance. However, at that time WS 81 which has the relatively low WIP and the negative difference is identified as bottleneck WS. In the period of B, WS 79 has no buffer inventory and less negative influence on throughput time performance, but it is still recognized as bottleneck WS. Moreover, in the period C the WIP of WS 77 suddenly decreases to under the WIPI$_{min}$ level; and the WIP of WS 79 suddenly increases and exceeds the WIPI$_{min}$ level, but the bottleneck shifts from WS 79 to WS 77 with no reason. On the contrary, WS 79 being bottleneck WS can be explained due to the highest WIP level and the maximum difference between WIP and WIPI$_{min}$ level in the period D.

4.3 Fault Detection and Isolation

A comparison of the performance measures of scenario 2 and 1 reveals that the system performance, especially throughput time performance, has been underachieved in scenario 2 than in scenario 1, though the bottleneck WSs have been already discovered and delegated the highest decision power to cooperate with each other. Now that improving system performance relies on bottleneck information, the most likely cause of the problem is the misestimation of bottleneck WSs in scenario 2. One question that needs to be asked, therefore, is whether the faults of bottleneck configuration can be detected and isolated, so as to further achieve short throughput time. However, the fault detection task is very tricky, on one hand, the error bottleneck situations are not supposed to be ignored, on the other hand, the unexplainable bottleneck situations caused by the nonlinear behaviour are not supposed to be detected as faults. So we only consider the extreme situations in which bottleneck situations may most possibly be misestimated.

Due to the highly dependent relation between throughput time, WIP and $WIPI_{min}$ level, the differences between actual WIP and $WIPI_{min}$ levels as reference variables are introduced into fault detection process based on the heuristic knowledge. I.e. when the value of the identified bottleneck WS is negative or the minimum value of the set of all the reference variables, this WS might be unsuitable as a bottleneck WS due to its less negative influence on throughput time performance. On the contrary, when the value of a WS is the maximum value of the set of all the reference variables, this WS is most likely to be a bottleneck WS. To find out an effective fault detection method, scenario 4, 5 and 6 are developed: In scenario 4, the identified bottleneck WSs with the minimum values are detected as faults; in scenario 5, the identified bottleneck WSs with negative values are detected as faults; in scenario 6, the identified bottleneck WSs with negative or the minimum values are detected as faults. In scenario 4, 5 and 6, the detected error bottleneck WSs are replaced by the WSs with the maximum values during simulation process. With the increasing of the fault isolation times, the order throughput time has been gradually reduced in scenario 4, 5 and 6 as shown in Table 2.

The mean (unweighted) order and operation throughput time have been significantly reduced in scenario 6. It indicates that the proposed fault detection and isolation method in scenario 6 is the most effective one. One of the more significant findings is that the delegation of unequal decision power based on bottleneck situations does improve system performance. Moreover, since the logistic potential is determined by not only the mean value but also the variance of throughput time,

Table 2 Comparison performance measures

Scenario no.	Mean order TTP (h)	Mean operation TTP (h)	Fault isolation times (−)
Scenario 1	9.47	7.77	−
Scenario 2	11.94	9.77	0
Scenario 4	9.83	8.07	174
Scenario 5	9.26	7.59	1,312
Scenario 6	9.22	7.56	1,369

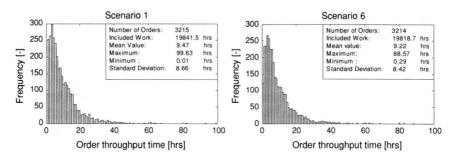

Fig. 5 Distribution of order throughput time

the throughput time variance is also decreased in scenario 6 by comparing to scenario 1 as shown in Fig. 5.

5 Conclusion

Aiming at enhancing global performance of distributed control systems, the proposed research didn't attempt to dig out new control system structure or to utilize the global information derived from the complicated scheduling algorithms, but instead to suggest taking the bottleneck information, as minimal global information, into account in designing distributed control systems. Since the simulation results proved that taking account the bottleneck information in decision-making process is helpful to enhance global system performance, the bottleneck entity should have the highest power to accept or deny an offer submitted by other entities in distributed control systems. In this way, the objective contradiction between bottleneck entity and other entities can be solved, and the globally coherent local decisions can be guaranteed. However, the current study has only examined throughput time oriented bottleneck so as to achieve the sole logistic objective of short throughput time. In future work the other bottlenecks (e.g. schedule reliability and utilisation oriented bottlenecks) as well as corresponding fault detection and isolation methods will be investigated as well. Based on the multi-bottlenecks information, a distributed control system will be developed to well trade off multiple logistic objectives and react adequately to the unexpected events.

References

Alexander, S. (1983): Dialectical Materialism. Progress Publishers, New York.

Brussel, H.V.; Wyns, J.; Valckernaers, P.; Bongaerts, L. (1998): Reference architecture for holonic manufacturing systems: PROSA. Computers & Industrial Engineering 37, pp. 255–274.

Brussel, H.V.; Wyns, J.; Valckernaers, P.; Bongaerts, L. (1998): Reference architecture for holonic manufacturing systems: PROSA. Computers & Industrial Engineering 37, pp. 69.

Bongaerts, L.; Brussel, H.V.; Valckenaers. P. (1998): Scheduling execution using perturbation analysis, IEEE International Conference on Robotics and Automation, pp. 2747–2752.

Dilts, D.M.; Boyd, N.P.; Whorms, H.H. (1991): The evolution of control architectures for automated manufacturing systems. Journal of Manufacturing Systems 10, pp. 79–93.

Duffie, N.A.; Prabhu, V.V. (1994): Real time distributed scheduling of heterarchical manufacturing systems. Journal of Manufacturing Systems 13, pp. 94–107.

Duffie, N.A.; Prabhu, V.V (1996): Heterarchical control of highly distributed manufacturing systems. Computers & Industrial Engineering 9, pp. 270–281.

Kim, J.H.; Duffie, N.A. (2005): Design and analysis of closed-loop capacity control for a multi-workstation production system. Annals of the CIRP, pp. 455–458.

Lin, G.Y.; Solberg, J.J. (1991): Effectiveness of flexible routing control. Journal of Flexible Manufacturing System 3, pp. 189–211.

Maturana, F.; Shen, W.; Norrie, D.H. (1999): An adaptive agent-based architecture for intelligent manufacturing. Journal of Production Research 37, pp. 2159–2173.

Nyhuis, P.; Wiendahl, H.-P. (2003): Logistische Kennlinien: Grundlagen, Werkzeuge und Anwendungen. Springer-Verlag, Berlin, Heidelberg, New York.

Ottaway, T.A. (2000): An adaptive production control system utilizing agent technology, Journal of Production Research 38, pp. 721–737.

Ramaswamy, S.E.; Joshi, S.B. (1995): Distributed Control of Automated Manufacturing Systems. 27th CIRP Seminar on Manufacturing Systems, pp. 411–420.

Scholz-Reiter, B.; Freitag, M. (2007): Autonomous processes in assembly systems. CIRP Annals, pp. 712–729.

Scholz-Reiter, B.; Windt, K.; Freitag, M. (2004): Autonomous Logistic Processes: New Demands and First Approaches. 37th CIRP International Seminar on Manufacturing Systems, pp. 357–362.

Schneider, M. (2004): Logistische Fertigungsbereichskennlinien. Fortschritt-Berichte VDI, Düsseldorf.

Wiendahl, H.-P.; Lutz, S. (2002): Production in networks. Annals of the CIRP, pp. 573–586.

Windt, K. (2000): Engpaßorientierte Fremdvergabe in Produktionsnetzwerken. Fortschritt-Berichte VDI, Düsseldorf.

Part III
Information, Communication, Autonomy, Adaption and Cognition

Synchronization of Material and Information Flows in Intermodal Freight Transport: An Industrial Case Study

Jannicke Baalsrud Hauge, Valentina Boschian and Paolo Pagenelli

1 Introduction

Today, globalization and technological progress lead to changes which suppliers, manufacturer, logistic service providers and customers need to react on in order to stay in the market. These changes as well as the trend toward complex products with short product life-cycle time cause a close collaboration among global distributed partners and finally to the evolvement of global supply chain networks (Thoben and Jagdev 2001) in order to increase the efficiency and flexibility.

The goal of these networks is the optimization of logistical and production processes (Pfohl 2002; Jüttner 2005), but due to their design and structure they are more vulnerable and the management is challenging due to the large number of different entities with different structure, objectives and cultures (Seiter 2006; Pfohl 2002).

The efficiency and operative excellence of the logistic processes is of predominant importance. The objective of the management of supply chain networks is therefore to manage both material and information flows throughout the entire network. Although much effort is undertaken in global enterprise networks for sourcing and fast and in-time delivering, and in several different strategies like JIT, Kan-ban, lean management, outsourcing of functions, stock reduction, etc., as well as the implementation of information and communication systems has been introduced in order to control the flows in a network, any statistics shows that there

J. Baalsrud Hauge (✉)
Bremer Institut für Produktion und Logistik GmbH, Bremen, Germany
e-mail: baa@biba.uni-bremen.de

V. Boschian
Department of Electronics, Electrical Engineering and Computer Science, University of Trieste, Trieste, Italy

P. Pagenelli
INSIEL S.p.A, Trieste, Italy

H.-J. Kreowski et al. (eds.), *Dynamics in Logistics*,
DOI: 10.1007/978-3-642-11996-5_20, © Springer-Verlag Berlin Heidelberg 2011

are several challenges which needs to be met before the suppliers and logistic companies manage to deliver the right goods in the right amount and quality and order at the right place on time (Sørensen 2005; Jüttner 2005; Pfohl 2002). The reason for this is manifold: the networks comprise several different entities like the suppliers, the producers, warehouses and distribution centers as well as all the logistic service providers carrying and handling the goods on its way from one location to another.

The Fig. 1 below illustrates a simple supply chain, leaving out the logistic service providers, but still it illustrates one of the main problems: the asynchronous information and material flows and the number of interfaces within these flows.

This article will analyze problems arising in intermodal freight transport due to the asynchronous flows as well as look at the possibilities and boundaries of the implementation of existing ICT for reducing the problems.

2 Problem Analysis

As a result of international competition, production processes become increasingly specialized. The implication is that an increasing part of the value added is placed outside the company's own production facilities. The trend in network collaboration is toward "distant collaboration", where ICT tools reduce the need for tight vertical integration (Richard and Devinney 2005). The result is decoupled and loosely integrated networks able to form new supply chains immediately. While as Fig. 1 shows the simplified information and material flow, Fig. 2 depicts an example of a small intermodal transport chain. The industrial case deals with the supply network of a product, sheet of glass, coming from China to an EU plant. The information and material flows in this industrial case are described below.

The product transported is sheets of glass used to produce solar panels. The flow of goods can be shortly described in the following paragraph. Firstly, the producer assigns the production of the goods to Chinese companies. Secondly, after completing the production, the sheets are sent with the cost and freight (C&F)

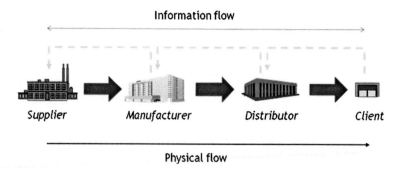

Fig. 1 Simplified supply chain (TSLU, VIU 2009)

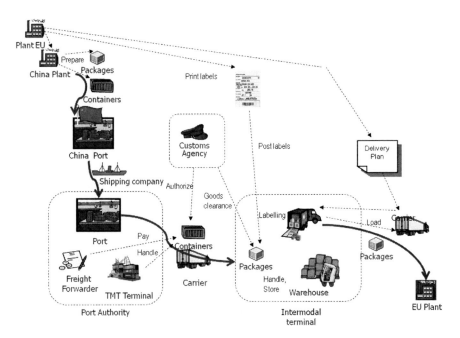

Fig. 2 Example of intermodal transport chain

procedure from the factories to a Chinese loading port, where it is shipped by the shipping company. At this point the shipper has the property of the goods and then goods arrive at the Port of Trieste where they are unloaded by the terminal operator after a while. This is the entry of the goods into the European Union. Trieste is a free port, so there are specific procedures for the customs and shipping documents and payments. At this moment, the carrier has the property of the sheets and hands over them to the shipping agency. Then the shipping agency receives the information regarding units and packages inside them about the packing list for loading, called "manifest", a copy of the bill in the express form and the bill of entry and the customs tariffs is paid and the customs clearance operations are carried out by the shipping agency. After completing this process, the goods are transported on trucks to the truck terminal of Gorizia where they are stored in the warehouse in Gorizia and managed by SDAG. In the warehouse SDAG personnel stick labels on packages. These are produced by Ilva and then posted from Verona before the goods arrival; finally, SDAG loads the goods on trucks depending on the packing list and then they are transported to destination by FERCAM. The flow of information starts much earlier than the flow of goods and is neither synchronic nor seamless nor does it have the same responsibilities and stakeholder. The freight forwarder is not only handling the goods for one customer, mostly it transports the goods of several. Therefore, it needs to be able to handle the different and fast changing requirements that his customers put on him (Zografos and Regan 2004; Caris et al. 2008). Thus, during the last years a lot of research has been carried out in the field on intermodal transportation (Crainic and Kim 2007; Macharis and Bontekoning 2004). Hence, in order to be efficient, an intermodal transportation system needs to synchronize the logistics operations and the information exchange among stakeholders. Solutions based on the application of Information and Communication Technologies (ICT) in logistics managements are the key tools to enhance logistics competitiveness (Giannopoulos 2004). The application of ICT in logistics managements is relatively recent, it lets real-time/ on-line information communication and data exchange through the entire operation chains become realistic speaking of time and cost (Feng and Yuan 2006), but it is still a challenge for SMEs like a lot of the freight forwarders to implement it.

Looking at the information and material flows, important considerations arise concerning the fact of lack of synchronism between the two flows and due to the large number of stakeholders, often having their own proprietary ERP systems, causing many interruptions in the flows. Thus, it can be stated that a major problem is that the flows of goods and information have not the same timing and organizations in charge of the goods and of the accompanying information are often not the same. Actually, logistic processes are characterized by limited sharing of information on goods movements, status and authorizations between the various transport chain actors. At a first sight, there seems to be two different approaches which might reduce the problems: the former concerns the reduction of the time for processing the information on the goods, so that the delay between the flows are reduced, while the latter is related to larger organisational changes, since it is to attach the needed information to the material. During the last decades, much effort was used on implementing the first approach. A main requirement for reducing the time is to increase the interoperability within the whole chain, so that it is possible to have a seamless information flow between the stakeholders hence several SCM-tools were developed for this purpose, but they are very costly, so that a lot of SME cannot afford this investment. Additionally, since hardly any company is only involved in one supply network, it cannot be expected that different SCM tools will be maintained and implemented. Secondly, looking in more detail at supply networks, other problems and barriers appear too. These can be divided in organizational barriers (e.g. lack of transparency and unclear responsibility in the transport chain), technical barriers (e.g. friction at transfer points, lack of standardization of equipment and loading units, missing information flows), financial and economical barriers (perceived high costs of investment in intermodal infrastructure), infrastructure barriers (e.g. different power supply systems, low speeds, different rail gauges), and logistical barriers (e.g. missing services, missing information about services, missing awareness of intermodal services). Additionally, especially for intermodal transport chains, inefficient handling of the cargo and accompanying documentation is a problem.

As mentioned above, it is expected that if it is possible to link the information to the material, so that they travel together through the chain, it is expected that it will be possible to reduce the lack of synchronicity. In the next section we will therefore discuss possible technical solutions, their advantages and disadvantages.

3 Available Identification and Communication Technologies

In the problem analysis, it was stated that if the information could be attached to the material/cargo or if a complete integration of the different proprietary systems at each individual entity, a lot of the problems caused by the lack of synchronicity would be effectively reduced. Therefore, the objective of this section is to give a short summary on available technologies and to show how these can contribute improve the synchronism. Figure 3 shows the information and material flow, but do also contain the function an information and communication system needs to handle.

A first step toward storing the information directly on the cargo item, is to use RFID technology. The advantage of RIFD tags is that the information is directly linked to the cargo, and the information stored in the RFID tags is always accessible at the different stages of the transport and also bulk reading is possible. If active RFID-transponders are in use, these can be coupled with sensors to gather data from the environment (temperature, humidity). However, the use of RFID is quite expensive, even though the tag itself is quite cheap (Riedel et al. 2007).

Waters and Rahman (2008) have done a literature survey on the benefits of using RIFD. Summarizing, it can be said that the main benefits of using RFID are the increased traceability, the information accuracy and sharing, which may lead to a faster information processing and thus a higher efficiency as well as to decreased stock level and improved customer service. In this sense, the use of RFID technology in a supply chain network do reduce the problem of the lack of synchronicity. The next step would now be to look at concepts and technologies offering additional advantages compared with just the use of RIFD. A concept deriving out

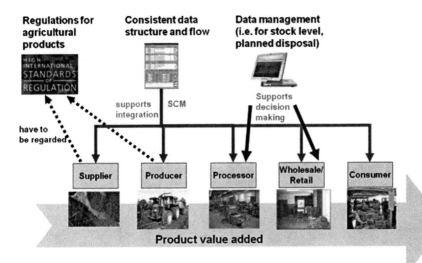

Fig. 3 Supply Chain networks' functionalities (Bemeleit and Thoben 2007)

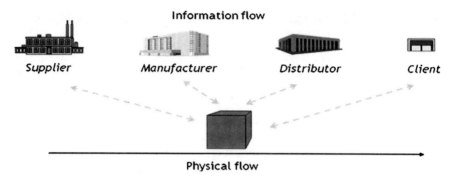

Fig. 4 Material and information flow applying the intelligent cargo concept

of the CRC 637, dealing with autonomous logistics, is to make the cargo item itself intelligent. Simplified it can be said that such an item needs to be able to process and execute decisions autonomously (Böse and Windt 2007). Secondly, this will only be helpful if the item itself can communicate its decision to its environment. The concept foresees not only the implementation of RFID, but also of sensors, so that product can gather additional information by the sensor during the transport. Other necessary basic technological components needed for the realization of intelligent item/cargo is "service oriented architectures" (SOA) and interoperability platforms for data interchange and collaboration between business partners mobile technologies and global positioning systems. The availability of standards, addressing all the aspects of cargo identification and management, is equally important from RFID tag specifications, to identification of individual items (EPC) and shipments (GTIN), to definition of logistic data and processes (e.g., EDI/EDIFACT, EDIFER, ebXML) (EURIDICE 2009; Hans et al. 2007; Smirnov et al. 2007). Up to now cargo intelligence is a concept on a prototype level, so that there is no statistic showing that this concept will revolutionize efficiency of supply chain networks during the next few years, but due to the possibility of the cargo to take their own decision based upon real-time information from its environment, it is very likely that this concept will improve the advantages of RFID, contribute to a reduction of delays and risks and therefore increase the resilience of transport chains and supply chain networks.

Figure 4 shows the improved synchronization of the information and material flow by the implementation of intelligent cargo concept.

4 Conclusions

Supply chain networks are operating in an increasingly dynamical environment and due to number of stakeholders (including suppliers, distributers, producers, logistic service providers and customers), each of them having their own information system, they are vulnerable to every unexpected event. Furthermore, the globalization, the increase of transport volume and political requirement of more environmental friendly transport do also increase the number of stakeholders being involved in the transport process. This article describes the problems caused by the lack of synchronicity in the material and information flows and looks at technologies and new concepts based on advanced ICT improving the synchronicity. It can be stated that even though the use of RFID neither solves the problem with the interoperability by its own nor does it actively communicate with the stakeholders' information system, it increases the synchronization of material and information flows throughout an intermodal supply network. It can be expected that a successful implementation of the intelligent cargo concept will increase the synchronism even more, and also contribute more to the resilience of such global intermodal supply networks, but still not solve the interoperability problem, since all stakeholders have to support the technology. But it will reduce any problem with network availability, since the communication does not depend on network availability, since all information is stored on the cargo.

RFID technology has been commercially available for quite a few years now, but even though it seems really to be able to solve serious problems within supply chain networks, it has still not penetrated every industrial sector. In the problem analysis different barriers were listed. Although some barriers can be reduced by the use of RFID, most of them remain as they are. In their work, Water and Rahman (2008) and Riedel et al. (2007) also look at organizational and economical challenges. For some of the organizational barriers like the willingness to share information, where it is a matter of organizational behavior, and not only of availability of technology, it will take time to change. Comparing benefits and disadvantages of only using RFID or introducing the intelligent cargo concept, it can be expected the costs are higher for the implementation, operation and maintenance of the intelligent cargo. Also the topics of RFID, intelligent cargo and security as well as consumer privacy will be more challenging to deal with, but it seems that the efficiency will further increase with the use of the intelligent cargo concept.

Acknowledgments We thank our project partners and the European Commission for all scientific, organisational and financial support. The project is funded in the 7.Framework programme, ICT. Please find more information under http://euridice-project.eu.

References

Bemeleit B., Thoben K -D.: Challenges for the integration of European and Chinese Vegetable Supply Chain Management. *In Proceedings of "The 12th International Symposium on Logistics (12th ISL)", 8–10 July 2007*, Budapest, Hungary, pp 75–81
Böse F., and Windt K.: "Catalogue of Criteria for Autonomous Control in Logistics." In *Understanding Autonomous Cooperation and Control in Logistics—The Impact on Management, Information and Communication and Material Flow*, by Michael Hülsmann and Katja Windt, 57–72. Berlin: Springer, 2007

Caris A., Macharis C., Janssens G.: "Planning Problems in Intermodal Freight Transport: Accomplishments and Prospects", *Transportation and Planning Technology*, vol. 31, i 3, 2008

Crainic T. G. and Kim K. H.: "Intermodal transportation", In: *C. Barnhart and G. Laporte, Editors, Transportation, Handbooks in Operations Research and Management Science vol. 14*, North-Holland, Amsterdam (2007), pp 467–537

EURIDICE: White paper. http://www.euridice-project.eu/index.php/web/pubdocs/58, 2009

Feng C.-M., Yuan C.- Y.: "The Impact of Information and Communication Technologies on Logistics Management", *International Journal of Management*, vol.23, No. 4, 2006

Giannopoulos G.A.: "The application of information and communication technologies in transport", European Journal Of Operational Research, vol. 152, pp. 302–320, 2004.

Hans C., Hribernik K. and Thoben K-D.: "An Approach for the Integration of Data within Complex Logistics Systems." *LDIC2007 Dynamics in Logistics: First International Conference. Proceedings.* Heidelberg: Springer, 2007. 381–389

Jüttner U.: "Supply chain risk management", The International Journal of Logistics Management, 16, 1, 2005, 120–141

Macharis, C. and Bontekoning, Y.M.: "Opportunities for OR in intermodal freight transport research: a review", European Journal Of Operational Research, vol. 153, pp. 400–416, 2004

Pfohl, H.C.: „Risiken und Chancen: Strategische Analyse in der Supply Chain", *in: Pfohl, H.C. (Hrsg.), Risiko- und Chancenmanagement in der Supply Chain, Darmstadt, 2002, Seite 1–50*

Riedel, J, Pawar K, Torrini S., Ferrari E.: "A Survey of RFID Awareness and use in teh UK Losgistics Industry", *LDIC2007 Dynamics in Logistics: First International Conference. Proceedings.* Heidelberg: Springer, 2007. 105–115

Richard, P, Devinney T.: "Modularity, Knowledge-flows and B2B Supply Exchanges", *California Management Review*, 47, 4, 2005

Seiter, M.: „Management von kooperationsspezifischen Risiken in Unternehmensnetzwerken", München, 2006

Sørensen L. B.: "How risk and uncertainty is used in supply chain management", *a literature study, Copenhagen, 2005*

Smirnov A, Levashova T., Shilov N.: "RFID-based Intelligent Logistics for Distributed Production Networks", *LDIC2007 Dynamics in Logistics: First International Conference. Proceedings.* Heidelberg: Springer, 2007. 117–124

Thoben K.-D., Jagdev H. S.: "Typological Issues in Enterprise Networks"; *Journal of Production Planning and Control*, Volume 12, Number 5, July–August 2001, p 421–436

TSLU, Venice International University, Basic course on logistic, EURIDICE tr.portal, 2009

Waters S., Rahman S.: "RFID in Supply Chains: Literature review and an agenda for future research". *In Pawar, K. S., Lalwani, C.S., Banomyong, R.: Conference Proceedings of the 13th intenational symposium on Logisics, ISL 2008*, Centre of concurrent Enterprise, Nottingham University Business School. Nottingham 2008, 418–425

Zografos K. G and Regan A. C.: "Current Challenges for Intermodal Freight Transport and Logistics in Europe and the United States", *Journal of Transportation Research Board*, No. 1873, pp.70–78, 2004

EURIDICE: Platform Architecture in Logistics for "The Internet of Things"

Jens Schumacher, Manfred Gschweidl and Mathias Rieder

1 Introduction on Logistics and "Internet of Things"

The Internet of things has seen a huge interest since more and more companies would like to adapt RFID technologies in order to improve their processes. The idea is to equip all objects of our daily life with identifying devices. Those devices are usually part of a self configuring wireless network where real-world objects seamless integrate themselves into computer systems. These concepts, originally elaborated by the AutoID-Labs (Architecting the Internet of Things. http://www.autoidlabs.org/) have great potentials not only in the logistics sector where the possibility to individually track and trace goods on their way through the supply chain promises much better control mechanisms as well as an improved quality assurance. This is obviously of particular interest for the logistics sector which, by definition, is concerned to physically move and store goods. Here the RFID technology promises a better transparency of the physical goods as they move through the different logistics operations. The RFID technology promises a better representation of these processes and their actual status also in the virtual world. The new possibilities that are enabled by the virtual representation of real-world goods in the Logistics sector will be investigated in this article.

J. Schumacher (✉), M. Gschweidl and M. Rieder
Vorarlberg University of Applied Sciences, Hochschulstrasse 1,
6850 Dornbirn, Austria
e-mail: jens.schumacher@fhv.at

M. Gschweidl
e-mail: manfred.gschweidl@fhv.at

M. Rieder
e-mail: mathias.rieder@fhv.at

H.-J. Kreowski et al. (eds.), *Dynamics in Logistics*,
DOI: 10.1007/978-3-642-11996-5_21, © Springer-Verlag Berlin Heidelberg 2011

1.1 Overview of EURIDICE and Issues Addressed

To identify the opportunities that in the transport sector are unleashed with the availability of cargo that is actively interconnected with the software applications managing and controlling it, the European Commission has launched the "EURIDICE"-project under the Framework Programme 7. The goal of this project is to realize, with the help of the technologies and services provided, a section of the Internet of Things and Services, which enables the long-term deployment of intelligent Cargo for the different stakeholders involved in the transport sector. This includes e.g. customs, ports, terminals, shippers, forwarders etc. Overall the EURIDICE Project consists of more than 20 partners from different business sectors and authorities, which ensure that the project will provide an open platform, where services from different stakeholders as well as service providers can be integrated. The platform will enable the service providers to combine different transport related services e.g. for dangerous or high value transport in an open and freely customizable manner. For the evaluation of the project companies and authorities from the transportation sector are integrated in the design process of the platform and pilots will be installed for them showing the applicability of the system.

1.2 The Vision of EURIDICE

"In 5 years time, most of the goods flowing through European freight corridors will be 'intelligent', i.e.: self-aware, context-aware and connected through a global telecommunication network to support a wide range of information services for logistic operators, industrial users and public authorities." states the vision of the EURIDICE project team. The difference to already available implementations of such information service providers for the logistics sector is the cargo centric approach of EURIDICE. In EURIDICE single cargo items show a pro-active behavior, provide information and offer suggestions to reach their goals (in our case most probably to reach their destination in time and in good shape).

The next chapters will discuss the EURIDICE architecture from the very top where business related services are offered to the most fundamental parts where mobile devices attached to cargo items handle their processes.

2 Overview of EURIDICE Architecture

The implementation and development of such a platform, for different business cases and with the objectives of EURIDICE, involves the definition of a clear and understandable architecture.

This part will discuss various aspects of the high level architecture and give an overview of it and the principles used for the realization. Therefore a short description of "Service-orientated Architecture" (SOA), "Business Process Execution Language" (BPEL) and "Business Process Modeling Notation" (BPMN) and their relevance for the project will be done, as also in more detail the concepts of intelligent, mobile agents within EURIDICE, which is a key technology, used for fulfilling the vision of intelligent goods and distributed intelligence in logistics chains.

2.1 Description of High Level Architecture

The whole framework and platform will consist of services in different levels and scopes. Roughly the services can be distinguished as shown in Fig. 1.

The common EURIDICE specific services allow the composition of applications specific services for the definition and orchestration of business processes on top of them. These application specific services or whole business processes can also be used by other application specific processes. On lower levels of the framework, the generic components level, EURIDICE will also provide services but only for internal use within the framework and not compromising them to the public, although they are based on open standards and if necessary, this manner can be easily changed in the future.

By using new information and communication infrastructures (e.g., Galileo, UMTS) it is possible to define a uniform Information triple item (time, place, status) for all transported goods in Europe. The use of these data limits itself

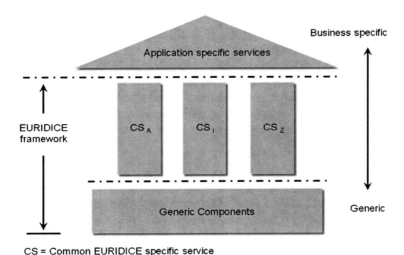

CS = Common EURIDICE specific service

Fig. 1 General levels and scopes of EURIDICE services

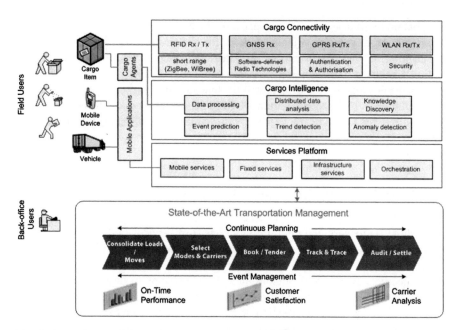

Fig. 2 State-of-the-Art transportation management and EURIDICE

currently, nevertheless, in primarily to easy "tracking and tracing"-functionalities. In the project EURIDICE, based on this information, value added services should be defined, allowing an individual control of the transported goods in the European home market. For an effective implementation the data acquisition and decision-making should result to a very great extent on mobile devices, being able to react on one hand without delays to logistic events and on the other hand to reduce communication expenditures. Figure 2 shows how State-of-the-Art transportation management will be extended by the EURIDICE platform.

As already mentioned, the key concepts used for the development of the platform are SOA, with the associated technologies e.g. BPEL and BPMN, mobile agent technology and the interaction between different services will be based on a common ontology used for the context model and also for the definition of rules and events. Therefore we will give a short overview on them and their usage within the EURIDICE project, before dealing in more details with the mobile agent technology.

2.2 "Service-orientated Architecture" and "Enterprise Service Bus"

There are many books and papers around describing principles, requirements and the application of SOA in projects, and this is not the intention for this article, but

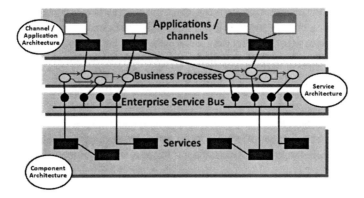

Fig. 3 EURIDICE service orchestration and business process integration

as SOA plays an important role in the overall concepts of EURIDICE a short look on how SOA fits in the platform and how it will be used should be provided.

The concept of a distributed and intelligent platform intends that every good has its own intelligence. On the one hand for the realization of an efficient tracking and tracing of the goods, on the other hand to also receive and evaluate other relevant information with regard to the transported goods on a real-time basis. Round these single intelligent goods an open platform is created to integrate the preserved information and offered services with existing (legacy) systems and to offer the services by means of SOA to other ICT systems, and interoperability with most different software products is thereby guaranteed on the basis of established standards and technologies (Fig. 3).

The communication between the single services is handled by an "Enterprise Service Bus" (ESB). On top of the ESB a Business Process Engine will be used for the orchestration of these services and the business processes needed for the business cases.

One key issue for the easy adoption of the system is the interoperability and adaptability not only for the implemented pilot applications, but also in the future by other companies and authorities involved in the transportation sector, first of all in Europe, but possible also elsewhere, therefore open standards are a must. The realized web services will be based on the available standards for them and for the modeling of the business processes BPEL and BPMN will be used, for the implementation on top of the platform.

2.3 BPMN and BPEL

The complexity of interactions and transactions between companies and their partners, suppliers and customers are increasing. It is becoming more and more evident that the further development and performance of business processes depend on close cooperation between different parties involved in the value creation. As one can easily see that the EURIDICE project has to achieve the goal to minimize the barriers for attending enterprises. To be successful a standardized process modeling language can help companies to describe their internal and external business processes transparently and flexibly. Companies must also be in a position to communicate the modeled processes to their partners appropriately, clearly and comprehensibly. All involved parties should speak the same language regarding the processes.

In the last years many effort has been spent on developing web service based XML execution languages for Business Process Management (BPM) systems. Such a language is the Business Process Execution Language for Web Services (BPEL4WS) which provides a formal mechanism for the definition of business processes, optimized for the operation and inter-operation of Business Process Management Systems (BPMS) and therefore not easy to understand by humans or to design, manage and monitor these processes. The close relation to formal mathematical models provides the foundation for handling the complex nature of both internal and Business-to-Business (B2B) interactions and takes advantage of the benefits of Web Services. This makes it possible to design complex business processes, which can be organized in a potentially complex, disjointed and distributed system, easy understandable for software systems and making it suitable for the EURIDICE platform in a technical view (Juric and Pant 2008).

Business people and business analysts are very comfortable with visualizing business processes as flow charts, but this creates a technical gap between the format of the process design and format of the execution language. This gap is addressed by the "Business Process Modeling Notation" (BPMN). The "Business Process Modeling Notation" (BPMN) offers an open standardized notation for the modeling of business processes. The development of BPMN is coordinated by the "Object Management Group" (OMG). It is a graphical notation system for describing business processes in a Business Process Diagram (BPD). The notation is intended to be easily understood by all involved users and makes it suitable for business people and developers to have a common understanding about the processes and the implementation. It is also designed to allow the visualization of XML-based languages for business process automation, e.g. Business Process Execution Language for Web Services (BPEL4WS). BPMN provides a formal mapping to BPEL4WS. Thus, BPMN is a standard visualization mechanism for business processes defined in an execution optimized business process language (Juric and Pant 2008). Therefore BPMN in conjunction with BPEL4WS allows easy deployment of designed processes on business process execution engines which will be included in the developed platform in EURIDICE. We believe that the above mentioned technologies offer the most promising opportunities to successfully very different enterprises, their services and their business rules into the EURIDICE system.

After elaborating the business-aligned parts of the architecture which enable the integration of different stakeholders of the European logistics sector we will not discuss the more fundamental, more technical parts of EURIDICE.

2.4 Intelligent Cargo

As already mentioned above one of the most important concepts of the EURIDICE platform is the concept of intelligent cargo. What we mean by the word *intelligent* must be defined carefully. We use the definition of Wooldridge (2002) which describes intelligent behavior by listing the kinds of capabilities that we might expect of such an *intelligent* cargo.

2.4.1 Reactivity

Intelligent cargo is able to perceive its environment. With the help of such information it is able to reason about its current state within the business-process, about its global position and about the validity of its handling. Based on its believes of the real-world it is able to reactively initiate new processes like invoking a re-routing of its transportation vehicle, alarming about its shape or informing about its early arrival.

2.4.2 Proactiveness

Intelligent cargo shows a goal-directed behavior. It pro-actively takes the initiative. One can easily imagine use cases like cargo that autonomously books some freight-space to be transported to its destination or cargo that independently negotiates about the most economical way of being transported.

2.4.3 Social Ability

Intelligent cargo does not only cooperate with the central part of the EURIDICE platform, it does also interact with other intelligent cargo. Cargo being shipped within the same vehicle shares their capabilities; cooperate in the meaning of sharing their knowledge, their processing power and their electrical power reserves.

2.4.4 Concepts about Embedding Sensing, Communication and Computation in Networked Physical Items

Since we are able to build microcomputers of the size of a mobile phone or even smaller, computers are no longer bound to any certain location. Small computers like mobile phones or other mobile devices can be relocated very easily. Today a mobile phone fits into every pocket, radios are equipped with high end micro controllers or MP3-players contain very sophisticated hard and software.

The paradigm of ubiquitous computing offers new possibilities to provide processing power wherever you want. The EURIDICE project takes advantages out of this new trend. Not only transport vehicles, but also single transportation goods can be equipped with quite sophisticated, mobile devices communicating with each other and other IT-systems. Every single part is responsible for a quite small amount of functionality where a combination of the single parts results in a quite complex and powerful system. If single nodes of the system need to communicate involving mobile devices, they need to use some wireless communication facilities like Bluetooth for short range and GPRS or UMTS for long range communication. The whole system becomes a distributed one which appears coherent and continuous to an end user (Greenfield 2006).

The EURIDICE system extends the paradigm of ubiquitous computing by adding some context awareness to the application. Context awareness means that an application is able to sense its environment in terms of detecting its global position, accessing several sensors or reacting on environmental events (Jia and Zhou 2004).

As mentioned before, the EURIDICE system combines both aspects. It implements mobile, intelligent agents hosted on mobiles devices, located within transportation vehicles or cargo items. By sensing the environment, the mobile and intelligent agents can react on changes or events appropriately by initiating counteractions or alarming a monitoring unit.

2.4.5 Multi Agent System

The field of multi agent systems (MAS) is a quite young discipline where researches date back to the early 1980s. The paradigm of MAS is meant to cover massively distributed system where multiple emancipated computing entities are interconnected. The MAS paradigm, which tends to be a very intuitive and human like way of designing a system, separates the system into a lot of, quite small and encapsulated entities called agents. Within the MAS single agents share services they offer, communicate and cooperate with each other to perform a task—their design goal. Intuitively the concept of a MAS perfectly fits to the vision of EURIDICE's intelligent cargo.

Agents are more or less just a piece of software which does one job, acts as a delegate for someone else and this without any user interaction. Agents and especially intelligent agents communicate with other agents, other processes and their environment. According to Wooldridge (2002) *intelligent agents* show the behavior discussed before. Because of the fact that one agent is designed to execute only one simple job there have to be other agents or communication partners to build up an agent network. With this network the agent system becomes intelligent (Winikoff and Padgham 2008) and flexible. Intelligent because the agents can proactively interact and share information, resources and computation power within this network of agents and other communication partners (e.g. sensors) and react on changes in it. Flexible because the agent network can be tailored to its exact needs. An agent has also the ability to decide autonomously how to reach its goal. They have the capability to take actions on their own to achieve their objectives. And an agent has the ability of adaption, which means they are able to adapt their strategies to react on changes inside of the agent network by learning from their experiences.

Within EURIDICE single cargo items receiver their virtual intelligent representation by an implementation of an intelligent agent. A MAS offers the flexibility needed for such a complex system. Different type of cargo with different requirements to the desired agents can simply be handled by using different combination of agents as long as agents stay small, encapsulated entities fulfilling a small amount of functionality. With the help of mobile agents, which allow taking the advantages of code-mobility, which means that agents dynamically migrate to different hosts at runtime, different configurations or orchestrations of intelligent-cargo behavior can easily be implemented. Imagine some cargo leaving a stock where according to its needs different agents simply migrate to the cargo's device. One can imagine some agents observing the temperature for sensitive-cargo like fish or agents continuously reporting the position of some very high value or expensive cargo.

3 Challenges and Solutions

The described platform offers an implementation of new services that are strictly good or "things" related, thus turning the current process oriented architectures around and placing the goods into the middle of the SOA. In principle this change can be compared with the change from functional to object oriented programming languages and this fundamental change will allow a much better usage and customization of services according to the real need of the users. However this changes also requires an adaption of the existing applications like ERP, SCM etc. since these are mainly process oriented and do not cover e.g. the usage of distributed decision making or de-centralized intelligence.

The described technological platform will obviously also affect the structure of the underlying business processes. On the one hand the architecture allows the development of new open services that can be exploited via the EURIDICE platform, on the other hand EURIDICE enables the free composition of transport related services on the bases of SOA. This new possibility to construct and use services around transport oriented functions will enable a new service composition that will lead to a better integration of the logistics market and subsequently to a better utilization of the different transport resources throughout Europe. At the same time the better transport utilization leads directly to the saving of fuel and helps to reduce the carbon footprint that transport operations carry with them. Thus the EURIDICE platform also contributes to the establishment of more sustainable processes.

4 Outlook on Project and Further Developments

In the next few month the platform will be further developed and the different services will be integrated into the platform. A demonstrator is currently in the development which illustrates the capabilities of the EURIDICE platform to the different stakeholders in the transport sector. The current trend of more and more decentralized intelligence backs up the concept of EURIDICE and shows that the required business process changes are imminent. The known controlled environments and planning methods used over the past few years are more and more replaced by chaotic schemes that require a planning that is not streamlined for the best utilization but for robustness. With the advent of the EURIDICE platform these planning schemes are completed by ad-hoc decision making provided by intelligent entities. While the current transport market is maybe not yet prepared for the full uptake of these new developments the demand for this development is clearly foreseeable.

Acknowledgments This work was supported by the Seventh Framework Programme of the European Commission through the EURIDICE Integrated Project contract number ICT 2007-216271

References

Greenfield Adam, Everyware—The dawning age of ubiquitous computing, New Riders Publishing, 2006
Jia Weijia & Zhou Wanlei, Distributed Network Systems, Springer, 2004
Juric Matjaz B. & Pant Kapil, Business Process Driven SOA using BPMN and BPEL, Packt Publishing, 2008
Winikoff, Padgham, Developing Intelligent Agent Systems: A Practical Guide, Wiley & Sons, 2004
Wooldridge Michael, An introduction to Multiagent Systems, Wiley, 2002

Initial Benefits of Using Intelligent Cargo Technology in Fresh Fishing Logistics Chain

Donatella Vedovato, Tatjana Bolic, Marco Della Puppa
and Marco Mazzarino

1 Introduction

The goal of the present work is to demonstrate how Intelligent Cargo (IC) solution produces benefits and satisfies the identified requirements of a distribution company, at the same time offering improvements (or not compromising) in the competitiveness, optimization of the logistics process (efficiency), and better implementation of safety and security of cargo.

Here, the presented case study deals with fresh fish, that is a high perishable product, subject to very strict laws and regulations requiring strict control, precise product history and status information (i.e. temperature during transport).

The company considered is the biggest fish distributor in Italy and the third one in Europe. Annually the company distributes and processes nearly 40 million kg of fish, working around the clock, 7 days a week. Activities of the company consist of receiving and distributing fresh fish coming from worldwide markets that are directed mainly to European markets.

The work is structured as follows: the first part is dedicated to the explanation of the concepts of the IC, the second part defines the methodology followed, briefly describes the logistics chain of the company, its critical issues and requirements and the third part lists the possible benefits from the use of IC, comparing them with the benefits coming from the current technologies.

D. Vedovato (✉), T. Bolic, M. D. Puppa and M. Mazzarino
Venice International University, Isola San Servolo, 30100 Venetia, Italy
e-mail: donatella.vedovato@univiu.org

T. Bolic
e-mail: tatjana.bolic@univiu.org

M. D. Puppa
e-mail: mdellapuppa@units.it

M. Mazzarino
e-mail: mazzarin@iuav.it

H.-J. Kreowski et al. (eds.), *Dynamics in Logistics*,
DOI: 10.1007/978-3-642-11996-5_22, © Springer-Verlag Berlin Heidelberg 2011

2 The Intelligent Cargo

The IC refers to a particular application of Information and Communication Technologies (ICT) in the freight transport and logistics. The IC is able to connect itself to the users in order to transfer transport related information.

Currently, the information about the transport is linked to the means of transport and consequently to the single organization that manages the transport. The approach for the communication of the information is organization to organization where the users exchange data through private platforms. In this way it frequently happens that the physical and the information flows are separated, needs to be transferred or translated. This communication approach leads to a not yet integrated freight transport system that needs a common architecture in order to develop new systems and applications (Giannopoulos 2004). In particular, it is needed a change on the approach, from organization to organization to thing to thing in order to go over the defined concerns.

In particular, this approach is followed in an European project, Euridice, that focuses in the IC. The project has started in 2008, aims to create the necessary concepts, technological solutions and business models to establish the most advanced information services for freight transportation in Europe. The IC concept that has these six capabilities (Euridice White Paper 2009):

- Self identification, that is the ability of the cargo to identify itself;
- Context detection, that is the cargo is connected to the environment and is able to get information from it;
- Access to services, the actors along the supply chain have the access to different services that give information about the cargo;
- Status monitoring and registering, the cargo is able to monitor its physical conditions such as temperature, humidity, etc.;
- Independent behaviour, the cargo is able to detect deviations from a previous plan;
- Autonomous decision, using the intelligence, the cargo can propose solutions.

The communication occurs between the items, and between the items and users (i.e. cargo owners, shippers) thanks to a DNS-like system for cargo that links users and cargo.

The IC will use an information service integrated platform centred on the single cargo item that communicates with it and allows cargo objects to perform basic interactions on their own and to involve the users' information systems. Different players along the supply chain can access the information they need on cargo at any point along its route through fixed and mobile web services infrastructures. The services used by the IC are (see Fig. 1):

- Business services; a set of fixed services that allow integration with user back-offices, third parties and legacy system;
- Mobile services that usually run on mobile devices that monitor the status of the cargo (reaching the location on time, etc.);

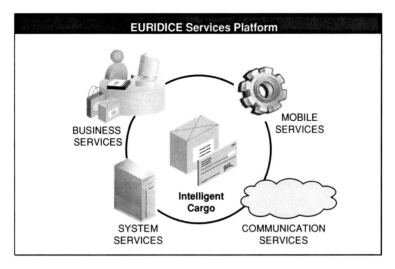

Fig. 1 Euridice service platform (Euridice White Paper 2009)

- System services provide cooperation between the users and the IC, users themselves and IC itself;
- Communication services support connection between users, vehicles and cargo and between the IC and sensors like RFID, antenna, GPS, temperature and humidity sensor types.

3 Methodology

The methodology used to understand the benefits that the IC can bring to the fresh fish logistics chain of the Italian distributor, consists of three steps:

1. Definition of the current business process; the business process has been detailed in terms of both physical flow (goods) and the information flow.
2. Identification of the critical issues and requirements; the critical issues are identified within the current business process and, consequently, the firm requirements that can be fulfilled with IC solutions.
3. Definition of the benefits coming from the use of the current technology that can fulfil the requirements; these benefits are compared with the ones coming from the use of the IC.

The main tool used in all the steps are repeated interviews and group meetings with the company employees, the goal of which was to exchange the information and fine tune all the findings. In this paper we focus on step two and step three, since we want to identify mainly the innovation possibilities coming from the use of IC solutions. Comparison between the current and the future business process scenarios allows for identification of potential benefits. Of course, the requirements identified in step two can be fulfilled using the state-of-the-art solutions and technology but the IC has a different approach. Section three will try to answer the question: "Why is IC better to use than the-state-of-the-art technology?"

Step 1 The logistics chain under study is the sourcing and distribution of perch fish. It has been chosen since it represents the most important flow for the company: it corresponds to one-third of the total quantity and value of the fish handled by the company. Here, we will shortly describe the physical flow only. The consequences of current information flow and its relation with the physical flow are reported in step two as critical issues.

The suppliers are located in Tanzania, from where the fish is transported by refrigerated trucks to Nairobi airport from where it is delivered to Paris airport. From Paris, the cargo travels by refrigerated trucks to the distributor warehouse where logistics activities such as handling and splitting take place. The fish is then transported by refrigerated trucks to end customers, like large-scale retailers and fish markets.

Step 2 The identified critical issues arising from the nature of existing information flow and its relation with the physical flow are: traceability, unreliability of information and waste of time in low value added activities.

First of all, the fresh fish distribution requires traceability, in terms of tracking the product origin, the operations done on it, the transport and logistics processes, etc. The traceability is required for all the foodstuffs, and the requirements are even stricter for the perishable goods. The company has no means of automatic tracking of the maintenance of the cold chain along the supply chain, especially in Africa, or any sort of automatic or semi-automatic tracking of where the shipment is. Currently both of these issues are being controlled by phone calls—calling all the actors involved in the process to find out where the shipment is, and/or if there was a problem with the cold chain maintenance. As can be imagined, this way of tracking and tracing involves a rather high workload of employees, it is time consuming, and not very reliable or timely.

The second critical issue is the (un)reliability of information. The information involved in this supply chain is entered and exchanged mainly manually. The main reason for it is a poor use of technology by some of the actors in this supply chain. This can lead to data elaboration errors, missing data, etc. The information reliability creates the problems in the following areas:

- Paper document accuracy. Paper documents that are required in the freight transport are filled by different actors in the supply chain. Most of these documents are still filled-out manually. Manual data entry implies a certain error rate. Some of the errors can even bring to a halt of the goods transport. Even more serious, if the paper documents that follow the goods are lost, then the shipment is stopped which is almost synonymous to the goods destruction when we are talking about the highly perishable goods.

- Order fulfilment by the supplier. The information about the product details, packing lists and the total amount of goods in a shipment sent by the supplier is manually entered after being received by email, or sms, or fax. The information can change a few times before the shipment leaves the supplier. For these reasons, shipment information is susceptible to mistakes that can be compounded when the "final" information is the result of data elaboration. This often results in discrepancies between the shipment content description sent ahead and the content of the shipment on its arrival. Often it is the case that the company discovers the discrepancies only at the arrival of goods during the physical inspection. If the discrepancies are substantial, the company is forced to improvise and change on the spot its distribution plan which can involve numerous actors—logistics operators, logistics platforms and the final customers. These on the spot changes often prevent the company to use the most economic and efficient solutions for the final distribution.

Order fulfilment to the customer. The customer often claims the faulty order fulfilment even when it is correct, but since the distributor has no means to ascertain the status of the order when it arrives to the customer, they have no choice but to accept the complaint. Thus, this issue can result in the distorted picture of the customer service levels provided by the distributor, making it difficult to measure and/or act on it and erroneous rates of perfect order fulfilment.

The last discovered issue is about the waste of time on activities, such as:

- Copies of paper labels that are attached to the boxes in the shipment are sent by fax to the distributor.
- Manual entry of shipment details (provided by fax by the supplier) in the distributor's ERP.
- Physical (manual) control of shipment contents by the distributor to ascertain the level of the order fulfilment.
- Shipment tracing and the customer delivery confirmation that are done by phone.

Given these critical issues, the fresh fish supply chain of the company has these requirements:

1. Upload and download of documents that are usually sent by fax even if the paper documents are still compulsory for the international transport of goods;
2. Cargo identification for the electronic information elaboration and exchange of the details between the different actors in the chain (such as product type, fishing and production place, etc.);
3. Recording of the shipment temperature, sealing history and the temperature breach notification;
4. Estimated time of arrival and notifications of the arrival to certain nodes of logistic network.

4 Results

Here we compare the state-of-the-art approach in the logistics of the fresh fish and the IC solution. First of all we describe the current technology and approach then we highlight the innovation that IC can bring to the whole SC.

4.1 Innovation Obtained Using the State of the Art Approach

In the fresh fish sector, different technologies are available for the traceability to guarantee the consumer a healthy and high quality end product. The traceability in the fresh fish sector relies mainly in the information availability and in the temperature monitoring. For the first type of traceability, the most innovative technologies that could overcome the identified critical issues and that are available rely mainly on RFID and bar code technologies. The monitoring of the temperature is usually performed with strip chart recorder placed into a limited number of boxes and dataloggers. There are pros and cons for using these technologies and here we describe them as they relate to our study case.

Pros RFID and bar code have replaced manual activities and have given the companies the access to more reliable and more or less real-time information about the cargo. The manual activities substitution with IT tools has provided benefits to the companies by reducing the amount of time consuming activities and reducing the amount of errors. Perishable goods markets (fresh food) are required to apply high quality control and keep track of it. The availability of real-time, reliable and sometimes new information has also improved the asset utilization and planning of the activities. It enabled the companies to have more accurate picture of stock levels, which in turn enabled application of lower inventory policies, thus lowering inventory costs and often at the same time lowering the probability of stock-outs.

Also the traceability of the temperature is today performed with different electronic systems that offers a precise and reliable monitoring of the temperature (Abad et al. 2009) respect to the manual one.

Cons the deficiency identified for the information and temperature traceability is the information availability, while other critical issues discovered for the temperature traceability are linked also with costs and hardware issues.

The first deficiency is that temperature tools, RFID and bar code are usually set up at a "firm level" and not at "cargo level". The communication/information exchange between different actors along the supply chain can be fragmented, can require a data translation or conversion. Consequently, the distributor has to translate, convert and elaborate the information coming from the supplier, assuming that the information is correct and that it does not get corrupted in the process.

The tools used for the traceability of the temperature are usually expensive and not automated (require manual inspection). For this reason it is performed only in the transport and storage phases, not along the entire supply chain (such as handling activities), even if there is a requirement for it (Abad et al. 2009).

4.2 Innovation Obtained Using the Intelligent Cargo

IC offers a new approach that bridges the problems presented in the state-of-the-art approach mentioned before. Firstly, IC can overcome the "firm level" issues since the approach is based on the cargo and not in the single organization. This is possible because the solution provides the communication from the cargo to the users also through a unique open platform (not proprietary) the access to which is available to each actor, in a specified form. It will enable access to the "same" cargo information for all the actors involved. This could diminish the integration problems since all the actors would be using the information presented in the same form. The standard use would reduce the number of instances and the need for information conversion, thus reducing the risk of errors. This is particularly important in the fresh fish chain where the number of actors involved in the supply chain is high.

Second, IC solution uses smart tags that record the temperature and devices that read it through the entire supply chain. In this way the hardware has not to be bought by each actor in the supply chain and can be re-used.

IC solution aims to be low cost and easy to use, thus reducing the information management related costs and easing the new entries into the supply chain. The users will have the option to start using IC gradually, depending on the needs and available resources. Thus avoiding the need for high initial investments, and easing the adoption of IC. Currently, in the fresh fish supply chain the players are not used to IT tools and they do not have the resources to sustain high investments.

4.3 Possible Benefits from Using the IC

On the base of the requirements identified, the IC solution can lead to following possible benefits.

1. Upload and download of paper based documents in an electronic form: which results in many time consuming activities, high rate of errors and the loss of documents. With the current technology it is possible to exchange the electronic versions of the documents, but it is still rather difficult to have them available for all the actors in the supply chain that need the access to them
2. Cargo identification: the traceability of the information in based on paper labels that can be lost and require great amount of time for the data entry and for the

translation and communication. The IC allows a more reliable information exchange between the different players that belong to the same supply chain since it does not require translations
3. Temperature recording: the control of the temperature is performed today through phone calls or with semi-automatic means, especially in the developing countries. The IC provides the actors with reliable, on time and precise information and gives the product a value added about the quality assurance
4. Estimated time of arrival and notifications: today the position of the cargo is checked by phone calls; the IC allows the different players to know on time the status and plan better the outbound activities.

5 Conclusions

The IC offers a pool of benefits that seems far better than the state of the art technology applied since it changes the communication approach about the transport related information. The concept of IC is based on the "thing to thing" strategy rather than "organization to organization". In this way actors belonging to different organizations are able to share information about the cargo since the solution is cargo-centric and able to communicate with the context, user and with the integrated centred platform.

The initial benefits highlighted in the paper are available only if the business models of the companies change since the IC concept modifies completely the work inside the company. The risk is that an innovative solution with high potential benefits cannot be adopted because the business models remain the same as in the past without take into consideration the changes given with a new solution.

On the other hand, a study about the assessment of an innovative solution is required, and for the IC is ongoing, in order to avoid the development of a great ICT product that does not match the business requirements. We think it could shed new light to the research in this field with an assessment study from a technical and user point view to assure an easy use of an innovation solution.

Acknowledgments We thank our partners at European Commission for all scientific, organisational and financial support. The project is financed in the 7. Framework Program ICT, http://euridice-project.eu.

References

Abad E., Palacio F., Nuin M., Gonzalez de Zarate A., Juarros A., Gomez J.M., Marco S. (2009). RFID smart tag for traceability and cold chain monitoring of foods: Demonstration in an intercontinental fresh fish logistics chain. *Journal of food Engineering*, 93, 394–399

Euridice White Paper (2009). http://euridice-project.eu
Giannopoulos G.A. (2004). The application of information and communication technologies in transport. *European Journal of Operational Research*, 152, 302–300

Autonomous Co-operation of "Smart-Parts": Contributions and Limitations to the Robustness of Complex Adaptive Logistics Systems

Michael Hülsmann, Benjamin Korsmeier, Christoph Illigen and Philip Cordes

1 Introduction

In times of globalization, the availability of standardized information technologies enables Logistic Service Providers to act on several worldwide markets leading to an increasing number of competitors, more price competition, and homogeneity of logistic services (Klaus and Kille 2006). Therefore, Logistic Service Providers are confronted with changing settings of requirements, caused by the phenomenon of hyper-competition, describing a fast moving business with a high competition in the fields of price-quality positioning (D'Aveni 1995). This demonstrates that the differentiation by offering lower price or higher quality services is one possible solution to gain competitive advantages (Müller-Stewens and Lechner 2005), whereas the standardized and homogenous services are unable to create added values for a customer leading to the possibility of substitution.

Due to the homogeneity of the offered services, the recognition of differences in logistic services quality is a challenging task (e.g. the transportation of goods). So, the need for another capable way for differentiation and gaining competitive advantages emerges. In this term, a significant increase of service quality might result from "Value Added Services" (Pfeiffer 2008). Every added service (e.g. packing of goods, mounting, and quality control) allows the Logistic Service

M. Hülsmann (✉), B. Korsmeier, C. Illigen and P. Cordes
Systems Management, School of Engineering and Science, Jacobs-University Bremen,
Campus Ring 1, 28759 Bremen, Germany
e-mail: m.huelsmann@jacobs-university.de

B. Korsmeier
e-mail: b.korsmeier@jacobs-university.de

C. Illigen
e-mail: c.illigen@jacobs-university.de

P. Cordes
e-mail: p.cordes@jacobs-university.de

H.-J. Kreowski et al. (eds.), *Dynamics in Logistics*,
DOI: 10.1007/978-3-642-11996-5_23, © Springer-Verlag Berlin Heidelberg 2011

Providers' customer to improve his own services and thereby constitutes a higher value to the customer. Moreover, if no competitor is able to imitate or substitute this special service, a "unique selling proposition" and thereby a competitive advantage for the Logistic Service Provider is created (Hülsmann and Grapp 2008). Since the Logistic Service Provider's long-term survivability essentially depends on their ability of creating competitive advantages by offering unique services (De Wit and Meyer 2005), companies should focus on that.

One way to create competitive advantages by offering these special services is the appliance of new Autonomous Co-operation technologies, like RFID or sensor-networks (Scholz-Reiter et al. 2004). These autonomous logistic processes are based on the use of interacting system elements, called "Smart-Parts". They can be described as intelligent machines or goods enabling a system to carry out a non-human based decision-making and problem solving (Wycisk et al. 2008). Thus, enhanced using and the further development of these new information and communication technologies leads by trend to smaller autonomously acting parts and therefore to an increasing number of logistic systems including "Smart-Parts". Hence, an increasing number of interrelations in logistic processes occur (Hülsmann et al. 2007), leading to a higher degree of complexity and dynamics, because of the increasing number of possible interactions and the increase in different possibilities of behavior. Due to the complexity observed and the ability of "Smart Parts" to act autonomously, such systems can be described as Complex Adaptive Logistics Systems (CALS) (Wycisk et al. 2008). They have to have a high adaptivity to cope with the increasing complexity and the resulting higher amount of information, due to the fact that they have to be flexible to get sufficient information as well as to be stable to avoid the incoming of too much information. In this context, the term "adaptivity" as the simultaneous requirement for system's stability and flexibility can be described as the system's ability to find the balance between absorbing the increasing amount of information from the environment (flexibility) and the necessity to keep the amount of needed information on a manageable level (stability) (Hülsmann et al. 2008). In other words, to assure an adequate information supply as well as to avoid an information overflow the system's success and survivability depends on its robustness. Robustness means on the one hand the ability to resist against a number of endangering environmental influences and on the other hand the ability to restore its operational reliability after being damaged (McKelvey et al. 2008). However, the autonomous interaction of the system's "Smart-Parts" leads to a higher degree of dynamic and thereby nearly to an impossibility to predict the system's behavior and future system states (Böse and Windt 2007), due to the fact that the "Smart-Parts" decision-making process depends on further behavior and decisions of other elements. Hence, the process of absorbing, selecting, and handling information depends on the behavior of the system's "Smart-Parts" and cannot be controlled by a higher institution.

Due to the direct interconnections between the "Smart-Parts", their behavior and the resulting dynamic and complexity of CALS, the system's robustness is not given per se. But to understand the effects and interdependencies on robustness, resulting from autonomous interacting "Smart-Parts" in CALS, and to show possibilities for achieving a high degree of robustness and assuring the system's long-term success the following question has to be answered: How does the robustness of CALS depend on the dynamics of autonomous cooperating "Smart-Parts"?

Hence, the aims of this paper are threefold: first, it intends to give a description of CALS as well as of the concept of Autonomous Co-operation. Second, it aims to analyze possible effects for CALS' robustness emerging from the using of Autonomous Co-operation. Third, implications to handle robustness as flexibility and stability in CALS will be outlined.

The paper includes three main parts plus an introduction and final conclusions. The first main section deals with the problem of robustness in "Smart-Parts" CALS. Therefore, a literature research is used to describe the future developments of logistic systems to the point of CALS and the resulting need for robustness. The second part is also based on research of current literature and technological developments to describe the implementation of the concept of Autonomous Co-operation in CALS. Thereby, the characteristics of Autonomous Co-operation are linked to the "Smart-Parts" to show the influences resulting from an implementation. The last section outlines the contributions and limitations for robustness resulting from the Autonomous Co-operation in CALS by discussing the possible effects on increasing or decreasing the system's flexibility and stability, resulting from the Autonomous Co-operation's characteristics in connection with the systems including "Smart-Parts". This allows a link between the CALS and the Autonomous Co-operation to show the possible effects of their robustness.

2 Problems of Robustness in "Smart-Parts" CALS

2.1 "Smart-Parts" in CALS

According to McKelvey et al. (2008), a paradigm shift in logistic research occurs in terms of an ongoing change from centralized control of non-intelligent elements in hierarchical structures to decentralized control of intelligent elements in heterarchical structures. Due to the parallels between logistic systems and Complex Adaptive Systems (CAS) concerning their properties, a further change regarding the understanding of logistic systems is rising, evolving from linear structures to complex systems and newly to CALS (Wycisk et al. 2008). The CAS concept comes from biology where it is mostly linked to living entities (Gell-Mann 2002). According to the CAS definition, it is a system that emerges into an adequate form by autonomously acting and co-evolving agents without being controlled or managed by a higher entity (Wycisk et al. 2008). The co-evolving and interacting agents, trying to reach own goals over time, enable the CAS to adapt to changing environments (Holland 1995).

Logistic Systems also consist of a high number of entities, the so-called "Smart-Parts", differing by their dimension (e.g. whole organization, departments, teams, machines, containers) (Wycisk et al. 2008). Thereby, the whole organization (e.g. company) can be a "Smart-Part" in logistic systems like worldwide Supply Chains, whereas the whole organization consists of own "Smart-Parts" like containers. These parts interact by exchanging resources like finances, products, services, or information (Surana et al. 2005). Furthermore, for sustaining their willingness to interact, the entities have to be heterogeneous in their characteristics and goals. Otherwise, there are no incentives, which motivate the individual element to participate in a co-operation. The heterogeneity is a pre-condition for the existence and functionality of CALS (Wycisk et al. 2008). These incentives are maintained through the essential information sharing caused by the different goals the entities are aiming for. Additionally, organizations and departments have to be able to learn to adapt their entities to changing environment requirements. But this ability is decreasing by looking to "Smart-Parts" like ships, containers or single goods (Wycisk et al. 2008). This leads to the point that all parts in logistic systems shall be able to interact autonomously and to make own decisions to cope with the mentioned problems. A solution is given by progress and recent developments in communication and information technologies (e.g. RFID and sensor networks). These new technologies can be used to enable non-living items to change their decision-rules and therefore to learn (Spekman and Sweeney 2006). Based on the learning ability they can act autonomously to a certain degree and render own decisions without the need to consult another entity on a superior level (Kappler 1992). Therewith, the "Smart-Parts" can be seen as co-evolving and interacting agents, rendering own decisions, and the system can be described as CALS.

2.2 The Need for Robustness in "Smart-Parts" CALS

Due to an increasing amount of inter-relations between the systems' elements, described as "Smart-Parts", as well as between the system and its environment (Patzak 1982), this intensification of interaction leads to an increase of the systems complexity. Furthermore, more elements ("Smart-Parts") in logistic systems can change the systems' states, causing an increase of dynamics (Conner 1998). In this term, dynamics describes the rate of modification of a system over a specific time (Coyle 1977) with a dependency of the system's state from its elements (Krieger 1998). Higher dynamics leads to an increasing amount of information about the system's environment as well as to an increasing change rate within the information the logistic management is confronted with (Hülsmann and Berry 2004). According to Ansoff (1984), the success and survivability of a system depends on its ability to reach and maintain a strategic fit between on the one hand the organization's positioning and competence-profiles and on the other hand its environmental requirements. This strategic fit is necessary for the efficient usage of resources, in terms of avoiding frictions in the internal processes (Scholz 1987). Hence, organizations have to absorb and process information from and about their environments to avoid a so called "information overload" (Hülsmann et al. 2007). Based on this lack of information or lack of information processing capacity, an inability to fulfill the organization's functions can occur (Schreyögg et al. 2003), because important information about logistic tasks is missing or cannot be processed adequately. Therefore, it is essential for an organization to find the balance between absorbing the increasing amount of information and keeping the amount of information that has to be processed at a manageable level (Hülsmann et al. 2008). The absorption of relevant and important information leads to the possibility of reacting to environmental changes and therefore to the organization's flexibility, whereas the second part assures the organization's stability by avoiding the information overflow (Luhmann 1994). This ability to balance the flexibility and stability of an organization is called adaptivity (Hülsmann et al. 2008), which is on the one hand needed to absorb enough information to avoid an undersupply and on the other hand important to impede the inflow of too much information. In addition, a system can be called robust, if the system itself and its including "Smart-Parts" are able to adapt to complexity and dynamics (Meepetchdee and Shah 2007) by balancing the systems' flexibility and stability itself without any intervention. Moreover, the system's robustness is a requirement to resist against influences (e.g. information overload), which endanger the fulfillment of logistic tasks (e.g. ensuring the material flow with regard to the time, quality and place). Furthermore, the robustness of CALS can be seen as the ability to restore itself after being damaged (Wycisk et al. 2008).

3 Implementation of CALS

3.1 The Concept of Autonomous Co-operation

As shown above, CALS including their processes are more and more characterized by an increasing complexity due to lots of part variants and a tremendous number of possible combinations and variations (Hülsmann and Windt 2007). That is caused by today's customers who expect shorter delivery times with a higher reliability, a global availability of desired goods and a broader variety of products and services (Hülsmann and Windt 2007). A high robustness of logistic systems as the balance between system's stability and flexibility (McKelvey et al. 2008) is desirable to cope with the complexity, to offer a feasible management of logistic systems and to assure the system's ability to fulfil its logistic tasks (Schreyögg et al. 2003; Hülsmann et al. 2008). Therefore, a concept is needed to create higher robustness for handling the tensions of an increasing complexity (Wycisk et al. 2008). One possible approach is using of the concept of Autonomous Co-operation for managing CALS' "Smart-Parts" in an appropriate and successful manner (Hülsmann and Windt 2007). Windt and Hülsmann (2007) defined Autonomous Co-operation as "(...) *processes of decentralized decision-making in heterarchical structures. It presumes interacting elements in non-deterministic systems, which possess the capability and possibility to render decisions. The objective of Autonomous Control [and Co-operation] is the achievement of increased robustness and positive emergence of the total system due to distributed and flexible coping with dynamics and complexity.*"

According to this definition, the main characteristics of Autonomous Co-operation are decentralized decision-making, autonomy, interaction, heterarchy, and non-determinism.

Decentralized decision-making means that the entities ("Smart-Parts") of a system are capable to make their own decisions about their actions based on the available information to them (Hülsmann et al. 2008). They are also able to reflect the reactions of other entities regarding their decisions and to adapt their own behavior (Kappler 1992). Accordingly, these entities act *autonomously* and, moreover, they establish *interaction* with other entities ("Smart-Parts") to receive relevant information (Surana et al. 2005). But the autonomous behavior and the interaction presume a suitable information supply to guarantee an efficient decision-making (Hülsmann and Windt 2007). Information supply and coordination in common logistic systems (e.g. central planning) is usually done by a superior entity, which initiates communication, coordination etc. whereas entities in CALS act autonomously. In autonomously controlled systems the need for a *heterarchical structure* arises (Goldammer 2002) to enable entities to share information and interaction (Böse and Windt 2007). Then if the system structure is homogeneous, all entities would be aiming for the same goals and the information flow would be constrained after achievement of objectives. As a result, the elements are on the one hand independent from any kind of control entity (Hülsmann and Windt 2007), but on the other hand the interdependencies within the systems are significantly higher. In conjunction with the autonomy of the system's entities, the predictability of the overall system's behavior is nearly impossible (Böse and Windt 2007). And even if all relevant data is available to measure the current system state (which is unlikely due to the common size of logistic systems), the behavior would still remain unpredictable because the combination of the relevant data with all decision alternatives makes the problem unmanageable again (Flämig 1998). This aspect is known as *non-determinism* (Böse and Windt 2007).

All the mentioned characteristics can be transferred either to a single "Smart-Part" or to a system of "Smart-Parts" whereas each "Smart-Part" or system of "Smart-Parts" have to have these characteristics as a requirement to apply the concept of Autonomous Co-operation. Therewith, the appliance of "Smart-Parts" is one vehicle to realize the concept of Autonomous Co-operation and they can be seen as a pre-condition. The more "Smart-Parts" are implemented, the more the characteristics of Autonomous Co-operation can be achieved and vice versa.

3.2 Achieving Autonomous Co-operation by "Smart-Parts"

An implementation of Autonomous Co-operation in logistic systems can be realized through the appliance of new information and communication technologies. RFID-chips or sensor networks are some examples, which enable entities in a logistic system ("Smart-Parts") not only to communicate with each other, but as well with their environment (Hülsmann and Windt 2007). Such entities may decide autonomously based on available information. Implementing these technologies also represents the other characteristics. Beside the decentralized and autonomous decision-making, entities can interact, a heterarchical structure is created and finally, the system's behavior becomes unpredictable (Hülsmann et al. 2008).

However, even if all of the mentioned characteristics seem to apply, there will be no logistic systems, which meet all of them to 100%, yet. The appliance of the concept of Autonomous Co-operation and the degree of Autonomous Co-operation always has to be regarded as a continuum between total external controls on the one hand and total Autonomous Co-operation on the other hand. To which extent the specific characteristics are matched has to be evaluated case by case. Therewith, the measuring problem of Autonomous Co-operation emerges (Hülsmann and Grapp 2006). The more complex and dynamic the system is, the more difficult is the measurement of the degree of Autonomous Co-operation. It has to reflect the interactions on the level of sub-systems, like manufacturers, suppliers and distributive trades, as well as interactions on the level of single elements within those sub-systems, like trucks, ships, planes, containers and single goods. The next section evaluates the effects of autonomous cooperating "Smart-Parts" on the robustness of CALS.

4 Contributions and Limitations of Autonomous Co-operation to the Robustness of CALS

According to the explanations above, it can be summarized, that a system's adaptivity, resulting from its stability as well as from its flexibility, leads to the robustness, which is a requirement for the system's success and its ability to handle logistic tasks. Moreover, it can be assumed that the characteristics of Autonomous Co-operation influence the robustness of CALS because the flexibility as well as the stability depends on the "Smart-Parts" behavior and decision-making process. To prove this assumption and detect the specific effects, the characteristics of the Autonomous Co-operation have to be analyzed regarding their contributions and limitations to the flexibility and stability of a logistic system. Since the characteristics of autonomously cooperating systems depend from each other, every specific characteristic with its own effects should be analyzed regarding the others. Figure 1 illustrates the interdependencies between the characteristics and a system's adaptivity.

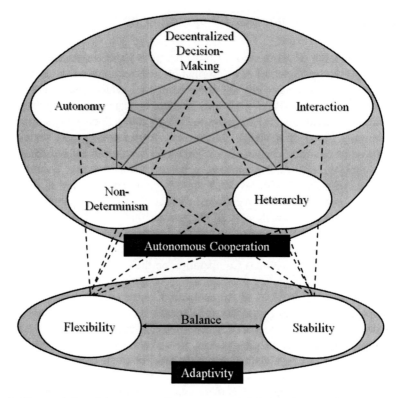

Fig. 1 Characteristics of Autonomous Co-operation and their interdependencies and effects on adaptivity

Figure 1 shows the characteristics of Autonomous Co-operation (decentralized decision-making, autonomy, interaction, non-determinism, heterarchy) as well as the adaptivity and the interrelations between on the one hand the several characteristics (solid lines) and on the other hand the Autonomous Co-operation and adaptivity (dashed lines). Thereby, it points out the requirement of balancing the system's flexibility as well as its stability to achieve the adaptivity, which is illustrated as equilibrium between them. By striving for the optimal balance between flexibility and stability higher system adaptivity and therewith a higher robustness might be provided. Moreover, it is shown that there is a high number of interrelations between the characteristics of Autonomous Co-operation and the stability and flexibility. This emphasizes the high amount of possible effects on robustness, resulting from the implementation of Autonomous Co-operation in CALS. Nevertheless, logistic systems will never meet the characteristics of living systems to 100% because of the existence of human elements managing parts of it. Hence, a "Smart-Part" element of the system usually does not interact with the others to a degree of 100% (Wycisk et al. 2008). Therewith, the degree of the system to act autonomously depends on the degree of interaction.

If a single "Smart-Part" changes its behavior, the change of other elements' behavior is caused automatically. Thus, the existing problem of measuring individual degrees of Autonomous Co-operation of the "Smart-Parts" (Hülsmann and Grapp 2006) leads to the related problem of measuring the number of relations, which means that the adaptivity of a system is not easy to identify without knowing the exact relations between the characteristics. For the following research, the effects are differentiated by their influence to increase or decrease the system's stability and to increase or decrease the system's flexibility. Between those two extremes, a negative trade-off can be assumed: less flexibility induces more stability and vice versa. That means an optimization of both stability and flexibility is always an optimization of the balance between them and not an isolated optimization of both, which is impossible, because of their *contradictionary* characteristics.

4.1 Influence of Autonomous Co-operation on Increasing System's Stability

Since "Smart-Parts" in CALS act autonomously without be controlled by a higher entity, the complexity the whole system is confronted with is distributed to its several elements, because it has to be absorbed by every single element. In consequence, that leads to a release of the supervising entity (Hülsmann et al. 2008), by *the distribution of complexity*. Moreover, the whole system's capacity to process information increases by the ability of the "Smart-Parts" to render their own decisions (Hülsmann et al. 2007). That means that the threat of an information-overload decreases by the *increase of the system's information processing capacity*. Because of the direct interaction of the "Smart-Parts", they are able to provide each other with relevant and target-oriented information (Hülsmann et al. 2007), reducing the *amount of information-flow* to get the same information standard. In addition, these effects might be able to stabilize the system by supporting the handling of incoming information, and thereby leading to an increase of the system's robustness.

4.2 Influence of Autonomous Co-operation on Decreasing System's Stability

For a high degree of interaction between the "Smart-Parts", an intensification of the interrelations is needed to create a more standardized and robust communication basis and enable the "Smart-Parts" to interact more efficient. Therewith, more complexity has to be absorbed, since a higher amount of information has to be processed. This leads in fact to an increase of the system's inherent complexity by the *additional complexity*. Moreover, an increase of the system's inherent dynamics can occur by the increase of its complexity (Conner 1998), caused by shorter decision-periods whereas the amount of information remains the same. This leads for example to an increasing amount of information about the environment, which is constantly changed by the *additional dynamics*. Furthermore, the interaction between the "Smart-Parts" leads to local bases of information, which are needed to make own decisions and reach the system's global optimum (Surana et al. 2005). Therefore, information about the system's optimum must be allocated to the elements by their interaction. Nonetheless, if the elements have the relevant information about the global optimum, it can although differ from the local optimum, if the elements decide not to go for it. This results from the fact, that one element does not know how the others will act, which leads to a significant relevance of expectations about the other elements' actions for the own decision-making process (Poundstone 1993). Therefore, it is possible, that the elements do not try to reach the goals of the whole system, which leads to the *risk of global inconsistencies of decisions*. Another effect results from the non-determinism of the autonomously controlled systems. Due to the *Non-Predictability* of the system's future states and the missing of an external control entity, no one can monitor the "Smart-Parts" and detect critical incidents regarding their behavior or decision-making. Because a control entity is not required, the autonomously adaption of the "Smart-Parts" behavior to environmental changes causes a decrease of the system's stability. In addition, the first two effects result from an increasing amount of information, while the last two result from a decreased information processing capability. So the Autonomous Co-operation can destabilize the system by decreasing the system's ability to keep information inflow at a manageable level and thereby decreasing the system's robustness.

4.3 Influence of Autonomous Co-operation on Increasing System's Flexibility

Owing to the system's non-determinism, the concept of Autonomous Co-operation enables the system to react flexible to environmental changes as well as to become capable and possess the right to develop its own paths. It is not determined to a certain future system state. The "Smart-Parts" in such a system are flexible in decision-making, since they do not have to consult a higher entity to render a decision, what finally leads to an *increase of the system's flexibility in decision making*. The elements heterogeneity is an important requirement for Autonomous Co-operation, because the individual decision-making process leads to a certain variance of decisions and therewith of their behavior. If too much elements would render the same decision, this could in turn lead to a damage of the system's reliability because of inefficiencies based on redundancies and a worse information flow caused by a decrease of interaction. So the decentralized decision making contributes to the heterogeneity of the system's "Smart-Parts" (Wycisk et al. 2008) and therewith act as *Assurance of sufficient heterogeneity*. In addition, these effects increase the system's flexibility and lead in fact to a higher robustness.

4.4 Influence of Autonomous Co-operation on Decreasing System's Flexibility

Especially in autonomous cooperating systems, where some of the "Smart-Parts" are human beings, the problem of *longer decision making times* can occur. As every single unit has to decide upon its goals and how to reach them, a longer decision-making process is the consequence, compared to the decision-making of a higher entity where the decisions are made at a central planning unit. So it can be outlined that on the one hand the Autonomous Co-operation can lead to a higher information processing capacity, whereas on the other hand this process can be decelerated by the Autonomous Co-operation. In consequence, this might reduce the system's ability to react on environmental changes and thereby decreases its flexibility, what finally leads to a decrease of the system's robustness.

In addition, it can be summarized that several effects, resulting from the concept of Autonomous Co-operation, influence the stability and flexibility and thereby the adaptivity and robustness of CALS. Nevertheless, the specific effect on the adaptivity cannot be described, because it depends on the balance between stability and flexibility in the current case and thereby on the current degree of system's stability and flexibility. Therefore, the adaptivity has to be evaluated in every specific scenario. On the one hand, the stability can be increased by a distribution of complexity on several "Smart-Parts", an increase of the system's information processing capacity and a lower amount of information flow. On the other hand, the stability can be decreased by additional complexity and dynamics, by a higher amount of elements, the risk of global inconsistencies by different goals and the non-predictability of the system's future states. Moreover, an increase of the system's flexibility is given by the system's flexibility in decision-making and the assurance of sufficient heterogeneity. Finally, a decrease of the flexibility can occur by a possible deceleration of information capacity. Furthermore, the outlined effects show the possible effects of increasing and decreasing the system's stability and flexibility, by the implementation of Autonomous Co-operation. Due to that, it points out possible activities to reach equilibrium between the stability and flexibility as well as keep the stability and flexibility in balance. The effects are summarized in Table 1.

Table 1 Effects of Autonomous Co-operation on a system's adaptivity

	Stability	Flexibility
+	Distribution of complexity	Increase of the system's flexibility in decision making
	Increase of the system's information processing capacity	Assurance of sufficient heterogeneity
	Lower amount of information flow	
−	Additional complexity	Possible deceleration of information processing (esp. for systems with human beings)
	Additional dynamics	
	Risk of global decision inconsistence	
	Non-predictability	

5 Conclusion

The overarching intention of this paper is to illustrate that robustness could be increased using "Smart-Parts" for implementing organization principles of Autonomous Co-operation. The robustness provides a logistic system with the essential capability for an optimal and successful handling of its logistic tasks. Hence, an evaluation of several effects of "Smart-Parts" to Autonomous Co-operation and finally on the robustness of CALS is accomplished by investigating the effects of the specific characteristics of Autonomous Co-operation towards the relation between flexibility and stability. As the main result, several possible effects of the characteristics of Autonomous Co-operation on the increase as well as on the decrease of the system's stability and flexibility and therewith on its robustness were outlined. It was illustrated that an optimization of the balance between stability and flexibility should be aimed for to achieve an optimal degree of Autonomous Co-operation associated with an optimal adaptivity. Furthermore, the existence of interrelations and the occurrence of effects caused by the characteristics of Autonomous Co-operation towards the adaptivity of logistic systems were demonstrated. However, the interdependencies of the Autonomous Co-operation characteristics as well as the not measureable degree of "Smart-Parts" interaction and unpredictable future system states lead to the impossibility of deriving an approach for the successful balancing of stability and flexibility in CALS. In addition, the explicit identification and measurement of the characteristics of Autonomous Co-operation as well as of the indicators stability and flexibility in common logistic scenarios is not quite easy, caused by the difficulties of determining an appropriate scaling. Because of these limitations, further research to develop an adequate measurement instrument should be performed. Moreover, relevant indicators should be defined to facilitate the process of identifying the characteristics of Autonomous Co-operation in CALS and the other mentioned values. This specific measurement results and the indicators could then be used to develop an efficient approach and to deviate general strategies for increasing the robustness of autonomous controlled CALS. For logistic systems practice an equilibrium between stability and flexibility should be aimed for, and the main focus should lie on the exploitation of effects that increase the system's flexibility and stability combined with a simultaneously avoidance of effects that decrease its flexibility and stability.

The main result of this research is a theoretical framework to deduce practical recommendations which are applicable to logistic systems, whereas the possible consequences have to be evaluated critically from case to case. In order to further enhance the findings, the next step should be the development of a measurement model to investigate and evaluate the interrelations between the relevant characteristics (Autonomous Co-operation, flexibility, stability, etc.).

Acknowledgments This research was supported by the German Research Foundation (DFG) as part of the Collaborative Research Centre 637 "Autonomous Cooperating Logistic Processes—A Paradigm Shift and its Limitations". Special Thanks to Elisabeth Meußdoerffer for reviewing and comments.

References

Ansoff I (1984) Implanting strategic management. Englewood Cliffs, New Jersey, pp. 10–28

Böse F, Windt K (2007) Catalogue of Criteria for Autonomous Control in Logistics. In: Hülsmann M, Windt K (eds.) Understanding Autonomous Cooperation & Control: The Impact of Autonomy on Management, Information, Communication, and Material Flow. Springer, Berlin, pp. 57–72

Conner D R (1998) Leading at the Edge of Chaos: How to Create the Nimble Organization. John Wiley & Sons, New York

Coyle R G (1977) Management system dynamics. John Wiley & Sons, London

D'Aveni R A (1995) Coping with hypercompetition; Utilizing the new 7S's framework. Academy of Management Executive 9(3):45–57

De Wit B, Meyer R (2005) Strategy Synthesis—Resolving Strategy Paradoxes to Create Competitive Advantage. South-Western College Pub, London

Flämig M (1998) Naturwissenschaftliche Weltbilder in Managementtheorien: Chaostheorie, Selbstorganisation, Autopoiesis. Campus Verlag, Frankfurt/Main

Gell-Mann M (2002) What is complexity. In: Alberto Q C, Marco F (eds.) Complexity and Industrial Clusters: Dynamics and Models in Theory and Practice. Physica-Verlag, Heidelberg, pp. 13–24

Goldammer E v (2002) Heterarchy and Hierarchy—Two Complementary Categories of Description. Vordenker October 2002

Holland J H (1995) Hidden Order. Perseus Books, Cambridge, MA

Hülsmann M, Berry A (2004) Strategic Management Dilemmas: Its Necessity in a World of Diversity and Change. In: Lundin R et al. (eds.) Proceedings of the SAM/IFSAM VIIth World Congress on Management in a World of Diversity and Change. Göteborg, Sweden, 2004, CD-Rom, 18 pages

Hülsmann M, Grapp J (2006) Monitoring of Autonomous Cooperating Logistic Processes in International Supply Networks. In: Pawar K S et al. (eds.) Conference Proceedings of 11th International Symposium on Logistics (11th ISL). Loughborough, United Kingdom, 2006, pp. 113–120

Hülsmann, et al. (2007), Hülsmann M, Scholz-Reiter B, Austerschulte L, de Beer C, Grapp J (2007) Autonomous Cooperation—A Capable Way to Cope with External Risiks in International Supply Networks?. In: Pawar K S, Lalwani C S, de Carvalho J C, Muffatto M (eds.) Proceedings of the 12th International Symposium on Logistics (12th ISL). Loughborough, United Kingdom, 2007, pp. 172–178

Hülsmann M, Windt K (2007) Changing Paradigms in Logistics. In: Hülsmann M, Windt K (eds.) Understanding Autonomous Cooperation & Control: The Impact of Autonomy on Management, Information, Communication, and Material Flow. Springer, Berlin, pp. 1–12

Hülsmann M, Grapp J (2008) Economic Success of Autonomous Cooperation in International Supply Networks?—Designing an Integrated Concept of Business Modelling and Service Engineering for Strategic Usage of Transponder-Technologies. In: Pawar K S, Lalwani C S, Banomyong R (eds.) Integrating the Global Supply Chain. Conference Proceedings of 13th International Symposium on Logistics (13th ISL), Loughborough, United Kingdom, 2008, CD-Rom, pp. 117–124

Hülsmann M, Grapp J, Ying L (2008) Strategic Adaptivity in Global Supply Chains—Competitive Advantage by Autonomous Cooperation. Special Edition of the International Journal of Production Economics 114(1):14–26

Kappler E (1992) Autonomie. In: Frese E (ed.) Handwörterbuch der Organisation, 3rd edn. Poeschel, Stuttgart, pp. 272–280

Klaus P, Kille C (2006) Die Top 100 der Logistik –Marktgrößen, Marktsegmente und Marktführer in der Logistik dienstleistungswirtschaft. Deutscher Verkehrs-Verlag, Hamburg

Krieger D J (1998) Einführung in die allgemeine Systemtheorie, 2nd edn. Fink, Munich

Luhmann N (1994) Soziale Systeme: Grundriss einer allgemeinen Theorie. Suhrkamp, Frankfurt/ Main

McKelvey B, Wycisk C, Hülsmann M (2008) Designing Learning Capabilities of Complex 'Smart Parts' Logistics Markets: Lessons from LeBaron's Stock Market Computational Model. Working paper, Unpublished

Meepetchdee Y, Shah N (2007) Logistical Network Design with Robustness and Complexity Considerations. International Journal of Physical Distribution and Logistics Management 37(3):201–222

Müller-Stewens G, Lechner C (2005) Strategisches Management—Wie strategische Initiativen zum Wandel führen. Schäffer-Poeschel, Stuttgart

Patzak G (1982) Systemtechnik. Springer, Berlin

Pfeiffer K (2008) Zuverlässigkeit zählt. Logistik heute 11

Poundstone W (1993) Prisoner' dilemma. In: von Neumann J (ed.) Game theory, and the puzzle of the bomb. Anchor Books, New York

Riedel J, et al. (2008) A Survey of RFID Awareness and Use in the UK Logistics Industry. In: Haasis H D et. al (eds.) Dynamics in Logistics. Springer, Berlin

Scholz C (1987) Strategisches Management—ein integrativer Ansatz. Berlin, New York

Schreyögg G, Sydow J, Koch J (2003) Organisatorische Pfade – Von der Pfadabhängigkeit zur Pfadkreation? In: Schreyögg G, Sydow J (eds.) Strategische Prozesse und Pfade, Managementforschung 13, Wiesbaden: Gabler

Scholz-Reiter B, Windt K, Freitag M (2004) Autonomous logistic processes: new demands and first approaches. In: Monostori L (ed.) Proceedings of the 37th CIRP International Seminar on Manufacturing Systems, Budapest, pp. 357–62

Spekman R E, Sweeney P J (2006) RFID: from concept to implementation. International Journal of Physical Distribution & Logistics Management 36(10):736–754

Surana et al. (2005) Supply-chain networks: a complex adaptive systems perspective. International Journal of Production Research 43(20):4235–4265

Wycisk C, McKelvey B, Hülsmann M (2008) 'Smart parts' logistics systems as complex adaptive systems. International Journal of Physical Distribution and Logistics Management 38(2):108–125

Decentralisation and Interaction Efficiency in Cooperating Autonomous Logistics Processes

Arne Schuldt

1 Introduction

The complexity of logistics processes has increased significantly in the last decades. Traditionally linear supply chains have evolved into complex supply networks. Participants in these networks are highly distributed, often even over multiple continents. Furthermore, they are highly interconnected and thus highly dependent on each other. Every customer has many suppliers and vice versa. Efficiency of conventional centralistic control of such supply networks is limited. Furthermore, it is often not applicable. The limitations of centralistic control are as follows:

(1) Complexity
(2) Dynamics
(3) Distribution.

Logistics processes usually comprise a high number of participating entities and parameters to be considered. However, already problems that seem to be rather simple at first glance, like the Transport Problem and the Travelling Salesman Problem, exhibit a high computational complexity. Optimal plans might thus already be outdated as soon as their generation is finished. This problem is even aggravated by the dynamics of logistics processes because changes in the environment require frequent re-generation of plans. Finally, the physical distribution of supply networks prevents relevant information from being available for centralised planning.

The paradigm of autonomous logistics (Hülsmann and Windt 2007) addresses these issues by delegating decision-making to the local logistics entities.

A. Schuldt (✉)
Centre for Computing and Communication Technologies (TZI), University of Bremen,
Am Fallturm 1, 28359 Bremen, Germany
e-mail: as@tzi.de

H.-J. Kreowski et al. (eds.), *Dynamics in Logistics*,
DOI: 10.1007/978-3-642-11996-5_24, © Springer-Verlag Berlin Heidelberg 2011

For instance, packages and shipping containers are expected to plan and schedule their way through the logistics network on their own (Sect. 2). In this approach, computational complexity can be reduced significantly because the number of parameters to be considered by each single container is limited. Robustness against dynamics is increased because re-planning involves only the affected entities instead of the whole system. Furthermore, it is no longer necessary to transmit all information to a centralistic entity. Decentralised control, however, requires delegating both the ability and the autonomy to make decisions to the participating entities. To this end, intelligent software agents are employed to represent logistics entities and to act on their behalf.

However, there are also limitations in autonomous logistics. Autonomous entities can rarely succeed in their objectives on their own (Schuldt and Werner 2007). Instead, it is necessary to cooperate (Wooldridge and Jennings 1999) which requires interaction with other entities. Obviously, the communication effort depends on the number of logistics entities involved. Therefore, it is important to find an appropriate granularity at which autonomous control is applied. Besides, communication complexity depends on the interaction mechanisms applied (Sect. 3). In general, there is a tradeoff between decentralisation and communication effort. The particular contributions of this paper are two interaction protocols (Section 4) for cooperating autonomous logistics processes: One with minimal communication effort (i.e., high interaction efficiency), the other one with a maximal degree of decentralisation. Which of them is appropriate depends on the concrete logistics task to be solved. A thorough examination (Sect. 5) helps choose the adequate protocol for implementing autonomous logistics processes. Both protocols have been implemented (Sect. 6) in an agent framework.

2 Cooperating Autonomous Logistics Processes

An application scenario is onward carriage in container logistics. Shipping containers arriving at a container terminal are expected to organise their transport into appropriate warehouses. This example involves three primary logistics functions: transport, handling, and storage. All of them require cooperation between containers for efficient process execution.

(1) A shipping container selects an appropriate warehouse. This decision depends on properties, capacity, and costs. Cooperation is necessary because it is desirable to receive similar goods at the same location. With regard to the subsequent distribution, this helps decreasing the number of truckloads by preventing empty vehicle running.

(2) Based on the warehouse chosen, the container selects a matching transport relation. Mass transport by barge or train is cheaper than transport by truck. Cooperation requires finding other containers with the same location and the same destination in order to share a train or barge.

(3) The container requests a time window for receiving at the warehouse. Cooperation is necessary to ensure that containers are received in accordance with their priority. Containers waiting at the same warehouse should therefore coordinate their demands.

These examples illustrate the necessity for cooperation in decentralised logistics control. Each of the above functions requires similar containers to form teams in order to succeed in their goals. Similarity is either defined by the goods loaded, by the current location, or by the scheduled destination. The task for the agents representing the logistics entities is therefore to find potential partners for cooperation. The teams formed may differ for different tasks addressed.

3 Problem Definition and Related Work

A multiagent system (MAS) is populated by a set of agents $A = \{\alpha_1, \ldots, \alpha_n\}$. Each agent represents an autonomous logistics entity, e.g., a shipping container. The task described in the preceding section covers two aspects: Representing agent properties for team formation and the process of team formation itself. The finite set of descriptors $D = \{\delta_1, \ldots, \delta_m\}$ describes relevant properties of the logistics entities. The choice of the concrete description depends on the logistics problem addressed. The description should follow some formal language, e.g., temporal or description logics (Schuldt and Werner 2007), so that agents can reason about it. The mapping

$$\text{description} : A \rightarrow D$$

maps agents to their descriptors. In the addressed scenario, agents with similar descriptors should form teams for successful cooperation. Note that this differs from applications where the capabilities of agents within one team supplement each other. The predicate

$$\text{match}(\delta_i, \delta_j)$$

indicates matching descriptors $\delta_i, \delta_j \in D$. A concrete specification of the predicate depends on the specific application in logistics as well.

The particular focus of this paper is therefore on agent interaction mechanisms for team formation. Previous work focused on formalising multiagent organisations (Fischer et al. 2003) and internal states of agents during team formation (Wooldridge and Jennings 1999). However, considerably less effort has been spent on actual *agent interaction* protocols for team formation. Distributed clustering approaches (Heinzelman et al. 2000), as applied in wireless sensor networks, are not applicable. They focus on clustering by quantitative (spatial) data. By contrast, in the task addressed here the partitioning already exists implicitly in the agent descriptions. The task is thus rather finding potential team members *without prior knowledge* about the other agents. Peer-to-peer approaches (Ogston and Vassiliadis 2001) provide each agent with an arbitrarily chosen set of other agents. Agents inform their peers about each other. Based on this foundation, they exchange their direct partners by others that are more similar. However, this setting is purely artificial for the addressed application in autonomous logistics. In particular, there is no meaningful choice for initial peers because the autonomous logistics entities are initially completely unaware of each other. A common approach to implement team formation is to apply the contract net (Smith 1977) interaction protocol. If an agent intends to form a team, it could use the contract net to announce the team description to all interested agents. It is, however, not applicable to the particular problem addressed here. Teams are usually not static but *dynamic* in autonomous logistics. That is, there is no distinguished point in time at which all members jointly establish the team. Instead, agents may join the team after it has been established. In the application scenario (Sect. 2) consider, for instance, containers that arrive after others.

An existing interaction protocol (Schuldt and Werner 2007) for this purpose involves a catalogue service for existing teams. This catalogue is queried by agents looking for potential partners. Subsequently, the agents send their description to the management agents of all existing teams. If one of them matches the description, the agent may join the team. Otherwise, it can itself register as a management agent for a new team with its description. This protocol leaves much autonomy regarding team formation to the agents and teams respectively. As a consequence, it has a comparatively high communication complexity. To prevent even higher communication efforts, the catalogue service still remains a centralistic entity. To summarise, the protocol is a tradeoff between decentralisation and interaction efficiency.

4 Team Formation Protocols for Autonomous Logistics

For reliable logistics processes, it is important to judge whether an interaction protocol is appropriate even before it is applied. To this end, boundary cases are of particular interest. Such cases include applications with a high demand for decentralisation or low communication effort respectively. The team formation protocol discussed in the preceding section balances these requirements. It has, however, shortcomings for boundary cases because it incorporates a centralistic entity and it exhibits a comparatively high communication complexity. Therefore, this paper introduces two derived protocols. The first one aims at minimising the communication effort (Sect. 4.1). The second one aims at maximising the degree of decentralisation (Sect. 4.2).

Agents that participate in the different team formation interaction protocols can take one or more of the following roles:

(1) Participant
(2) Manager
(3) Broker.

A participant agent aims at finding potential partners for cooperation in a specific logistics objective. A manager agent manages a team of agents that share a specific logistics objective. A broker agent administers a set of currently existing teams and their descriptors.

4.1 Minimising the Communication Effort by Broker

As a first step, this section aims at minimising the communication effort for team formation. This is important whenever communication should to avoid because it is expensive. As discussed in Sect. 3, there is a tradeoff between the interaction efficiency and the degree of decentralisation. That is, in order to minimise communication effort, one has to accept a decrease in the degree of decentralisation. The original protocol (Schuldt and Werner 2007) already incorporates a centralistic entity, the catalogue service. To recapitulate, the catalogue only administers the list of the teams established. The most significant part of the communication effort arises from the fact that all teams must be contacted by agents looking for cooperation partners. It is therefore promising to delegate more responsibility to the catalogue service. In particular, it should be able to decide directly which team descriptions match the description of the agent. This turns the former catalogue into a broker agent.

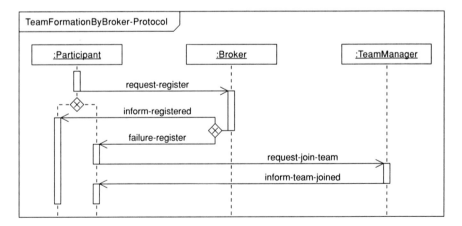

Fig. 1 Agent interaction protocol for team formation by broker. The notation is in accordance with the Agent Unified Modeling Language, in short AUML (Odell et al. 2000). Note that exceptional messages are omitted for the sake of readability

The protocol flow is as follows (Fig. 1). The protocol is initiated by a participant $\alpha_i \in A$ that is interested in team formation. This agent acts optimistically in that it initially assumes that no existing team matches its properties. In that case, it can itself register as the manager of a newly established team with its properties. To this end, the agent transmits its description$(\alpha_i) \in D$ to the respective broker agent in order to register itself as a team manager. The broker then compares this descriptor with those of all teams that are stored in its database. If there is no match, the agent itself is indeed registered as a new team manager. If the descriptions of α_i and an existing team manager $\alpha_j \in A$ resemble each other, i.e.,

$$\text{match}(\text{description}(\alpha_i), \text{description}(\alpha_j))$$

the registration of α_i fails. It may instead join the existing team of α_j.

4.2 Maximising the Degree of Decentralisation by Multicasting

Applying a broker agent significantly reduces the number of messages to be sent in the multiagent system. However, it also decreases the degree of decentralisation because all agents must contact this centralistic entity. The broker is thus a potential bottleneck of the system. In order to increase robustness, it is thus desirable to abolish centralistic entities. Of course, this also includes the catalogue service of the previous protocol (Schuldt and Werner 2007). To recapitulate, the catalogue is employed in order to administer the list of existing teams. That is, agents looking for cooperation partners can directly contact all team managers. Without this information, they would have to send a broadcast message to all agents because the intended recipients of the message are not known in advance. However, even if communication is affordable one should aim at reducing the number of messages sent. In general, a broadcast message is therefore not acceptable. As an alternative, multicasting can be applied. Computer network reference models like OSI or TCP/IP implement multicasting on the network layer. Hence, the application layer (which corresponds to agents) is disburdened from this task.

The protocol flow incorporating multicast messages is as follows (Fig. 2). Like in the broker-based protocol (Sect. 4.1), the participant α_i acts optimistically, i.e., it assumes that it may establish a new team. Therefore, it contacts the message transfer service (MTS) of its multiagent platform in order to receive future multicast messages on team formation. It may thus happen that multiple agents with the same description form new teams in parallel. However, this is not desirable for the application intended. The teams should clearly distinguish from each other at least during team formation. Whenever smaller teams are preferable, the partners may split up into multiple teams, e.g., during plan formation (Wooldridge and Jennings 1999).

In order to resolve potential conflicts between similar teams, the participant α_i sends its own properties to the respective multicast address. Therewith, it reaches all managers of existing teams to request a team match. The decision whether the

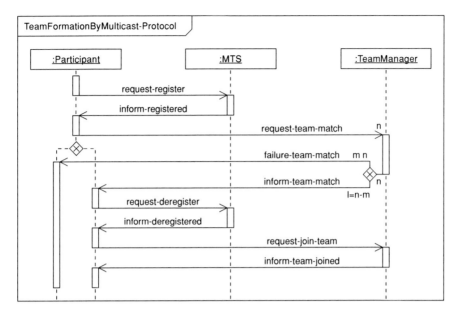

Fig. 2 AUML interaction protocol for team formation by multicast message service. Exceptional messages are omitted for the sake of readability

description of α_i matches the one of a particular team is made by the managing agent itself. That is, all autonomy regarding team formation is left to the teams. If two descriptions match, α_i is informed that it must deregister and that it may join the older team of α_j. Otherwise, α_i has successfully established its own team.

Note that the optimistic behaviour of the participant distinguishes this protocol from the original catalogue-based approach (Schuldt and Werner 2007). The original protocol comprises an additional iteration of matchmaking before a participant registers itself as a team manager. Due to the concurrent execution of multiagent systems, this does not suffice to prevent redundant teams. Abolishing this step reduces the number of messages to be exchanged. In turn, it increases the number registrations and deregistrations for multicast addresses.

5 Protocol Analysis and Discussion

The protocols introduced in the preceding section address different boundary cases in autonomous logistics. In order to enable developers to judge which protocol is appropriate for a concrete logistics application, it is necessary to examine them thoroughly. Several attributes for the categorisation of agent interaction protocols can be found in the literature (e.g., Rosenschein and Zlotkin 1994; Sandholm 1999). The following attributes are considered here in order to compare the protocols with respect to their particular advantages and drawbacks (Fig. 3):

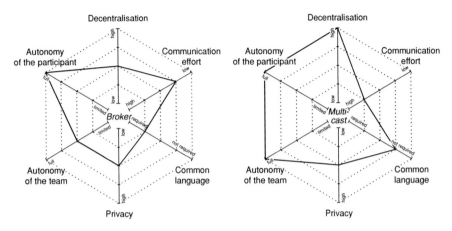

Fig. 3 Comparison of the team formation interaction protocols with respect to decentralisation, communication effort, common language, privacy, and autonomy

(1) Decentralisation
(2) Autonomy of the participant
(3) Autonomy of the team
(4) Communication effort
(5) Common language
(6) Privacy.

The degree of decentralisation indicates whether centralistic entities are required for protocol execution. The degree of autonomy measures how much decision-making is left to the participant and to the team respectively. The asymptotic communication complexity is an indicator for the communication effort, i.e., the interaction efficiency of the protocol. Furthermore, the extent to which a common language is required as well as the privacy is examined.

The team formation protocol based on multicasting has a high degree of decentralisation because no centralistic entity exists. All coordination is performed by the participating logistics entities themselves. The decentralisation of the broker-based counterpart is limited because the broker is a potential bottleneck. All entities looking for cooperation partners have to contact the broker agent. This is only acceptable in systems where the degree of decentralisation is less important than other attributes. In both protocols, the autonomy granted to the participants is high. Participants can deliberately decide whether they want to find partners and which team they want to join. Likewise, the multicasting-based approach does not restrict the autonomy of teams. By contrast, teams only have limited autonomy if a broker is applied. Decision-making whether a candidate matches the team description is delegated to the broker. However, the subsequent question whether a participant is actually accepted as a team member is again left to the team manager. If a broker is applied it must be able to understand all agent descriptions in order to decide whether they match. Hence, there is a demand for a common language all agents agree upon. Without a broker, only agents with similar goals must share a common language. Agents may then skip messages they do not understand. This also permits introducing descriptions for special purpose applications. Although for different reasons, both approaches are constricted regarding privacy. If a broker is applied, it is necessary to disclose decision processes about team matching to this centralistic entity. If multicasting is applied, agent descriptions are sent to all agents that claim to be team managers. However, virtually every agent can subscribe to this multicast channel. If there is a demand for security it is necessary to certify eligible agents.

The communication complexity of the broker-based protocol is as follows. Each agent exchanges one message with the broker. Agents that are not themselves team managers exchange two additional messages with their team manager. The complexity for every single agent is thus constant, either $O(2)$ or $O(4)$ respectively. The complexity for the whole system is thus linear, $O(4n - 2m) = O(n)$ with n being the number of all agents and m being the number of the team managers. In the multicasting-based approach each agent communicates with all m team managers. The communication complexity ranges between $O(2m + 2)$ and $O(2m + 6)$ and is thus linear for single agents. The complexity of the whole system is thus quadratic for all n participating agents, $O(nm) = O(n^2)$. That is, the increased degree of decentralisation also increases the communication complexity.

Despite of their differences, the outcome of the introduced team formation interaction protocols equals. They are interchangeable because they all result in unique teams that can be flexibly extended. Moreover, the result is even equal to that of the catalogue-based protocol (Schuldt and Werner 2007). The catalogue-based approach can act as a fallback solution if multicasting is not available because it resembles its multicasting-based counterpart in most attributes.

6 Implementation and Application

The protocols introduced in this paper have been implemented in JADE (Bellifemine et al. 2007), the Java agent development framework. The roles of agents (participant, manager, and broker) in the interaction protocols have been implemented as agent behaviours in this framework. Since JADE version 3.5, agents can subscribe to multicast topics. It is thus possible to benefit from multicast messages as demanded by the second protocol. The protocol implementation is generic in that it only incorporates agent interaction. Agent developers can add concrete agent descriptions demanded for the application intended. The protocols are currently applied in PlaSMA (Schuldt et al. 2008) in order to evaluate strategies for autonomous logistics. PlaSMA is a middleware that enhances JADE for parallel and distributed event-driven simulations.

7 Conclusion

This paper contributes two interaction protocols for software agents representing autonomous logistics entities. These protocols allow forming dynamic teams of agents sharing the same goals without any prior knowledge. A thorough examination helps agent developers choose the right protocol based on the logistics application intended. One protocol maximises the interaction efficiency, thereby also limiting the degree of decentralisation. The other one maximises the degree of decentralisation, thereby requiring higher communication efforts. Both protocols are generic regarding the descriptions for logistics entities.

Acknowledgments This research is funded by the German Research Foundation (DFG) within the Collaborative Research Centre 637 "Autonomous Cooperating Logistic Processes: A Paradigm Shift and its Limitations" (SFB 637) at the University of Bremen, Germany.

References

F. Bellifemine, G. Caire, and D. Greenwood. Developing Multi-Agent Systems with JADE. John Wiley & Sons, Chichester, UK, 2007.

K. Fischer, M. Schillo, and J. H. Siekmann. Holonic Multiagent Systems: A Foundation for the Organisation of Multiagent Systems. In HoloMAS 2003, pages 71–80, Prague, Czech Republic, 2003. Springer-Verlag.

W. R. Heinzelman, A. Chandrakasan, and H. Balakrishnan. Energy-Efficient Communication Protocol for Wireless Microsensor Networks. In HICSS 2000, volume 8, pages 8020–8029, 2000.

M. Hülsmann and K. Windt, editors. Understanding Autonomous Cooperation and Control in Logistics. Springer-Verlag, Heidelberg, Germany, 2007.

J. Odell, H. Van Dyke Parunak, and B. Bauer. Representing Agent Interaction Protocols in UML. In AOSE 2000, pages 121–140, Limerick, Ireland, 2000. Springer-Verlag.

E. Ogston and S. Vassiliadis. Matchmaking Among Minimal Agents Without a Facilitator. In Agents 2001, pages 608–615, Montreal, Canada, 2001. ACM Press.

J. S. Rosenschein and G. Zlotkin. Rules of Encounter: Designing Conventions for Automated Negotiation Among Computers. MIT Press, Cambridge, MA, USA, 1994.

T. W. Sandholm. Distributed Rational Decision Making. In G. Weiss, editor, Multiagent Systems. A Modern Approach to Distributed Artificial Intelligence, pages 201–258. MIT Press, Cambridge, MA, USA, 1999.

A. Schuldt, J. D. Gehrke, and S.Werner. Designing a Simulation Middleware for FIPA Multiagent Systems. In WI-IAT 2008, pages 109–113, Sydney, Australia, 2008. IEEE Computer Society Press.

A. Schuldt and S. Werner. Towards Autonomous Logistics: Conceptual, Spatial, and Temporal Criteria for Container Cooperation. In LDIC 2007, pages 311–319, Bremen, Germany, 2007. Springer-Verlag.

R. G. Smith. The Contract Net: A Formalism for the Control of Distributed Problem Solving. In IJCAI 1977, page 472, Cambridge, MA, USA, 1977. William Kaufmann.

M. Wooldridge and N. R. Jennings. The Cooperative Problem Solving Process. Journal of Logic & Computation, 9(4):563–592, 1999.

Design Aspects of Cognitive Logistic Systems

Carsten Beth, Jens Kamenik, Dennis Ommen and Axel Hahn

1 Introduction

Nowadays, it isn't possible to build a material flow system for a transfer station that will be efficiently productive for more than ten years without substantial rebuilds. Such a system has to be reconfigurable by design. It has to be possible to change the layout with a minimum effort in time and cost to be able to react to changing requirements (Windt 2006). Two major challenges have to be addressed to achieve these goals: (1) Centralization challenge: the centralized control structure of state of the art systems has proved to be too inflexible. A reconfiguration or just an extension of an operational system can lead to an expense equal to a complete new installation. (2) The conveyer gap challenge: continuous conveyers aren't flexible enough. It is very expensive to change their structure especially when they are organized hierarchically with central control. Discontinuous conveyors on the other hand are applicable for many applications and substitutable in case of damage. Regrettably, they don't achieve the required throughput in some scenarios (Heinecker 2006).

C. Beth (✉), J. Kamenik and D. Ommen
OFFIS Institute for Information Technology, Escherweg 2, 26121 Oldenburg, Germany
e-mail: carsten.beth@offis.de

J. Kamenik
e-mail: jens.kamenik@offis.de

D. Ommen
e-mail: dennis.ommen@offis.de

A. Hahn
Department of Computing Science, Carl v. Ossietzky University, Ammerländer Heerstr. 114-118, 26129 Oldenburg, Germany
e-mail: hahn@wi-ol.de

H.-J. Kreowski et al. (eds.), *Dynamics in Logistics*,
DOI: 10.1007/978-3-642-11996-5_25, © Springer-Verlag Berlin Heidelberg 2011

To overcome these limitations we describe an approach to combine the best qualities of both continuous and discontinuous conveyers into one system with respect to flexibility and throughput.

2 The Cognitive Logistic System

In cognitive logistic systems the intelligence is implemented in the modular robotic conveyers and also the surrounding environment. In our approach (see Fig 1) the cognitive transportation units (CTU) are modular and they can interact with other basic CTU devices to form a swarm. Each swarm has the minimum abilities that the transportation order requires. According to these requirements CTUs can act as continuous conveyors or discontinuous conveyors (see Figs 1 and 2).

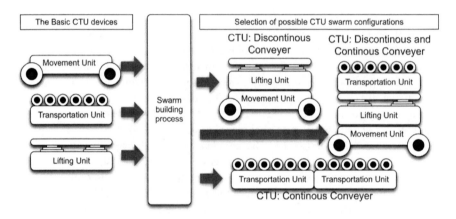

Fig. 1 Basic CTU devices and their possible configurations

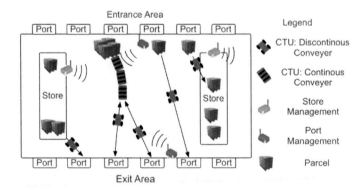

Fig. 2 Transfer station scenario

In general the CTU swarm can be composed of three basic devices to address the specific load situation. Each of those three CTU's basic devices has its own control unit to enable the cooperation with other CTUs and to permit autonomous behavior.

This approach is more flexible, adaptable and reconfigurable than a monolithic transportation system (Günthner et al. 2008b). The control paradigm of smart, independent, autonomous modules shall lead to positive emergence with the promise to cope with the high dynamic of logistic systems (Windt 2006).

3 The Scenario

Although the proposed cognitive logistic system is applicable to a wide range of intra-logistic scenarios, we focus on a specific one in our actual work. In order to evaluate the approach with a well known real life scenario we have selected a transfer station were cross docking is performed.

An example transfer station working with CTUs is drafted in Fig 2. Here, the station consists of entrance and exit areas, a storing area and a working area between the entrance and the exit. Goods (represented as parcels in the figure) enter the station at the ports of the entrance area and leave it at the ports of the exit area. The CTUs are responsible for good transportation between entrance end exit ports. If the delivery time of a good is too far in the future and exceeds the entrance port's storing capacity, it will temporarily be placed in the store area. Goods are assumed to be heterogeneous, i.e. they are different in size and weight and there are no standardized carriers.

Two classes of approaches can be separated depending on where decisions about the transportation process are made: (1) In good driven systems (Scholz-Reiter et al. 2006, 2007b) embedded devices attached to the goods escort the goods to their destinations. During the transportation process the embedded devices cooperate with the environment to achieve their goal. (2) In transportation system driven approaches the decision are made by the environment surrounding the goods. In our scenario we use the transportation system driven approach because in real life goods aren't equipped with embedded devices, even with the heterogeneous goods.

The distribution of control and planning tasks is not specified in all detail yet, although we want to describe the system roughly in this section.

The planning task of good transportation is distributed across autonomously acting components in the transfer station, the ports, stores and CTUs which are acting supported by sensor information about the environment. The planning task can be divided in two sections, (1) placement of order and (2) execution of transportation. (1) is performed, when a good is identified at an entrance port. The port receives additional information about the good from the ERP-system, e.g. weight, dimension, destination port and time window of delivery, and creates a transportation order. These orders are broadcasted as an auction. CTUs can participate in the auction in dependence of their actual schedule. To calculate the transportation cost, they plan routes to drive dependant on the traffic situation and transmit their quote back to the port. After receiving the quotes the port assigns the order to a CTU.

If the delivery time window is too far in the future, the port can decide to temporarily store the good. In this case a storing auction is performed before the transportation order action. Good information together with the time window of storage is transmitted to the stores, which reply with warehousing costs (e.g. to-bin transfer and stock removal times), if they have the wanted capacity in the time window. This information, together with an estimated transportation cost to the stock, is the basis for the port's decision on which stock to take.

In times where a lot of orders are offered by ports, CTUs can try to reduce the transportation costs by reconfiguration and cooperation. If many CTUs build a continuous conveyer at a heavily used track, the transportation capacity and the flow-rate of goods on the track can be increased. The main challenge will be to define an award-system to give CTUs the motivation to reconfigure and form a swarm of discontinuous conveyors. These CTUs can profit by being involved in the transportation of more goods as if they acting alone as discontinuous conveyers. CTUs not involved in the swarm can profit by having shorter ways to deliver goods to the destination, which leads to lower transportation costs and furthermore better chances to get orders in auctions. If the transport volume decreases at the discontinuous conveyor swarm, the motivation of the involved CTUs to maintain the swarm also will decrease. At the end the swarm will dispense.

After the placement of a transportation order, the responsible CTU performs the necessary actions to execute it (2). It drives to the port, loads the good, drives to the destination and puts it down. While the CTU drives through the transfer station, it observes the environment with its own and external intelligent sensors. They can register for sensor events affecting specific regions of their course. If events occur, e.g. detected obstacles or bottlenecks, the CTU can decide to reorganize its plan by changing the route or it can decrease the velocity to avoid a potential collision.

4 Related Work

In Ommen et al. (2009) a classification scheme for robotic intra-logistic systems was proposed. This scheme is depicted in Fig. 3 and categorizes the related work of ongoing research and existing products of these systems into three categories: (1) Central control and autonomous behavior: these material flow systems are controlled via a central instance, where all the decisions regarding the transportation order are scheduled. These robotic systems are usually application specific and, hence there is no need for cooperation between the robots to fulfill a goal. (2) Local control with autonomous behavior: because of local control the presented systems are scalable, flexible and failsafe. Thus, the installation and reconfiguration costs are lower in comparison to central controlled systems. The robots act

Fig. 3 Classification scheme for autonomic transportation systems

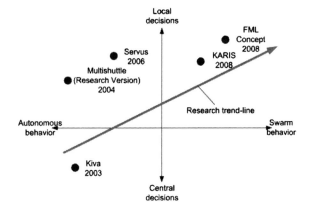

autonomously and don't cooperate. (3) Local control with swarm behavior: robots cooperate with each other to achieve a common goal. This requires the ability to communicate with each other and to make local decisions.

4.1 Category I: Central Decision with Autonomous Behavior

The Kiva warehouse management system by Kiva Systems (Guizzo 2008; Wurman et al. 2008; Kiva Systems 2009) is a commercial system for commissioning of products in stocks with small parts. The stock consists of many adjustable shelves with a matrix like structure on a flat ground. Small autonomous robots are able to drive under a shelf, lift it up and bring it autonomously to other locations, e.g. a picking station. Orders are accepted from a warehouse management system by a central computer, which is responsible for the scheduling of the autonomous robots and picking stations, as well as the shelf space at the station. Because of the agent based architecture, the system is highly scalable and can grow with the requirements, where the centralized Job Manager is a limiting part. Another disadvantage is the limited field of application domains. The system is mainly useful for order picking processes that have a high degree of manual work.

4.2 Category II: Local Decision with Autonomous Behavior

Multishuttle is a product by Siemens Dematic AG developed in cooperation with the Fraunhofer-Institute for Material Flow and Logistics in Dortmund, Germany. In contrast to traditional warehouse systems like a shelf access equipment with constant throughput capacity, the Multishuttle system is scalable. The system consists of autonomous vehicles that drive horizontally into a shelf and are able to autonomously load and unload product carrier (at the same time). Rails are laid in several stacked levels and guide the vehicles and deliver energy at the same time. Movement of the vertical direction is achieved by lifts. Transport orders are communicated to the vehicles by WLAN.

The system Servus form the Austrian company Servus Robotics (Servus Robotics 2005, 2006) is intended for intra-logistics assembly automation. Like the Multishuttle, the Servus system is also rail bounded. The vehicles are able to act autonomously. They accept transportation orders through an infrared or WLAN interface. Additional information on the goods, like necessary processing steps, is stored at the vehicle. Therefore, the goods themselves don't need to be intelligent. Additional actuators can be build upon the vehicles, e.g. to be able to perform processing steps while the goods are carried. The rails, unlike within the Multi-shuttle system, do not supply energy. Instead each vehicle has its own fast rechargeable energy supply.

4.3 Category III: Local Decision with Swarm Behavior

Another project investigating in robotic conveyers is the ARMARDA project of the Institute for Conveying Technology and Logistics (IFL) at the University of Karlsruhe, Germany. They have presented a robotic transportation system (Baur et al. 2008) called BINE, that consists of homogeneous transportation units that are able to drive on the floor or stand on the floor while acting as a conveyer. The wheels thereby are rotatable by 360 degrees providing free movement at the horizontal plane. A BINE unit is able to carry payload by its own or, if the charge is too large or too heavy, as a swarm with many BINE units together. If high throughput is required, a continuous conveyer with sorter function can also be build by a BINE swarm. Swarm building and acting is the actual research work at IFL.

The institute for Materials Handling, Material Flow, Logistics (FML) in Munich (Germany) proposed a concept for future material handling systems (Günthner et al. 2008a, b) consisting of low-scale autonomic transportation units. All transportation vehicles are small and have a simple and basic design at a low price due to high volume production. For special roles they can be equipped with manipulators like a lift fork, roller or belt conveyer. They are autonomous with their own intelligence and communication options. If a task can't be achieved by a single vehicle, more of them can form a swarm and act together.

4.4 Conclusion on the Related Work

As it is depicted in Fig 3 there is clear trend line towards autonomous robotic systems that can act in a swarm to achieve common goals in material flow systems. The discussed research project ARMARDA is an elaborated robotic approach that shows that these systems can act in two ways: as a discontinuous or continuous conveyer. Nevertheless, this system has no flexibility regarding the transported goods. The concept of the FML has this ability because of its changeable manipulators. Thus, it can pick up different types of goods, like pallets and mixed cargo. But in the FML concept is still some discussion on the realization of such an approach missing. This paper will close this gap and will show how modular material flow system can be designed and controlled.

5 The Cognitive Environment and its Sensors

Sensors are devices that provide physical data from the environment in an electrical presentation. They are the basis of our system model and all decisions made by the strategies need to rely on them. In former logistic systems like forklifts the only sensory information comes from human perception. Every action the human does is based on his limited perception of the environment. The same problems have autonomous logistic systems e.g. driver-less systems that rely on built-in sensor information. Their view of the environment is limited to the perception ability of the integrated sensors.

In our cognitive logistic system sensors do not "belong" to a system. They are usable by everyone and provide their information to everyone who needs them. This expands the view of the environment to the whole "logistic scenario" world. Every autonomous system is able to get this view and is therefore able to raise the correctness of their decisions. As an example consider an autonomous robot trying to pass an intersection where it is not able to sense if another robot is crossing it. To prevent a collision, it is slowing down and advances step by step until its integrated sensors are able to provide the "free" information. In the cognitive way—external sensors attached to the walls (able to detect movement of non-cooperative systems) or even another autonomous system trying to cross the track (cooperative system) can give the free/non-free information.

Riedmaier (2008) shows rudimentarily the possible leverage of such an approach to the logistics. So lead the additional sensor information of the condition of the track to a optimization of the speed of forklifts that thereby increase the handling of palettes about 5%.

Further usage of the network of sensors can be as a communication relay for the autonomous robots and sensors. Due to limited propagation of radio waves in logistic in-door facilities and the limiting governmental regulation regarding maximum transmission power of radio devices, the multi-hop capability of modern wireless sensor networks is a good way to extend the range of the sensors and the autonomous robots communication system. For harsh industrial environments WirelessHART (HART Communication Foundation 2007) is a standardized protocol for reliable wireless communication. It provides robust self-organizing and self-healing mechanisms to encounter communication failures, e.g. when fixed or mobile obstacles interrupt a communication path.

In a cognitive system networked sensors can be used as distributed observers. That means a sensor stores a snapshot of the past inside its memory. This kind of sensors can answer questions like "Did robot x passed intersection z?" or "Is there already a robot/somebody in this area?" The answer to these questions is useful for coordination and optimization purposes. Instead of a central observer that knows everything (every state of the system) a fixed distributed sensor has an area it observes. For example, a fixed sensor knows about the robot traffic in his area and can therefore give a usage estimation of the path belonging to his observation area. Technically, sensors now have to store their data instead of just sending real-time data to the autonomous robots. The robots then ask the sensors for certain events in their stored history snapshot. For fault-tolerance reasons, sensors are allowed to replicate their data to other fixed or mobile sensors.

6 Design Methodology

The presented scenario contains numerous of heterogeneous devices like sensors and actuators (S&A) and also different platforms. Each of these devices have different capabilities and therefore different software (SW) interfaces that need to be addressed during the run of the cognitive material flow systems. Therefore an abstraction layer for the automation domain is needed that copes with the heterogeneity of S&A and their platforms in a right way.

The integration problem of heterogeneous devices during the engineering process is not only related to the automation industry but has also been identified as an issue by the automotive industry. There the AUTOSAR (AUTOSAR Specifications Release 2009) standard was introduced in 2003, driven by Daimler, Volkswagen and also by other OEMs and suppliers. Some of the goals of this standard are: (1) improve the reuse of SW artifacts, between heterogeneous platforms with the use of a newly introduced abstraction layers that hides the platform specific SW components, (2) improve the maintenance of existing SW artifacts, and (3) define a common SW development process that can be used by all industry participants.

Nevertheless the automation industry has also made some endeavor to ease the situation of heterogeneous devices. Promising approaches here were CORFU (Thramboulidis 2002) and IDA (IDA Organization website). Like AUTOSAR both also proposed a new abstraction layer to hide the device complexity from the application. But what was success in the automotive industry, leads not to the same acceptance in the automation industry. One of the major reason is that the automation industry is not driven by a couple of OEMs but either of the many heterogeneous plant and machines vendors. This situation makes the agreement on a common standard unlikely. Therefore a new approach is needed that deals with the interoperability of heterogeneous interfaces/standards itself and which not tries to introduce a new standard.

We propose therefore a semantic driven approach to achieve interoperability in the automation domain. Such an approach is easily extendable to new interfaces just by semantically describing the S&A interface that has to be integrated into the automation network and by describing the behavior of the S&A interface in a formal way. A design process would look like that: (1) The engineer describes the state model of a S&A interfaces in a formal way for instance in UML. Further he annotates the S&A data structures, as input and output fields to which he also adds literality their description, like the physical unit. (2) With this information a semantic mapping of these S&A interfaces to an existing real-time middleware, which is written in a platform independent manner can be created. (3) The engineer selects now the desired platform and the design process automatically create the middleware in which the S&A SW interfaces are integrated for the selected platform. The middleware itself abstracts now complexity of the devices.

7 Conclusion and Future Work

This paper proposed a robotic construction kit for material flow systems that raises the modularity and flexibility. Furthermore, a unified sensor integration scheme was proposed that raises the cognitive perception ability of the whole logistic system and a sensor data concept that enables the idea of a distributed observer was shown.

In a next step the proposed models will be simulated. The goal of the simulation is to find the best granularity of the modularization and to find the best cooperating strategies for autonomous logistic systems. After this achieve a test bed implementing Fig. 2 for validation of the chosen strategies will be created.

Acknowledgments This contribution was supported by the German federal state of Lower Saxony with funds of the European Regional Development Fund (ERDF) within the scope of the research project "Cognitive Logistic Networks (CogniLog)"

References

AUTOSAR Specifications Release 3.1. http://www.autosar.org. Accessed 4 April 2009
Baur T., Schönung F., Stoll T., Furmans K. (2008) Formationsfahrt von mobilen, autonomen und kooperierenden Materialflusselementen zum Transport eines Ladungsträgers. 4. Fachkolloquium der Wissenschaftlichen Gesellschaft für Technische Logistik e. V
Guizzo E. (2008) Three Engineers, Hundreds of Robots, One Warehouse. IEEE Spectrum, pp. 26–34
Günthner W. A. et al. (2008a) Vom Prozess zum Ereignis—ein neuer Denkansatz in der Logistik. Jahrbuch Logistik, pp. 224–228
Günthner W. A., Durchholz J., Kraul R., Schneider O. (2008b) Technologie für die Logistik des 21. Jahrhunderts. Kongressband zum 25. Deutschen Logistik-Kongress, pp. 360–393
HART Communication Foundation (2007) WirelessHART Technical Datasheet, Revision 1.0B

Heinecker M. (2006) Methodik zur Gestaltung und Bewertung wandelbarer Materialflusssysteme, Dissertation, Technische Universität München

IDA Organization website. http://www.ida-group.org. Accessed 4 April 2009

Kiva Systems website http://www.kivasystems.com. Accessed 4 April 2009

Ommen D., Beth C., Kamenik J. (2009) A system architecture for robotic movement of goods—approaches towards a cognitive material flow system, International Conference on Information in Control, Automation and Robotics

Riedmaier S. (2008) Neue Möglichkeiten für Schmalgangstapler. Hebezeuge Fördermittel 12, pp 762–763

Scholz-Reiter B., Rekersbrink H., Freitag M. (2006) Kooperierende Routingprotokolle zur Selbssteuerung von Transportprozessen. Industrie Management, p. 22

Scholz-Reiter B., Jagalski T., Bendul J. (2007b) Bienenalgorithmen zur Selbststeuerung logistischer Prozesse. Industrie Management, pp. 7–10

Servus Robotics (2005) Na, Servus! Automation 6

Servus Robotics (2006) Wie von Geisterhand. Automation 4

Thramboulidis K. (2002), "Development of Distributed Industrial Control Application: The CORFU Approach." Vasteras, Sweden

Windt K. (2006), Selbststeuerung intelligenter Objekte in der Logistik. Selbstorganisation—Ein Denksystem für Natur und Gesellschaft, Böhlau Verlag

Wurman P. R., D'Andrea R., Mountz M. (2008) Coordinating Hundreds of Cooperative, Autonomous Vehicles in Warehouses. AI Magazine 29/1:9–19

Autonomous Units for Solving the Traveling Salesperson Problem Based on Ant Colony Optimization

Sabine Kuske, Melanie Luderer and Hauke Tönnies

1 Introduction

In logistics, one is often faced with optimization problems that are NP-hard. One successful way towards efficient solutions of complex optimization problems is swarm intelligence that makes use of the emergent behavior of a set of autonomously and independently acting individuals. Ant colony optimization (ACO) algorithms are famous representatives of this type. ACO algorithms are inspired by the way how ants find short routes between food and their formicary. They have been shown to be well-suited not only for the solving of shortest path problems, but for a series of more complex problems, typically occurring in logistics (cf. (Dorigo and Stützle 2004)). In particular, the Traveling Salesperson Problem (TSP) plays an important role in ACO algorithms. It not only occurs in many logistic applications, it also serves to illustrate the basic features of ACO algorithms.

This paper proposes to use communities of autonomous units (see e.g., Hölscher et al. 2007; Hölscher et al. 2009) as a formal graph-transformational and rule-based framework for modeling ACO algorithms. A community of autonomous units consists of a set of autonomous units, a global control condition, and an overall goal. Autonomous units are equipped with a set of rules, a set of auxiliary units, a control condition, and a goal. Independently from each other, they try to reach their individual goals by applying their rules and auxiliary units according to their control conditions. Applications of rules and auxiliary units transform the

S. Kuske (✉), M. Luderer and H. Tönnies
Department of Computer Science, University of Bremen, P.O.Box 330440,
28334 Bremen, Germany
e-mail: kuske@informatik.uni-bremen.de

M. Luderer
e-mail: melu@informatik.uni-bremen.de

H. Tönnies
e-mail: hatoe@informatik.uni-bremen.de

H.-J. Kreowski et al. (eds.), *Dynamics in Logistics*,
DOI: 10.1007/978-3-642-11996-5_26, © Springer-Verlag Berlin Heidelberg 2011

common environment, and autonomous units can react to these transformations. Hence, the autonomous units interact and act in a dynamically changing common environment. For solving the TSP, the ants are realized as autonomous units, the complete graph forms the common environment and the overall goal consists in finding the shortest Hamiltonian cycle in the graph.

The use of communities of autonomous units for ACO algorithms has the following advantages:

- It is not uncommon that realistic logistic problems, when solved by an ACO algorithm, require a great number of ants performing some non-trivial actions. Especially for complex problem solution strategies, it quickly becomes difficult to see whether the algorithm actually solves the problem. Community of autonomous units have a well-defined operational semantics which opens possibilities to prove interesting properties, like termination, by induction on the length of the transformation sequences and thus increasing the probability that the algorithm is actually performing the task that it is supposed to perform.
- The graph- and rule-based representation allows to visualize and specify the ACO algorithms more naturally and can help to gain a better understanding of the algorithm. Furthermore, there exist tools that are suitable to provide an implementation platform for communities of autonomous units; in particular, in (Hölscher 2008) it is shown how an algorithm based on autonomous units can be implemented with the tool GrGEN (Geiß and Kroll 2008) for finding shortest paths in graphs.
- Communities of autonomous units are suitable to integrate autonomous control into the model of the logistic processes and algorithms (cf. Hölscher et al. 2007; Hölscher et al. 2008; Hölscher 2008). The advantage of autonomous control in logistics is the focus of the Collaborative Research Centre 637 *Autonomous Cooperating Logistic Processes: A Paradigm Shift and Its Limitations*.

The paper is organized as follows. In Sect. 2, ant colony systems for the heuristic solving of optimization problems are briefly introduced. In Sect. 3, autonomous units and communities of autonomous units are presented. In Sect. 4, it is shown how ant colony systems solving the TSP can be modeled by communities of autonomous units in a formal and visual way. The conclusion is given in Sect. 5.

2 Ant Colony Systems

ACO is an algorithmic framework for the heuristic solving of combinatorial optimization problems. The idea is based on the observation how ants find short routes between food and their formicary. A moving ant leaves a chemical substance, called pheromone, on the ground which can be sensed by other ants. The higher the concentration of pheromone along a way, the more probable it is that ants will choose this way. Since on short ways ants leave their pheromone in a shorter time than on longer ways the concentration of pheromone will be higher the shorter a way is. The more ants follow a specific route, the more attractive it becomes for other ants, thus resulting in a positive feedback loop. Furthermore, pheromone evaporates with time, so a too fast convergence to a short route is avoided.

ACO imitates this behavior for solving combinatorial optimization problems which often have a complete graph as input. A problem solution is encoded as an ordered sequence of edges in the graph. The artificial ants autonomously construct solutions by exploring the graphs. Since the ants start their work at a randomly chosen node in the graph, the crucial point consists in the decision which node is to visit next.

A common method is to assign a probability $prob_{ij}$ to every possible decision, where i denotes the node currently visited by the ant and j is a node that can be reached from i by traversing an edge. The value of $prob_{ij}$ is calculated as follows:

$$prob_{ij} = \frac{p_{ij}^{\alpha} \cdot x_{ij}^{\beta}}{\sum_{k \in J^i} p_{ik}^{\alpha} x_{ik}^{\beta}} \quad \forall i \in \{1, \ldots, n\}, \ j \in J^i.$$

The value p_{ij} simulates the pheromone intensity of the edge between the nodes i and j. Consequently, every edge of the graph has its own p value. The value x_{ij} is a heuristic value describing an estimated probability that the solution includes the way between the nodes i and j. Like the pheromone intensity, every edge has its own x value. The value x_{ij} is often called the *desirability* of the edge between i and j. The exponents α and β are problem-dependent parameters to control the influence of p and x. The set J^i consists of all nodes that are reachable from the node i via a single edge. In other words, $prob_{ij}$ is a value in the closed interval $[0 \ldots 1]$ giving a procentual estimation of how worthy it is to visit the node j being at node i.

With the help of this estimation, every artificial ant searches its way through the graph. If an ant succeeds in constructing a solution, e.g. having found a way from the start node to the goal node, it updates the pheromone value of all the edges by some value that reflects the quality of the solution. In the case of the TSP for example this value could be the reciprocal distance of the complete route. Afterwards the artificial ants will use different $prob_{ij}$ values according to the routes already found. Since the $prob_{ij}$ values of routes will be higher the shorter the route is, artificial ants will converge slowly to the shortest route found. This basic idea has been extended and modified in some ways to improve the performance. Details can be found for example in (Dorigo and Stützle 2004).

3 Communities of Autonomous Units

In this section, we briefly introduce communities of autonomous units. For more detailed introductions see (Kreowski and Kuske 2008) and (Hölscher et al. 2009).

Communities are composed of autonomous units that act and interact in a common environment which is typically a graph. The ingredients of communities are given by an underlying graph transformation approach (cf. (Rozenberg 1997) for an overview of graph transformation approaches).

Definition *(Graph transformation approach)* A graph transformation approach is a system $\mathcal{G}, \mathcal{R}, \mathcal{X}, \mathcal{C}$ where $\mathcal{G}$ is a class $\mathcal{G}$ of *graphs*, $\mathcal{R}$ is a class of *rules* such that every rule specifies a binary relation on $\mathcal{G}$, $\mathcal{X}$ is a class of *graph class expressions* each of which specifies a set of graphs in $\mathcal{G}$, and $\mathcal{C}$ is a class of control conditions each of which specifies a set of sequences of graphs.

Every autonomous unit consists of a set of pairs of rules, a set of auxiliary transformation units, a control condition, a specification of initial private states, and a goal. When a rule pair (r_1, r_2) is applied, the first component r_1 transforms the current common environment and the second component r_2 modifies the current private state of the unit. Every auxiliary transformation unit encapsulates a set of rule pairs, a set of further auxiliary transformation units, and a control condition.

Definition *(Units)* An auxiliary transformation unit is a system $\text{tu} = (P, U, C)$ where $P \subseteq \mathcal{R} \times \mathcal{R}$, U is a set of auxiliary units, and $C \in \mathcal{C}$. An *autonomous unit* is a system $\text{aut} = (I, U, P, C, G)$ where $I \in \mathcal{X}$ is the *initial private state specification*, U is a set of *auxiliary transformation units*, $P \subseteq \mathcal{R} \times \mathcal{R}$, $C \in \mathcal{C}$, and $G \in \mathcal{X}$ is the *goal*.

In the following, we use auxiliary transformation units with a hierarchical import structure. This means that every auxiliary transformation unit of import depth zero does not contain any auxiliary transformation unit, and every auxiliary transformation unit of import depth $n + 1$ can use only auxiliary transformation units of import depth at most n.

Every community is composed of a set of autonomous units, a specification of a set of initial environments, a global control condition, and a goal.

Definition *(Community)* A *community* is a system *(Init, Aut, Cond, Goal)* where Init, $Goal \in \mathcal{X}$ are graph class expressions called the *initial environment specification* and the *overall goal*, respectively, *Aut* is a set of autonomous units, and $Cond \in \mathcal{C}$ is a *global control condition*.

The semantics of communities for modeling ACO algorithms is a parallel one, i.e. in every computation step, several autonomous units may perform their actions simultaneously. More precisely, the parallel semantics of a community of autonomous units *COM = (Init, Aut, Cond, Goal)* consists of a set of state sequences that represent the transformation processes. Every state in such a sequence is composed of a common environment and a private state for every autonomous unit. The initial state of every transformation process s in the semantics of *COM* must be composed of a common environment specified by *Init* and an initial private state for each $aut \in AUT$. Moreover, s must be allowed by *Cond* as well as by the control conditions of all units in *Aut*.

4 Modeling Ant Colony Systems by Communities

In this section, we demonstrate how ant colony systems can be modeled with communities of autonomous units by translating an ACO algorithm for the TSP into a community.

The ACO algorithm for the TSP gets as input a complete graph without multiple edges in which every edge is labeled with a natural number denoting the distance between its two attached nodes, and a real number denoting an initial pheromone quantity. An optimal solution of the TSP is a Hamiltonian cycle in G whose distance is minimal. Basically, the TSP is solved by ACO systems according to the following procedure. First, every ant chooses nondeterministically a start node. Second, every ant traverses the graph by going along a Hamiltonian cycle. In each step it passes through exactly one edge which is chosen according to the probability rule given in Sect. 2. Moreover, it stores the cycle and its distance in its memory. Third, pheromone evaporation takes place and every ant traverses the cycle again while augmenting the pheromone quantity of every passed edge by $1/s$ where s is the distance of the cycle traversed by the ant. Then the next iteration starts.

For modeling the described algorithm as a community, we employ as underlying graph class undirected edge-labeled graphs. A *graph transformation rule r* is defined as $N \supseteq L \supseteq K \subseteq R$, where N, L, K, and R are graphs such that L is a subgraph of N, and K is a subgraph of L and R. The rule r is applied to a graph G according to the following steps (cf. also Habel et al. 1996; Corradini et al. 1997). (1) Choose a homomorphic image of L in G. (2) If L is a proper subset of N, make sure that the image of L cannot be extended to an image of N. (3) Delete the image of L up to the common part K of L and R, provided that the result is a graph again, i.e., no dangling edges should be produced. (4) Add a copy of R to the resulting graph such that K is identified with its image.

A rule $N \supseteq L \supseteq K \subseteq R$ is depicted as $N \to R$ where the parts of N not belonging to L are crossed out. The common part K consists of the nodes and edges that have the same forms, labels, and relative positions in N and R. A node with a loop is often depicted as a node with the loop label inside. We assume the existence of a special label *unlabelled* that is omitted in graph drawings. A pair of rules $(N_1 \supseteq L_1 \supseteq K_1 \subseteq R_1, N_2 \supseteq L_2 \supseteq K_2 \subseteq R_2)$ is depicted as $N_1|N_2 \to R_1|R_2$, where again the items of $N_i - L_i$ ($i = 1, 2$) are crossed out.

As graph class expressions we use *all* denoting all graphs and complete(A) where A is some set of labels. The expression complete(A) characterizes all complete graphs without multiple edges in which every edge is labeled with an element of A. Moreover, every graph G is a graph class expression which specifies itself. The control conditions of our approach consist of regular expressions over rules and auxiliary units equipped with the operator ! standing for *as long as possible*. Moreover, as global control conditions, we use regular expressions over sets of autonomous units, where a set of units means that they should run in parallel. For reasons of space limitations we do not introduce a formal semantics of control conditions but explain their meaning when they are used.

The community of autonomous units that models the presented ACO algorithm
for the TSP is equal to

$$(\text{complete}(\mathbb{N} \times \mathbb{R}), \{\text{Ant}_1, \ldots, \text{Ant}_k, \text{Update}(\rho)\}, (\{\text{Ant}_1, \ldots, \text{Ant}_k\};$$
$$\text{Update}(\rho))^*, \text{all})$$

where $k \in \mathbb{N}$, $0 < \rho \leq 1$, and the condition $(\{\text{Ant}_1, \ldots, \text{Ant}_k\}; \text{Update}(\rho))^*$ means
that in each iteration the autonomous units $\text{Ant}_1, \ldots, \text{Ant}_k$ run in parallel and then
the autonomous unit $\text{Update}(\rho)$ becomes active, where ρ is a pheromone decay
parameter.

For $j = 1, \ldots, k$, the autonomous unit Ant_j is equal to

$$\left(M_j^\bullet, \{\text{traverse}_j, \text{putpher}_j\}, \emptyset, \text{traverse}_j; \text{putpher}_j, \text{all} \right)$$

where $M_j^\bullet$ denotes the graph consisting of a node with an M_j-labeled loop (representing the initial memory of Ant_j), and the control condition requires that the
auxiliary units traverse_j and putpher_j, be executed sequentially in this order.

The unit traverse_j in Fig. 1 searches for a Hamiltonian cycle, guided by the present
pheromone trails. It uses the auxiliary unit prob_j and contains the three rule pairs
start, *go*, and *stop*. With the first component of every rule pair, the common environment is transformed whereas the second component updates the memory of Ant_j.

The rule pair *start* puts the ant on a randomly chosen node in the common
environment. In its memory, the ant inserts a node for the distance of the cycle

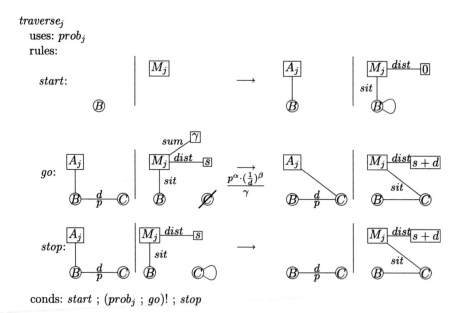

Fig. 1 The auxiliary unit traverse_j

found so far (which initially is equal to zero) and it stores its current location. The unlabeled loop at the *B*-node serves to remember the start node of the cycle. The rule pair *go* models a movement of the ant to a neighbor node. The crossed node guarantees that the ant only moves to a neighbor node *C* if the node is not stored in the memory, yet. Moreover, in the memory, the cycle distance is updated, the node *C* (together with the passed edge) is added to the visited path, and the current location of the ant is changed to node *C* by redirecting the *sit*-edge. The rule pair *go* is applied according to the probability presented in Sect. 2 and depicted under the arrow of the rule. The value of γ in the formula is inserted in the memory by the auxiliary unit prob$_j$, which for reasons of space limitations is not depicted. The rule pair *stop* closes the cycle. According to the control condition of traverse$_j$, the pair *start* is applied first. Second the unit prob$_j$ and *go* are applied as long as possible in this order, and finally, the pair *stop* is applied.

It can be shown that the unit traverse$_j$ finds a Hamiltonian cycle in the underlying graph.

The auxiliary unit putpher$_j$ in Fig. 2 leaves a pheromone trail on the passed cycle with the intensity $1/s$, where s is the distance of the cycle. According to the cycle stored in the memory, putpher$_j$ places additional "pheromone-edges", labeled with $1/s$, in parallel to the edges where trail update should take place. During this procedure the cycle is deleted from the local memory. It should be noted that the fact that each ant inserts separate edges for its individual pheromone trail allows the ants to put pheromone in parallel.

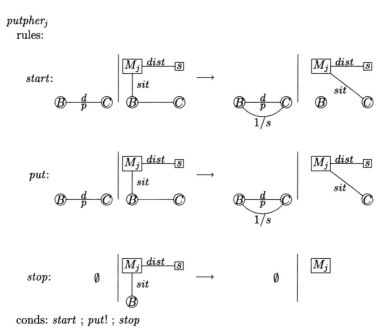

Fig. 2 The auxiliary unit putpher$_j$

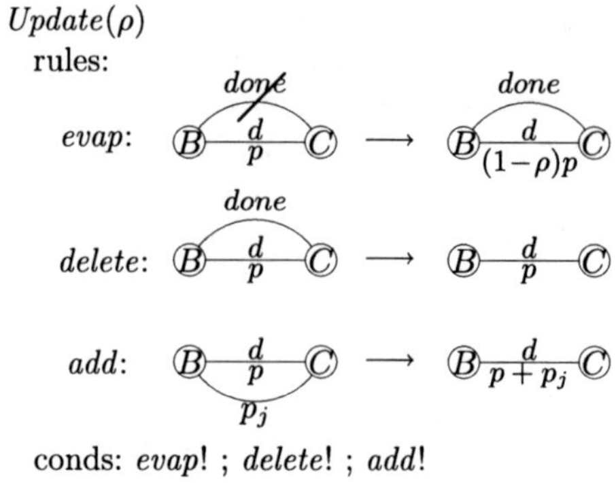

Fig. 3 The autonomous unit
Update(ρ)

$Update(\rho)$
rules:

evap:

delete:

add:

conds: *evap*! ; *delete*! ; *add*!

It can be shown that all transformation sequences of putpher$_j$ are finite and that the parallel edges are inserted along the Hamiltonian cycle in the memory of the ant.

The autonomous unit Update(ρ) in Fig. 3 is responsible for the evaporation and the summation of the pheromone updates made by the ant units. According to its control condition the rule *evap* is executed as long as possible to lessen the amount of pheromone at every edge. To remember where the evaporation has taken place, a parallel edge with label *done* is inserted.

When the evaporation for all edges is completed, *delete*! removes all the *done*-edges. Now for each edge e with a label in $\mathbb{N} \times \mathbb{R}$, *add*! adds up the values of the "pheromone-edges" placed by the ant units and updates the pheromone amount of e with the respective result.

Since Update(ρ) needs no local memory, the second components of the rule pairs of Update(ρ) are not depicted. Formally these second components consist of empty graphs, only.

It can be shown that all transformation processes of Update(ρ) terminate and model the update of pheromone trails in the required way.

5 Conclusion

In this paper, we have proposed communities of autonomous units as a modeling framework for ACO algorithms. In particular, we have modeled an ACO algorithm for the TSP as a community of autonomous units.

It has turned out that rule-based graph transformation allows to specify ant algorithms in a very natural way because (artificial) ants modify graphs and move along edges. This yields the following advantages:

- The specification of ants as autonomous units provides the ants with a well-defined operational semantics so that correctness results can be proved.
- The graph transformation rules of autonomous units allow for a visual specification of ants behavior instead of string-based pseudo code as it is often used in the literature.

- The existing graph transformation systems (cf. e.g. Ermel et al. 1999; Geiß and Kroll 2008) facilitate the visual simulation of ant colonies in a straightforward way (see also Hölscher (2008)).
- Implementing ACO algorithms with graph transformational systems is useful for verification purposes, i.e., to check whether the algorithms behave properly for specific cases.

In the future, this and further case studies should be implemented with one of the existing graph transformation systems so that (1) the emerging behavior of ant colonies can be visually simulated by representing transformations of the common environment and transformations of the private states on different visualization levels, and (2) ACO algorithms can be verified. For the implementation purpose we plan to use GrGen (Geiß and Kroll 2008) because it is one of the fastest and most flexible graph transformation systems.

Further case studies should take into account local search, elitist and rank-based ant systems as well as dynamic aspects (cf. Eyckelhof and Snoek 2002; Dorigo and Stützle 2004; Montemanni et al. 2005; Reimann et al. 2004; Rizzoli et al. 2007) to mention only a few examples).

Acknowledgments Sabine Kuske and Hauke Tönnies would like to acknowledge that their research is partially supported by the Collaborative Research Centre 637 (Autonomous Cooperating Logistic Processes: A Paradigm Shift and Its Limitations) funded by the German Research Foundation (DFG). We are grateful to Hans-Jörg Kreowski and the anonymous referees for their valuable comments.

References

Corradini, A., Ehrig, H., Heckel, R., Löwe, M., Montanari, U., and Rossi, F. (1997). Algebraic approaches to graph transformation part I: Basic concepts and double pushout approach. In: G. Rozenberg (ed), Handbook of Graph Grammars and Computing by Graph Transformation, Vol. 1: Foundations. World Scientific, Singapore, pp. 163–245.

Dorigo, M., and Stützle, T. (2004). *Ant Colony Optimization.* MIT-Press.

Ermel, C., Rudolf, M., and Taentzer, G. (1999). The AGG-approach: Language and environment. In: H. Ehrig, G. Engels, H.-J. Kreowski, and G. Rozenberg (eds) Handbook of Graph Grammars and Computing by Graph Transformation, Vol. 2: Applications, Languages and Tools. World Scientific, Singapore, pp. 551–603.

Eyckelhof, C.J., and Snoek, M. (2002). Ant systems for a dynamic TSP - Ants caught in a traffic jam. In M. Dorigo, G. Di Caro, and M. Sampels (eds), *Ant Algorithms - Third International Workshop, ANTS 2002, Volume 2462 of Lecture Notes in Computer Science*, pp. 88–98.

Geiß, R., and Kroll, M. (2008). GrGen.NET: A fast, expressive, and general purpose graph rewrite tool. In A. Schürr, M. Nagl, and A. Zündorf (eds), *Proc. 3rd Intl. Workshop on Applications of Graph Transformation with Industrial Relevance (AGTIVE '07), Volume 5088 of Lecture Notes in Computer Science*, pp. 568–569.

Habel, A., Heckel, R., and Taentzer, G. (1996). Graph grammars with negative application conditions. Fundamenta Informaticae 26(34):287–313.

Hölscher, K. (2008). Autonomous Units as a Rule-based Concept for the Modeling of Autonomous and Cooperating Processes. Logos Verlag. PhD thesis.

Hölscher, K., Klempien-Hinrichs, R., Knirsch, P., Kreowski, H.-J., and Kuske, S. (2007). Autonomous units: Basic concepts and semantic foundation. In M. Hülsmann, and K. Windt (eds) *Understanding Autonomous Cooperation and Control in Logistics. The Impact on Management, Information and Communication and Material Flow.* Springer, Berlin Heidelberg New York, USA.

Hölscher, K., Knirsch, P., and Luderer, M. (2008). Autonomous units for communication-based dynamic scheduling. In H.-D. Haasis, H.-J. Kreowski, and B. Scholz-Reiter, *Dynamics in Logistics, Proceedings of the First International Conference on Dynamics in Logistics (LDIC 2007)* (pp. 331–339). Springer.

Hölscher, K., Kreowski, H.-J., and Kuske, S. (2009). Autonomous Units to Model Interacting Sequential and Parallel Processes. Fundamenta Informaticae, 92(3):233–257.

Kreowski, H.-J., and Kuske, S. (2008). Communities of autonomous units for pickup and delivery vehicle routing. In A. Schürr, M. Nagl, and A. Zündorf (eds), *Proc. 3rd Intl. Workshop on Applications of Graph Transformation with Industrial Relevance (AGTIVE '07), Volume 5088 of Lecture Notes in Computer Science*, pp. 281–296.

Montemanni, R., Gambardella, L., Rizzoli, A., and Donati, A. (2005). Ant colony system for a dynamic vehicle routing problem. *Journal of Combinatorial Optimization* 10(4):327–343.

Reimann, M., Doerner, K., and Hartl, R. F. (2004). D-Ants: Savings based ants divide and conquer the vehicle routing problem. *Computers and Operations Research* 31(4):563–591.

Rizzoli, A., Montemanni, R., Lucibello, E., and Gambardella, L. (2007). Ant colony optimization for real-world vehicle routing problems. *Swarm Intelligence* 1(2):135–151.

Rozenberg, G. (1997). Handbook of Graph Grammars and Computing by Graph Transformation, Vol. 1: Foundations. World Scientific, Singapore.

Part IV
Radio Frequency Identification

Dynamic Management of Adaptor Threads for Supporting Scalable RFID Middleware

Chungkyu Park, Junho Lee, Wooseok Ryu, Bonghee Hong and Heung Seok Chae

1 Introduction

RFID (Radio Frequency Identification) technology has been regarded as a core technology of ubiquitous computing by providing a management mechanism using network (Floerkemeier and Lampe 2004). RFID technology has many application areas such as retail, healthcare, logistics, automotive, food industry, etc. In addition, many efforts have been actively performed to standardize various components of RFID technology.

An RFID system consists of RFID tags, RFID readers and RFID middleware. RFID tags are attached on each product in RFID environment. RFID middleware is software designed to process the streams of tag coming from one or more reader devices. RFID middleware performs filtering, aggregation, and counting of tag data, reducing the volume of data prior to sending to RFID Applications. In addition, because a lot of readers are usually installed in different locations, RFID middleware should handle extremely large quantities of data from many readers. Therefore, RFID middleware should provide scalability for readers in order to maintain its performance with the increasing number of readers.

C. Park (✉), J. Lee, W. Ryu, B. Hong and H. S. Chae
Department of Computer Engineering, Pusan National University, Pusan, Korea (REP.)
e-mail: allan@pusan.ac.kr

J. Lee
e-mail: junos52@pusan.ac.kr

W. Ryu
e-mail: wsryu@pusan.ac.kr

B. Hong
e-mail: bhhong@pusan.ac.kr

H. S. Chae
e-mail: hschae@pusan.ac.kr

H.-J. Kreowski et al. (eds.), *Dynamics in Logistics*,
DOI: 10.1007/978-3-642-11996-5_27, © Springer-Verlag Berlin Heidelberg 2011

To this goal, we define Reader Framework, which consists of adaptors and an adaptor manager. Each adaptor communicates with a physical reader and the adaptor manager manages one or more adaptors. The adaptor manager can use one or more threads to support many adaptors. To server data from each reader, the adaptor manager can utilize one thread for each reader. However, as the number of readers is increasing, the overhead for managing multiple threads is also increasing. Such overhead can adversely affect the performance of the middleware. Therefore, we need to maintain the number of threads at an appropriate level.

This paper proposes a technique to dynamic thread management for providing reader scalability. When dynamic thread management detects the occurrence of delay in processing tags, it adjusts the number of threads to maintain the performance. To dynamically manage the number of threads, we propose two techniques: group merging and group splitting. Group merging is performed to decrease the number of threads by merging two adaptor groups into one group. In contrast, group splitting is applied to create new adaptor groups by splitting adaptors. Theses dynamic thread management can improve the performance of RFID middleware by utilizing an optimal number of threads for serving multiple readers.

In addition, through an experiment we compare the dynamic thread management with conventional techniques: multi-threads method and single-thread method. According to the experiment, our dynamic management technique shows more scalability than those two conventional ones.

The rest of this paper is organized as follows. Section 2 introduces related work and Sect. 3 describes problem definition and target environment. The dynamic thread management is described in Sect. 4. Section 5 presents the result of the experiment. Section 6 describes conclusion.

2 Related Work

RFID middleware system (Nova Ahmed et al. 2007) is related to research of reader scalability in RFID middleware. Some of adjacent physical readers are mapped into one of virtual reader. Virtual readers can communicate each other. RFID middleware system uses virtual path to notify predictable tag data to another readers. Virtual path is set a same path of tag movement. When a new virtual reader is added, it is configured by load of virtual path. However, RFID middleware system has limitations in reader scalability. First, when new virtual reader is added, RF^2ID middleware system doesn't try to decrease a load and tries to avoid load. Second, virtual reader considers that data collection and process using distributed nodes are separated. Because we can not apply to distributed environment, it is not worthy to be considered reader scalability of RFID middleware.

Sun Java System Message Queue (Sun Microsystems 2007) of Sun Micro Systems supports two types of operations by registration information. It manages connections of client using by multi-threads (Table 1).

Table 1 Two types of thread management

Model	Description
Dedicated model	Each connection uses two threads for sending and receiving
Shared model	In case of sending and receiving messages, connection is processed in shared threads. This model can increase the available number of connections, but a decrease of performance occurs by additional overhead for managing threads

System manager can configure a number of limiting the number of threads in Sun Java System Message Queue. And if overhead of threads is exceeded, all of connections are denied after that time.

3 Problem Definition

Reader scalability is considered an important quality attribute of RFID middleware. Reader scalability is an essential requirement in RFID middleware architecture (Sharyn Leaver et al. 2004; Vlad Krotov 2006; Floerkemeier and Lampe 2005) and is used as an important criterion of performance evaluation (Sharyn Leaver et al. 2004; Gi Oug Oh et al. 2006; Zongwei Luo et al. 2006).

RFID middleware should provide aggregation and filtering of tag data from readers. To perform such functions, Reader Framework of RFID middleware consists of two components: adaptor and adaptor manager. Table 2 briefly describes these two components.

Adaptor manager receives commands and tag events from middleware engine and lots of adaptors in Fig. 1. So, RFID middleware stores the received messages in a queue and retrieves them from the queue which is dealing with buffer in Fig. 2. In case tags stay in available reader range for a long time, tag smoothing (EPCglobal Inc 2008a, 2008b) is proposed for avoiding duplications. Smoothing removes the duplicated tag event and decrease processing load of RFID middleware.

Adaptors and the adaptor manager use threads to efficiently process received tag data from readers. We can define a single-thread model and a multi-thread model for adaptor management.

- Multi-thread model use a separate thread for each adaptor. Because a thread is dedicated to one adaptor, each adaptor can quickly perform smoothing

Table 2 Components of reader framework

Component	Description
Adaptor	A component which simplifies reader interfaces and protocols for various vendors
Adaptor manager	A component which manages multiple adaptors to server multiple readers

Fig. 1 RFID middleware

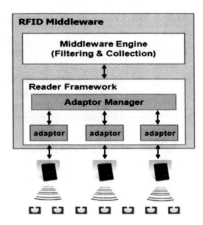

Fig. 2 Adaptor manager

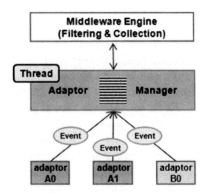

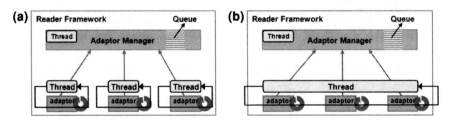

Fig. 3 a Multi-thread model. **b** Single-thread model

(see Fig. 3a). However, in case the number of readers increases, performance of RFID middleware can be degraded by excessive number of threads.

- Single thread model: On the other hand, in single-thread model, only one thread handles all adaptors (see Fig. 3b). Because one thread is used for multiple adaptors, it need to serve each adaptor in some order such as round-robin.

4 Dynamic Thread Management

We propose a dynamic thread management to solve these problems with the increase of the number of readers. This technique adjusts the number of adaptors that are managed by each thread. The dynamic thread management consists of two steps:

- Determine triggering condition of dynamic thread management. This step is for determine when to execute dynamic thread management. The triggering condition is based on two types of delays in processing tag data.
- Execute dynamic thread management. Depending on the type of delays, two types of thread managements can be applied: group merging and group splitting.

4.1 Triggering Condition of Dynamic Thread Management

We consider two delays which lead to a decrease of performance. One is queue processing delay due to excessive number of threads. The other is smoothing delay due to excessive number of adaptors managed by a thread. Load monitor detects delay of queue processing in the adaptor manager and smoothing delay in adaptor using following mechanism. As seen in Fig. 4, the load monitor monitors the operations of the adaptor manager and the adaptors and determines whether two types of delays take place.

4.1.1 Queue Processing Delay

Delay in queue processing is detected by monitoring queue utilization in the adaptor manager. In other words, queue processing delay is considered to occur when the queue has fewer spaces than a particular threshold.

4.1.2 Smoothing Delay

Smoothing is a process of filtering duplicated tags. When each reader cycle is finished and new tags are received, if a relevant adaptor is smoothing received tag

Fig. 4 Load monitor

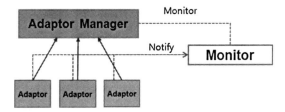

data in former reader cycle or is not smoothing, load monitor determines that smoothing delay take places.

4.2 Technique of Dynamic Thread Management

This section presents two dynamic thread management methods: Group Merging and a Group Splitting.

4.2.1 Group Merging

Group merging can decrease the number of threads by merging two groups into one group. Group merging starts, if queue processing delay is detected by load monitor in the adapter manager. In addition, addition of new reader also triggers group merging. By decreasing the number of threads, the adaptor manager can be allocated more time to process tag data.

For group merging, the adaptor manager chooses target groups with a minimum load. Load of adaptor is evaluated based on relation between reader cycles and the number of received tag data. Smoothing processing has the complexity of $O(n^2)$. So load is calculated that a power of the number of received data divided by reader cycle. And a load of a group is calculated by average of all adaptors in a same group.

For example, we can suppose that adaptors and threads exist in Fig. 5. If we assume that readers have a same reader cycle (100 ms), load of each adaptor is proportional to a power of volume of tag data. Therefore group merging is executed between an adaptor 1 and an adaptor 3.

A group with have minimum load is chosen because the number of tags each group processes cannot be predicted in advance. So we consider the current load to reduce possibility of additional load resulting from the merging.

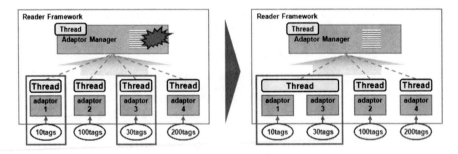

Fig. 5 Example of group merging

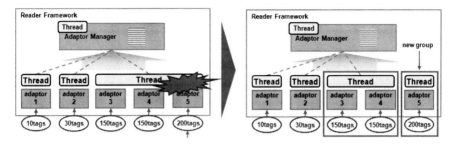

Fig. 6 Example of group splitting

4.2.2 Group Splitting

Group splitting is executed, when smoothing delay occurs. That is, it starts when too many adaptors exist in a group or huge data is received from relevant adaptors. In case smoothing delay is detected in one of groups, the adaptor manager determines an over-loaded group with too many adaptors. And the adaptor manager creates a new thread to create a new group for a target adaptor.

The adaptor manager chooses an adaptor to be moved to new group. It should calculate load to select adaptors of a delayed group, because a smoothing delay of adaptor does not mean the maximum load. Load calculation uses the same method in group merging.

Figure 6 shows an example and we assume that reader cycles are same (100 ms). If smoothing delay occurs in the third group, then the adaptor 5 is selected as an adaptor with a maximum load. The adaptor 5 is allocated to a newly created group.

5 Performance Evaluation

This section describes the result of performance evaluation for evaluating reader scalability of RFID middleware. An experiment compares the proposed dynamic thread management technique with two conventional ones: multi-thread model and single-thread model.

Criterion of evaluation is processing volume of dynamic thread management with measuring changed the number of threads and throughput of the adaptor manager queue in each model.

5.1 Experiment Environment

The proposed dynamic thread management technique has been implemented using Sun Java 2 standard edition development kit 1.6. An experiment has been performed on Windows XP and PC with 2 GB main memory and CPU Intel Dual core

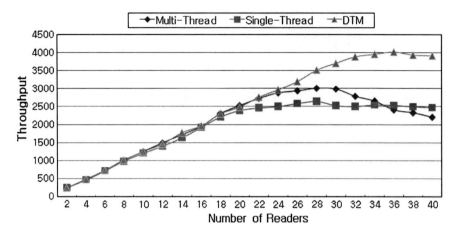

Fig. 7 Measurement of processing volume

1.86 GHz. Because of the non-availability of many physical readers, virtual readers are used for simulating them. The virtual readers run on different systems to send tag data to an RFID middleware. The number of readers and tags are controlled. We used forty virtual readers and arbitrary number of tags between 10 and 30 were set for each virtual reader for generating different loads. Reader cycle was set 200 ms.

5.2 Experiment Result

Throughput of adaptor manager is measured using three techniques: multi-thread, single-thread and dynamic thread management. Figure 5 shows the throughput with the change in the number of readers. Throughput is defined to be the number of processed tag that reach the adaptor manager queue.

As seen in Fig. 7, when the number of readers is over twenty, three techniques begin to show difference in their throughput. We can see that the dynamic thread management shows the best performance among them. According to the multi-thread model, the number of threads has increased as the number of readers increases. When about 30 readers are connected, performance degradation was observed. With the proposed dynamic thread management, we can provide better performance by adjusting the number of threads.

6 Conclusion

For large scale applications, RFID middleware need to provide reader scalability. Existing reader management techniques are using many threads and system resources. So RFID middleware require more effective reader management techniques. In this paper, we proposed dynamic thread management to efficiently manage multiple readers by adjusting the number of threads. Dynamic thread management includes group merging to decrease the number of threads and group splitting to increase the number of thread. Through an experiment, we compared the proposed method with two conventional methods and found that our method can have better scalability than them.

The main contribution of our work is that these techniques can reduce the load of reader management in runtime, so the RFID Middleware using these techniques becomes capable of connecting more readers without decrease of the performance. Future research includes techniques of enhanced reader scalability among one or more RFID middleware.

Acknowledgments This work was supported by the Grant of the Korean Ministry of Education, Science and Technology (The Regional Core Research Program/Institute of Logistics Information Technology).

References

EPCglobal Inc (2008a) Application Level Events Specification Version 1.1. http://www.epcglobalinc.org/standars/ale/ale_1_1-standard-core-20080227.pdf Accessed 7 March 2008

EPCglobal Inc (2008b) Reader Protocol Standard Version 1.1. http://www.epcglobalinc.org/standars/rp/rp_1_1-standard-core-20060621.pdf Accessed 24 May 2008

C Floerkemeier, M Lampe (2004) Issues with RFID Usage in Ubiquitous Computing Applications. PERVASIVE 2004 LNCS 3001:188-193

Christian Floerkemeier, Matthias Lampe (2005) RFID Middleware design: addressing application requirements and RFID constraints. ACM International Conference Proceeding Series vol 121:219-224.

Gi Oug Oh, Doo Yeon Kim, Sang Il Kim, Sung Yul Rhew (2006) A Quality Evaluation Technique of RFID Middleware in Ubiquitous Computing. ICHIT '06 International Conference vol 2:730-735, 9-11

IBM (2001) Design for Scalability. http://www.ibm.com/developerworks/websphere/library/techarticles/hipods/scalability.html. Accessed 16 October 2008

Nova Ahmed, Rajnish Kumar, Robert Steven French, Umakishore Ramachandran (2007) RF^{2}ID: A Reliable Middleware Framework for RFID Deployment. IEEE IPDPS:1~10 IEEE Computer Society Press

Sharyn Leaver, Tamara Mendelson, Christine Spivey Overby, and Esther H.Yuen (2004) Evaluating RFID Middleware. Forrester Research Inc

Sun Microsystems (2007) Sun Java System Message Queue 3.7 UR1 Administration Guide. Sun Microsystems Inc

Vlad Krotov (2006) RFID Middleware http://www.bauer.uh.edu/rfid/RFID%20Middleware.ppt. Accessed 5 September 2008

Zongwei Luo, Ed Wong, S.C. Cheung, Lionel M. Ni, W.K. Chan (2006) RFID Middleware Benchmarking. the 3rd RFID Academic Convocation in conjunction with the China International RFID Technology Development Conference & Exposition.

Tag-to-Tag Mesh Network Using Dual-Radio RFID System for Port Logistics

Jinhwan Kim, Hyocheol Jeong, Myungjae Kim, Haosong Gou, Munseok Choi and Younghwan Yoo

1 Introduction

In recent days, the remarkable increase in production and consumption extensively requests the adoption of the Radio Frequency IDentification (RFID) system in various fields. The RFID is the technology to identify and track human, animal, and goods via wireless communication, and it is currently used for logistics, traffic management, inventory management, and health care system.

The representative RFID usage in ports is the location and management of containers and equipments. Attaching RFID/RTLS (Real-Time Location System) tags to containers, they can either locate specific containers or collect the information related to goods through wireless communication. However, RFID readers in most current applications may not cover the entire area due to the limitation of places where readers can be deployed and its high cost. Even if there are enough readers to cover the entire area, sometimes the radio signal may not reach tags unexpectedly due to the interference by containers and large equipments such as

J. Kim (✉), H. Jeong, M. Kim, H. Gou, M. Choi and Y. Yoo
School of Computer Science and Engineering, Pusan National University, San-30,
Jangjeon-dong, Geumjeong-gu, Busan 609-735, Republic of Korea
e-mail: compunix@pusan.ac.kr

H. Jeong
e-mail: ketalong@gmail.com

M. Kim
e-mail: mjhero@naver.com

H. Gou
e-mail: gouhaosong@pusan.ac.kr

M. Choi
e-mail: bakas81@naver.com

Y. Yoo
e-mail: ymomo@pusan.ac.kr

H.-J. Kreowski et al. (eds.), *Dynamics in Logistics*,
DOI: 10.1007/978-3-642-11996-5_28, © Springer-Verlag Berlin Heidelberg 2011

crane and yard tractors. Containers are densely piled in the yard up to five layers and are made of metal. Besides, equipments in ports are mostly made of metal too, which gives bad influence on the propagation of radio frequency. The area where the RF of readers cannot reach is called *dead-zone*. When a tag enters a dead-zone, it cannot directly communicate with readers.

Although all applications do not need the 100% of tag ID collection, the more complete information is more helpful for the load and the shipping of containers in ports. Therefore, we propose the tag-to-tag mesh network (T2T-MN) to resolve the dead-zone problem, where tags in a dead-zone can communicate with the reader in a multi-hop manner through neighbor tags. However, the multi-hop communication naturally increases the number of transmission of tags, resulting in the high collision probability. This is why dual-radio RFID system is utilized in this research. The dual-radio system operates in two frequency bands: 433 MHz for reader-to-tag communication and 2.4 GHz for tag-to-tag communication. The separation of the two types of communications into different bands reduces the collision probability.

This paper is organized as follows. Sections 2 and 3 introduce various solutions to the dead-zone problem and the signal interference, respectively. Sections 4 and 5 suggest how to support the reader-to-tag multi-hop communication and the tag-to-tag mesh networking. In Sects. 6 and 7, the communication using the dual-radio and how to implement it are explained. Section 8 concludes this paper with future works.

2 Possible Solution for *Dead-Zone*

In this paper, we assume the tag ID is collected under the situation that containers are already deployed. For the dead-zone problem, there are three possible solutions: deployment of additional static readers, development of mobile readers, and multi-hop tag-to-tag communication using the tags with our tag mesh networking engine.

2.1 Deployment of Additional Readers

The easiest way is to deploy additional static readers near the dead-zone. But, some problems exist here: (1) Static readers cannot be installed freely anywhere in ports. The readers can be set up only in some limited places like light towers, because much space is already reserved for lots of containers and equipments, and the route for moving vehicles such as tractors and forklifts is also needed. (2) Still, a dead-zone can be made instantaneously and irregularly due to the signal interference by metallic equipments and densely piled containers. Thus it is impossible to exactly guess the appearance and the location of the dead-zone.

2.2 Development of Mobile Readers

The second possible solution is to develop a mobile RFID reader. Attached to working vehicles like yard tractor and forklift, mobile readers can approach dead-zones and collect tag IDs. This method has the following problems: (1) Some places still cannot be approached because the main purpose of moving vehicles is not the collection of tag information. The route of the vehicles is not optimized for the tag ID collection. Thus the RF signal may not reach the places farthest from the route of moving vehicles and the tags on the containers in the highest layer of a pile. (2) Reader mobility causes the frequent changes of the network topology, resulting in message overhead and network instability. (3) The price of a mobile reader, which is expected as two or three times as expensive as a static reader is another problem.

2.3 Multi-Hop Tag-to-Tag Mesh Network

The final and proposed method in this paper is to allow multi-hop communication between RFID reader and tags. When they are within the RF range from each other, they talk to each other directly. Otherwise, if a tag is in a dead-zone and cannot receive any signal from the reader directly, it communicates with the reader via neighbor tags in the multi-hop manner. This is the most economical solution because neither additional static readers nor expensive mobile readers are needed. Generally a static reader costs 200 times more than a tag, and mobile reader is 2–3 times as expensive as a static reader. However, one problem should be resolved first. The 433 MHz RFID active tags (e.g., E-Seal tag) are currently used for port logistics and they follow the standard ISO/IEC (2004; 18000-7) for the air inter-face. Then this standard specifies only the direct communication between reader and tags, thus a new architecture must be designed for multi-hop communication. The proposed architecture is shown in Fig. 1, which will be described in detail later.

3 Solution to the Radio Frequency Interference

In the existing RFID network, readers communicate with tags directly using the slotted ALOHA protocol. As mentioned earlier, however, RFID tags in ports might not receive reader commands sometimes due to the heavy interference by containers or equipments. Thus the proposed architecture enables tags in a dead-zone to send their packets through neighbor tags in the multi-hop manner. However, this multi-hop communication increases the total number of transmission in a network, resulting in lots of collisions caused by the radio frequency interference.

The collisions can be reduced through two ways: multi-channel or multi-radio.

Fig. 1 Tag-to-tag communication in a dead-zone

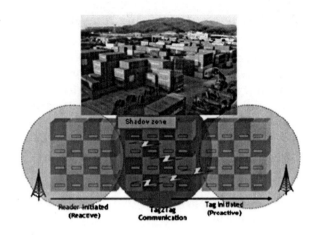

3.1 RFID Radio Frequency

Table 1 summarizes the radio frequencies adopted for current RFID systems. Particularly, 433.92 MHz is used for the active tag, and 860–960 MHz is available for both active and passive tags. The higher the range is, the more sensitive to its environment it is; the faster the read speed is, though (Finkenzeller 2003).

3.2 Use of Multi-Channel

The use of multiple channels may reduce the number of collisions intuitively. However, 433 MHz RFID systems currently used in ports does not provide multiple channels.

3.3 Use of Multi-Radio

Multiple radio bands can be used to improve the collision probability. In this paper, dual radios are adopted for different parts of the network: tag-to-reader and tag-to-tag.

Table 1 Frequency bands for RFID systems

Classification band	Frequency range	Read distance
Low freq. (LF)	125–134 kHz	~60 cm
High freq. (HF)	13.56 MHz	60 cm
Ultra high freq. (UHF)	433.92 MHz	50–100 m
	860–960 MHz	3.5–10 m (passive)
Microwave	2.45 GHz	30 m

4 Reader-Tag Communication

4.1 Standard for Reader-Tag Communication

ISO/IEC (2004; 18000-7) is the standard for the air interface of the 433.92 MHz active tag, defining the features of PHY and MAC like modulation schemes and packet formats.

4.2 Reader Command

The reader-tag communication in the standard is always initiated by a reader in the master–slave manner. A reader sends first one of two types of commands, peer-to-peer and broadcast commands, and tags reply to them. The reader must send a Wake-up signal to wake up all tags before transmitting any command to tags, since they are mostly in the sleep mode to save their energy. The Wake up signal is a 30 kHz sub-carrier tone for 2.5–2.7 s. Figure 2 depicts the processes for a P2P command and a broadcast command.

5 Tag-to-Tag Mesh Network

Several container ports in the United States and Europe have adopted the 433 MHz active RFID tags to locate or load/unload containers efficiently. The read distance of the 433 MHz active tag is known as 50–100 m, but the actual distance is shorter than this due to the signal interference by a lot of containers and equipments in the yard. Therefore, the existing readers deployed with the expected read distance of 50–100 m, may cause the dead-zone problem.

This paper proposes the tag-to-tag mesh network (T2T-MN) as the most economical solution to the dead-zone problem. In the proposed mesh network, the tags

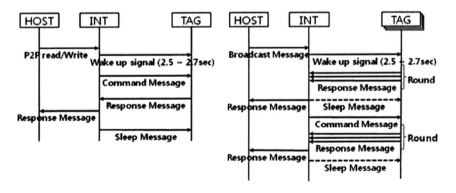

Fig. 2 Processes for peer-to-peer command and broadcast command

located in dead-zones can send readers their packets through neighbor tags in the multi-hop manner. Our T2T-MN software engine has the following features.

5.1 IEEE 802.15.4 and ZigBee

The 2.4 GHz ISM band is utilized for the tag-to-tag communication in the proposed network. The PHY and MAC layer in our network is based on IEEE 802.15.4 (2006): 16 channels are provided in 2.4–2.4335 GHz band and the transmission rate is 250 kbps. The O-QPSK (Offset Quadrature Phase-Shift Keying) modulation technique is used. Meanwhile, the upper layer follows the ZigBee standard (Ergen and ZigBee Alliance 2004). The ZigBee can provide a very economical and flexible wireless network solution. It is cheaper than the Bluetooth, consumes less energy, and supports the mesh network. Figure 3 illustrates the relation between the 802.15.4 and ZigBee protocol stacks.

5.2 Routing

A routing protocol is needed to find out paths between two tags in a mesh network. The AODV (Ad hoc On-demand Distance Vector) protocol is used in this research.

6 Dual-Radio RFID System

6.1 Dual-Radio Tag Hardware

Figure 4 shows the dual-radio tag architecture. It has two Micro Controller Units (MCUs), AT91SAM7S256 (Atmel 2005) and ATmega128L (Atmel 2005), and

Fig. 3 IEEE 802.15.4 and ZigBee stacks

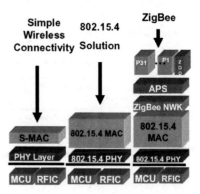

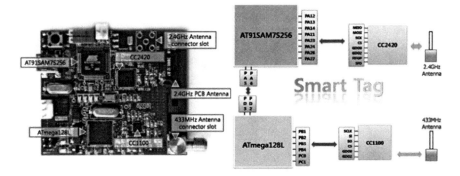

Fig. 4 Dual-radio tag architecture

two RF chips, CC2420 (Texas Instruments 2006) and CC1100 (Texas Instruments 2007), for dual-band frequencies, 2.4 GHz and 433 MHz. The right figure depicts the connection between components. Data IO is performed through parallel IO (PIO) ports.

6.2 Dataflow in the System

Figure 5 illustrates the dataflow in the system.

- *Dataflow #1*: Reader → CC1100 → ATmega128L → AT91SAM7S256 → CC2420 → Other tags
- *Dataflow #2*: Another tag → CC2420 → AT91SAM7S256 → ATmega128L → CC1100 → Reader
- *Dataflow #3*: Another tag → CC2420 → AT91SAM7S256 → CC2420 → The same tag or others
- *Dataflow #4*: Reader → CC1100 → ATmega128L → CC1100 → The same reader

6.2.1 Reader-Tag Communication Module

The above Dataflow #4 is the reader-to-tag communication using the ATmega128L MCU and the CC1100 RF chip. The reader command received by CC1100 is delivered to ATmega128L. If the command can be handled by ATmega128L by itself, the response message is made by ATmega128L and transmitted to the reader via CC1100. Otherwise, the response message is made by AT91SAM7S256. ATmega128L gets the response from the shared memory in AT91SAM7S256 and sends it to the reader.

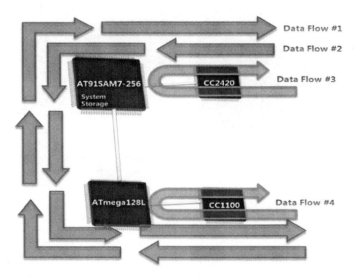

Fig. 5 Dataflow in the dual-radio tag

6.2.2 Tag-Tag Communication Module

The above Dataflow #3 is the tag-to-tag communication using the AT91SAM7S256 MCU and the CC2420 RF chip. A packet from another tag is received by CC2420. Being processed by AT91SAM7S256, the packet or the response message is transmitted to other tags or the tag that sent the packet before.

6.3 Embedded OS Porting

The μC/OS-II was adopted as the operating system of our dual-radio tag.

6.3.1 AT91SAM7S256

AT91SAM7S256 is the main MCU of the dual-radio tag. It has tasks to control CC2420 and to communicate with ATmega128L. It also manages the shared memory and supports the UART serial port.

6.3.2 ATmega128L

ATmega128L is the sub-MCU of the tag, managing the reader-tag communication. It has tasks to control CC1100 and to access the shared memory in AT91SAM7S256. It can also utilize the UART port through AT91SAM7S256.

6.4 Management System

Figure 6 shows the components of the management system. This system can send the *Collection* command to collect tag information through the reader connected to the UART port. The received information is stored into the database.

- *Controller*. It delivers messages from the manager to the reader via UART. The information collected by the reader is stored into the database.
- *Manager*. It sends commands to the controller and provides users with the information stored in the database.

7 Experiment

The experiment was performed using five tags to evaluate the performance of the proposed mesh network and the dual-radio tags. Dead-zone was made in our experimental network by adjusting transmission range of reader and tags. Then the ratio of the number of ID collected tags to the number of total tags in the network. This experiment substantiated that the multi-hop reader-to-tag and tag-to-tag communication work correctly.

Figure 7 is the actual develop environment of the proposed RFID system. The lower figure shows the deployment of tags and the upper figure shows that string messages arrived at the reader in real. This message is handed over to the management system.

Figure 8 illustrates that the management system displays tag information collected by a reader. The reader is connected to the UART port on the PC. In the figure, only Tag 0x00F3 is within the RF range of the reader. Tag 0x00F3 and the reader communicate directly in the 433 MHz band. On the other hand, others must

Fig. 6 Management system architecture

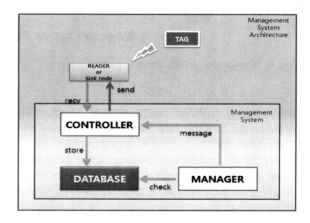

Fig. 7 Experiment environment

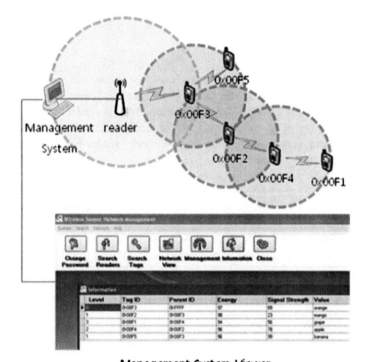

Management System Viewer

Fig. 8 Management system viewer

communicate with the reader through neighbor tags in the multi-hop manner, utilizing the 2.4 GHz ISM band. The experiment shows that all tag information is received correctly.

As for the collision probability, many collisions did not occur in the experiment. It looks like the network size is too small for any meaningful result to be acquired on the collision. Currently, we have only five tags. The experiment will be performed again using more tags in the future.

8 Conclusion

This paper suggests the tag mesh network to resolve the dead-zone problem in the RFID networks. Tags located in the dead-zone can communicate with readers or other tags through neighbors in the multi-hop manner, while they have to talk to each other directly in the previous networks. The multi-hop communication, however, increases the total number of transmission in the network, resulting in many collisions. In order to improve the collision probability, the overall network function was separated into the two parts, reader-tag communication and tag-tag communication, and a different frequency band was assigned to each of them using the dual-radio RFID system.

The proposed T2T-MN is an economical approach to the dead-zone problem, compared to the deployment of additional readers. T2T-MN is compatible to existing systems without any changes. Moreover, the T2T-MN can be easily set up where no network infrastructure is established, playing the role of the main network temporarily. Also, this solution can be applied to other wireless networks such as Ubiquitous Sensor Network (USN) and HomeRF where any two nodes cannot communicate directly.

The multi-hop communication not only increases the number of collisions but also makes tags consume more energy, resulting in the short lifespan of the tags. Thus various low-power techniques in routing and topology control should be supplemented in the future. Additionally, some security mechanisms are also needed to develop a secure RFID network.

The proposed solution needs a little much cost since the tags have two MCUs and two RF chips. These tags are still too expensive to be deployed at all containers and equipments in the port right now. However, the technological advancement will make the price of the modules cheaper and cheaper, and then the proposed solution will practically be able to be applied to the port and other network fields.

Acknowledgments This work was supported by the Grant of the Korean Ministry of Education, Science and Technology (The Regional Core Research Program/Institute of Logistics Information Technology).

References

Atmel, AT91SAM7S256 Datasheet (2005)
Atmel, ATmega128(L) Datasheet (2005)
Ergen, S.C.: ZigBee/IEEE 802.15.4 Summary. ZigBee Alliance (2004)
Finkenzeller, K.: RFID Handbook: Fundamentals and Applications in Contactless Smart Cards and Identification (2003)
IEEE: Part 15.4: Wireless Medium Access Control (MAC) and Physical (PHY) Layer Specifications for Low-Rate Wireless Personal Area Networks (WPANs) (2006)

ISO/IEC: 18000-7 International Standard—Parameters for Active Air Interface Communications
    at 433 MHz (2004)
Texas Instruments, CC2420 Datasheet (2006)
Texas Instruments, CC1100 Datasheet (2007)

Automation of Logistic Processes by Means of Locating and Analysing RFID-Transponder Data

Bernd Scholz-Reiter, Wolfgang Echelmeyer, Harry Halfar and Anne Schweizer

1 Introduction

The turnover of packaged goods is increasing constantly. Goods are generally transported in standardised packaging, e.g., cardboard boxes stacked on pallets inside standardised carriers. Typical standardised carriers are containers for overseas transportation, Unit Load Devices (ULD) for air transportation and swap trailers for national transfer. Today, nearly all unloading procedures are carried out manually. Especially in high-wage countries this is linked with equivalent costs. It is also a high physical strain for the employees, which leads to corresponding health risks.

In future, due to the expected demographic development, the costs for such labour will increase and the availability of capable manual labour will decrease. Therefore, the development of automatic solutions is a desirable target in order to improve the reliability of delivery chains and meet the increasing cost pressure.

The automatic unloading of carriers and the automatic transfer of packaged goods in logistic systems is a great technical challenge. Unlike other conventional automation systems the place of deposition, size and form of the objects is unknown beforehand. Therefore, the system has to be able to gather corresponding

B. Scholz-Reiter (✉), W. Echelmeyer, H. Halfar and A. Schweizer
BIBA-Bremer Institut für Produktion und Logistik GmbH, University of Bremen, Bremen Germany
e-mail: bsr@biba.uni-bremen.de

W. Echelmeyer
e-mail: ech@biba.uni-bremen.de

H. Halfar
e-mail: hal@biba.uni-bremen.de

A. Schweizer
e-mail: vir@biba.uni-bremen.de

H.-J. Kreowski et al. (eds), *Dynamics in Logistics,*
DOI: 10.1007/978-3-642-11996-5_29, © Springer-Verlag Berlin Heidelberg 2011

information regarding its surroundings, analyse this information and make autonomous decisions for the next action.

A first step for the automatic unloading of cubic packaged goods of a defined size spectrum has been achieved with the ParcelRobot (Scholz-Reiter et al. 2008) developed at BIBA. This robot can unload cubic packaged goods of a defined size spectrum from a container or swap trailer and place them on a conveyor belt. The necessary information regarding the surroundings is gathered by means of a laser scanner and followed by an image processing procedure. At this point, the system recognises the possible risks of collision and the deposition location of the object.

For many tasks robot systems are commonly supported by sensors. Automatic tasks that handle undefined packaged goods (undefined size, weight, position and location) require 3-dimensional data in order to calculate the gripping coordinates. These are generally acquired via Laser Measurement Systems (LMS). These LMS have to meet especially high demands concerning reach, measuring radius and measuring speed of tasks in logistical systems. The mentioned demands are accomplished with a low angular resolution (1°, corresponds to approx. 5 cm in a distance of 3 m) (Sick AG 2009). This can lead to stacking situations of packaged goods that are predicted incorrectly. Furthermore, the measurement systems do not provide information regarding size, weight, content, profound stacking situations (subsurface), etc. which are helpful to the reliable handling procedure. Therefore, the state of technology specifies systems that have a relatively high positioning accuracy (compared to robot systems in industrial production) but also an ignorance or insecurity regarding the individual composition and surroundings of the packaged goods.

2 Problem

Particularly the example of the ParcelRobot developed at BIBA shows that the ignorance of its surroundings leads to process insecurity and a low process dynamic. Thus, determined parcel stacking situations cannot be recognised correctly by the image processing system. In a staircase shaped stack of boxes, e.g., the system tends to choose a box from beneath another box as its first gripping target (see Fig. 1). This could be avoided if the system would have information regarding the depth of the packaged goods.

These difficulties result in errors or interruptions of the robot function or, in the worst case, can cause damage to the packaged goods. This example shows that there is a significant need for improvement of the construction and conception of sensor systems. In logistics, the use of information from RFID-transponders is a good solution.

In future, robot supported sensor systems need to consist in a combination of optical sensors that interact with each other and in data containing technical information. The vision for the future is to work with optical sensor data containing information regarding type, composition, weight, volume, location and

Fig. 1 Problematic stacking situation of parcels

position of the parcel in a universal and robot based coordinate system, etc., which will be attached to the packaged goods. This would increase the security, diversity and dynamic of the process considerably.

There is no existing research approach for processing and analysing the data that is attached to robot handled logistic objects, e.g., on RFID-transponders. There is a demand to develop analysis methods, which are capable of real time analysis and give suitable results in order to be integrated into the control of a robot or superpose the control.

3 Methods and Approach

Figure 2 shows an overview of the systems complete concept. The target of such a research project is to develop a procedure which can recognise, identify and locate objects provided with RFID-transponders in an industrial environment. With the help of additional sensor technology (laser scanner, image recognition) it should provide sufficient location information in order to deliver a complete automation system. Additionally, information should be supplied, e.g., regarding the mechanical characteristics or geometric forms of the object. This information can also be used by the automation system.

For this purpose it is necessary to find the appropriate alignment of antenna and RFID-laser devices as well as the appropriate analyses procedure, which can find the correct location of the transponder. The critical factors when locating or positioning transmitters such as RFID-transponders are the interferences and reflections. In this context, known methods will be examined in order to develop a suitable procedure. An overview of known methods is shown in (Jing Shi and Junyi Zhou 2008).

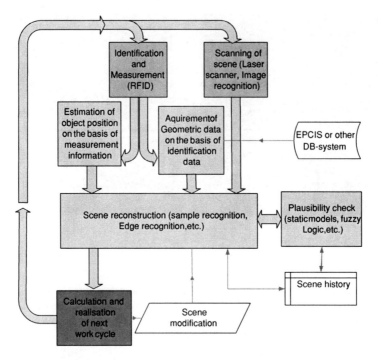

Fig. 2 Block scheme

Secondly, it is essential to develop a procedure, which acquires necessary information, e.g., the geometrical dimension of the object on the basis of identity information of the located objects. On the one hand this data could be saved additionally on the transponders. On the other hand it could also be possible to retrieve this information via the so called "internet of things" or via access to an own data bank (Thiesse and Fleisch 2008). Furthermore, the procedure for information acquisition contains a suitable code regulation for relevant data.

Thirdly, it is required to develop a procedure, which evaluates the gathered information in order to provide the subsequent control of the automation system with the necessary data. It is especially important to compare gathered location and geometric information with existing image and laser scanned data in order to give the automation system enough precise localisation information. Time is the main critical factor. Theoretically it is necessary to rerecord and evaluate the whole scene after every work cycle, which would lead to a significant restriction concerning the work velocity of the automation system. At this point it is imaginable to solely detect and evaluate the modifications of a scene after each work cycle.

A further task of the data evaluation consists in the development of a concept which handles uncertainties. Especially the localisation information can be afflicted with errors. In this case it is necessary to develop concepts that enable a plausibility check and create redundancies in the database in order to obtain reliable results.

The results of the complete research project will be gathered in a comprehensive scientific documentation of accumulated knowledge. The fundamental results of this research work will be published in suitable scientific conferences and trade magazines. On the basis of this study, recommendations for a possible serial production shall be specified and also recommendations for further research work, if applicable.

4 Summary and Prospects

The present article describes the research approach of automation of logistic processes with the help of localisation and evaluation of RFID transponder data. The existing problems were described concerning the reliability and accuracy when gathering surrounding information for the control of automation systems, particularly in logistics. A system concept was suggested, which could possibly solve these problems. Research demand was determined on the basis of this system model, which is necessary to realise the present concept. This research demand mainly consists in finding, testing and evaluating a suitable localisation procedure. There is also an existing requirement to find suitable methods of evaluation and analysis of gathered information, especially regarding the expected uncertainties and errors concerning the localisation procedure as well as the high demands set on the time interval of such a method.

The next target based on the present project is an interconnection of this system with an automation unit in order to create a complete demonstrator.

References

Jing Shi, Junyi Zhou. Rfid localization algorithms and applications -a review. *Journal of Intelligent Manufacturing*, 8 2008.

B. Scholz-Reiter, W. Echelmeyer, and E. Wellbrock. Development of a robot-based system for automated unloading of variable packages out of transport units and containers. *Automation and Logistics, 2008. ICAL 2008. IEEE International Conference on*, pages 2766–2770, Sept. 2008.

Sick AG, Nimbuger Strasse 1, 79276 Reute, Germany. *Technische Beschreibung, Lasermesssysteme LMS 200/211/221/291*, 2009

Thiesse, F., Fleisch, E. *On the value of location information to lot scheduling in complex manufacturing processes*. Int. J. Production Economics 112 (2008), pages 532–547.

Auto-Triggering of RFID-based Logistic Process in Inter-Workflow Using Complex Event

Hyerim Bae and Yeong-Woong Yu

1 Introduction

In rapidly revolutionizing business environments, collaboration with partners is considered to be an essential element of success (Liu et al. 2007; Rhee et al. 2007b), because the competitiveness of a company is derived from the entire scope of business activity that delivers products to end users. Such collaboration is especially required in logistic environments (Jung et al. 2008), since a logistic process inherently involves multiple participants. Collaborative success is achieved by means of systematic interfaces among partners, to the overall end of enhancing customer satisfaction (Gunasekaran and Ngai 2004; Liu et al. 2007; Rhee et al. 2007b).

For the efficient management of supply chains, supply chain management (SCM) has been introduced to plan, implement, and control collaboration among partners (Gunasekaran and Ngai 2004). However, execution issues such as the execution of inter-organizational processes have rarely been examined by SCM researchers. Their execution can be achieved by means of a systematic logistic process support.

Business process management (BPM) has been widely accepted as an effective and integrated way of managing and executing business processes (Basu and Kumar 2002; WfMC 1995). The BPM system is considered to be a general methodology for increasing a company's productivity through the systematic design, management, integration, and improvement of business processes (Basu and Kumar 2002; Rhee et al. 2007a, b). However, whereas logistic processes

H. Bae (✉) and Y.-W. Yu
Business and Service Computing Lab, Industrial Engineering, Pusan National
University, 30-san Jangjeon-dong Geumjeong-gu, Busan 609-735, South Korea
e-mail: hrbae@pusan.ac.kr

Y.-W. Yu
e-mail: hero@pusan.ac.kr

H.-J. Kreowski et al. (eds.), *Dynamics in Logistics*,
DOI: 10.1007/978-3-642-11996-5_30, © Springer-Verlag Berlin Heidelberg 2011

pursue inter-organizational optimization through the effective sharing of information, BPM, in its basic functionality, cannot be applied to the management of multi-organizational business processes.

In the present research, we discovered inter-workflow patterns to support logistic processes managed by multiple organizations. Accordingly, these patterns are converted into ECA rules, which enable a logistic process to be executed by triggering the action of another process without requiring any separate process engine.

The main objective of this research is to develop a systematic method for managing inter-workflow and to extend ECA rules for RFID environment. We also established the following sub-objectives, which are, at the same time, the three steps necessary for attaining the final goal (Bae et al. 2004).

- Employing inter-workflow patterns for multi-organizational processes in logistic environment: We employ inter-workflow patterns to represent the logistics of business processes among companies.
- Applying complex event with RFID event and business process event (BPE): We apply a complex event to represent the relations among the processes in an environment, where materials and products flow among different partners.
- Extending RFID-based ECA rules for auto-triggering of logistic process: We extend the rules for mapping between RFID logistic processes for auto-triggering in run-time environment.

To achieve these goals, we first discover inter-workflow patterns that occur over logistic processes, and then derive rules to execute the patterns. We represent the rules and provide RFID-based ECA rules to utilize complex event for auto-triggering in ubiquitous environment. We expect that our relation patterns can contribute to the systematic management of the relations between two or among three or more independent processes.

2 Background

2.1 Related Previous Work

The relevant previous research falls into one or another category: research on workflow patterns or on rule-based workflow execution. Certainly, a great amount of research has been conducted to model business processes using predefined patterns. Workflow Management Coalition (WfMC), an international standard organization on workflow, defines several types of workflow modeling semantics (WfMC 1995). van der Aalst et al. (2000) extends this specification defining 20 advanced workflow patterns. Workflow and BPM researchers have adopted these standard patterns, and thus a basis for workflow interoperability has been established.

Several research efforts have been undertaken to apply rule-based approaches to the execution of business processes (Bae et al. 2004; Casati et al. 1996; Chen et al. 2006; Liu and Kumar 2005; Lucia et al. 2003). Bae et al. (2004) proposed automatic business process execution achieved by replacing a workflow engine with an active database enabling ECA rules. In order to generate ECA rules for process execution, they converted the predefined-pattern process structure into block structures and generated ACTA formalism. Chen et al. (2006) devised a set of ECA rules as well as a method for their execution required for service composition.

Although the previous research, in sum, provides a sound foundation for our research, our method clearly is unique. Whereas the previous studies developed process patterns and ECA rules for a single process, our research focused on the relationships between multiple processes. Additionally we provide RFID-based ECA rule for inter-workflow. Thereby, we provide ECA rule-based execution of multi-organizational processes in logistic environments, where the business process management system (BPMS) is already installed.

2.2 Basic Process Model and ECA Rule

Prior to describing the relations among processes, we need to treat the concept of a basic process model. Our approach assumes that all partner companies participating in a logistic process have their process management system. Based on that assumption, we provide a simple definition of a process model that represents the process of a participatory company (Bae et al. 2004; Rhee et al. 2007a).

Definition 1: Process Model. A process structure is defined as a directed graph $P = (T, L)$ and a labeling function $f(\cdot)$ for split or merge types, such that

- $T = \{t_i \mid i = 1,\ldots, I\}$ is the set of tasks, where t_i is the i-th task, and I is the total number of tasks in P.
- $L \subseteq \{(t_i, t_j) \mid t_i, t_j \in T$ and $i \neq j\}$ is the set of links, where an element (t_i, t_j) indicates that t_i immediately precedes t_j.
- For a split task t_j, such that $|S| > 1$, where $S = \{t_k \mid (t_j, t_k) \in L\}, f(t_j) = $ 'AND' if all t_ks should be executed; otherwise, $f(t_j) = $ 'OR'.
- For a merge task t_j, such that $|M| > 1$, where $M = \{t_i \mid (t_i, t_j) \in L\}$, $f(t_j) = $ 'AND' if all t_is should be executed; otherwise, $f(t_j) = $ 'OR'.

Definition 2: Event–Condition–Action rule. A global logistic process also requires an execution mechanism for seamless interoperation between participants. In the present study, ECA rules were used. Even though previous research on the control of processes using ECA has been actively conducted, it has not yet been applied to the relations between processes. In general, ECA rules observe the syntax below (Goh et al. 2001; Tan and Goh 1999).

RULE (rule_name)
  ON (object).(event)
  IF conditionSet = { c_i | c_i is a condition }
  DO actionSet = { a_i | a_i is a condition }
ENDRULE

3 Inter-Workflow Patterns

In our research, we introduce inter-workflow patterns that have been discovered through our several years of research on logistic process modeling. In this chapter, as CSM is important pattern and used frequently in logistic environment, we use one this pattern to explain RFID-ECA rule for auto-triggering between RFID-based logistic processes. Other patterns are listed in van der Aalst et al. (2000).

3.1 Pattern 1: Chained Service Model

A chained service model (CSM) is the simplest case, which was originally introduced by WfMC's standard specification (WfMC 1995). In the CSM, once a process completes, another process is triggered and commences its execution. We classify the CSM into two sub-patterns according to its triggering object. If the succeeding process is triggered by the preceding process, we call it CSM-I. Otherwise, if the succeeding process is triggered by a task in the preceding process, we call it CSM-II. Figure 1a shows that P_B is triggered by P_A and initiates after P_A completes. On the other hands, in Fig. 1b, P_B is triggered by t_{A8}.

Generally, these two patterns are somewhat similar. However, there are some differences in the flow of state of process and tasks. When the process P_A in CSM-I initiates, its state can be changed from 'READY' to 'DOING', and first task can be started at this time. All tasks should be executed in 'DOING' state of process. After last task finishes in 'DOING' state of process, the process state can be changed from 'DOING' to 'FINISHED'. We can find the difference between CSM-I and CSM-II in Fig. 2.

3.2 Logistic Process Example with Inter-Workflow Pattern

Using the inter-workflow patterns, we can establish relations between processes. Let us consider the supply chain illustrated in Fig. 3. There are three organizations, each of which has its own process. In order to handle the manufacturers-to-customers logistics, all the partners need to interoperate systematically. We use

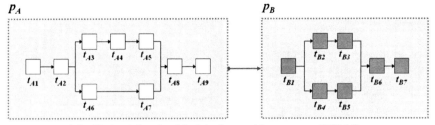

(a) Chained Service Model triggered by process (CSM-I)

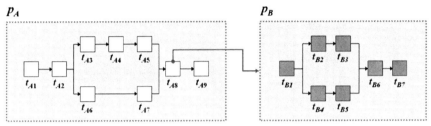

(b) Chained Service Model triggered by task (CSM-II)

Fig. 1 Chained service model (CSM)

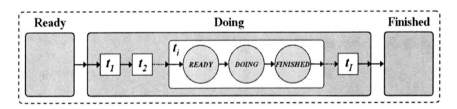

Fig. 2 Flow of state of process and task

inter-workflow patterns to represent the relations between processes and to auto-mate their interoperation among them.

As shown in the figure, five processes are interrelated by inter-workflow patterns. For example, the 'Manufacturing process (P_{A1})' and the 'Warehousing process (P_{C1})' are inter-related by the CSM-I pattern. According to the definition of the CSM-I pattern, when P_{A1} completes, P_{C1} is triggered and t_{B8} in P_{B1} trigger P_{C2} by CSM-II pattern.

4 Event Definitions

In this research, in order to develop an auto-triggering for inter-workflow patterns in logistic environments, we suggested two types of event; the RFID event (RFE) and BPE. We also defined a complex event (CPE), which is an event generated by

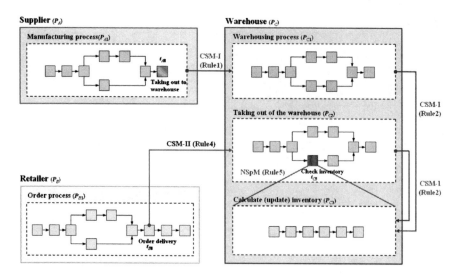

Fig. 3 Logistic processes example

combining the two event types. In the RFID-based logistic environment, where diverse RFID devices are used, process triggering between/among inter-workflow patterns should be based on not only BPE but also RFE. For this reason, we applied the concept of CPE to ECA rules.

4.1 RFID Event and Business Process Event

An RFID event is generated when RFID devices detect tag information, and a BPE is generated through process execution by workflow engine. These two events are defined as follows.

Definition 3: RFID Event

$$e_{RF} = <tagID, readerID, timestamp>$$

An RFID event (RFE) is a set of events that represent occurrences originated from RFID devices such as RFID tags and readers. An element of RFE is e_{RF}, which has three components: a tag identifier, a reader identifier, and a time stamp. According to the definition 3 above, an event $e_1(\in RFE) = <tag\_001,\ reader\_001,\ 2009.03.14.15:30:45>$ represents that a reader, *reader_001*, has read the data from a tag, *tag_001*, at time 2009.03.14.15:30:45.

Definition 4: Business Process Event

$$e_{BP} = <objectID,\ eventType,\ timestamp>$$

Table 1 Major events of business process

Object	Event		
	Event name	Corresponding function	Description (when to be generated)
Process	Create	*createInstance()*	When a process instance is created.
	Initiate	*initiate()*	When a process begins.
	Finish	*finish()*	When a process completes.
	Fail	*fail()*	When a process fails.
	Abort	*abort()*	When a process is aborted by a user.
Task	Assign	*assign()*	When a task is assigned to a user.
	Reject	*reject()*	When a user rejects a task
	Begin	*begin()*	When a task begins
	Suspend	*suspend()*	When execution of a task is suspended.
	Resume	*resume()*	When a suspended task resumes its execution.
	Finish	*finish()*	When a task completes normally.
	Fail	*fail()*	When a task fails.
	Abort	*abort()*	When a task is aborted by a user.

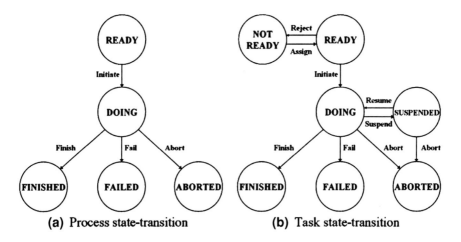

(a) Process state-transition   **(b)** Task state-transition

Fig. 4 State transition model for process objects

A BP Event (BPE) is a set of events that are generated in the flow of business process execution. An element of BPE is e_{BP}, which has three components: an object identifier, an event type identifier, and a time stamp. By the definition 4, we can represent an event $e_2(\in BPE) = < P_{A1}, Finish, 2009.03.14.15:30:45 >$.

Events in a business process take place according to the objects used in the BPMS, which are process, task, user, application, process variable and others. The present research uses events required only to interoperate among processes, which events are summarized in Table 1.

Process and task, the principal BPMS-managed objects, change their states while business processes are executed. Figure 4a, b illustrate the state transitions of process and task, respectively. A process, after being designed and deployed, remains in the 'READY' state, and an authorized user can execute it generating the 'Initiate' event and transitioning the process into the 'DOING' state. If the process completes, it enters the 'FINISHED' state. Or as a result of errors, it can be 'FAILED' or 'ABORTED' by users.

The state model of a task is more complex than that of a process. Let us consider a task in a process. When all of its preceding tasks complete, it is assigned to a user and a corresponding event, and 'Assign' is initiated. The state thereby changes from 'NOT READY' to 'READY'. If the user does not accept the task, the state will revert to the previous one, and if the user accepts the task, the 'DOING' state is entered. After the execution, the task state model enters one of three completed states, which are the same as those of the process.

4.2 Complex Event with RFE and BPE

In order to represent the CPE that indicates the relation of RFE and BPE, we introduce the following four basic operators in a symbolic format:

- **And** ($\wedge$): $e_1 \wedge e_2$ represents that both e_1 and e_2 occur.
- **Or** ($\vee$): $e_1 \vee e_2$ represents that either e_1 or e_2 occurs.
- **Precedence** (>): $e_1 > e_2$ represents that e_2 occurs after e_1 has occurred.
- **Negation** ($\neg$): $\neg e_1$ represents that e_1 does not occur.

Definition 5: Complex Event

$$e_{CP} = <e_t, start\_time, end\_time >$$

A complex event (CPE) $e_{CP} = <e_t, start\_time, end\_time>$ where e_t is the event semantics, *start_time* is the time stamp of the first event, and *end_time* is the time stamp of the last event. Based on those four basic operator symbols, we have the following four instances for e_{CP}:

- **AND** ($e_1, e_2,..., e_i,..., e_n$): A complex event that indicates $e_1 \wedge e_2 \wedge ... \wedge e_i \wedge ... \wedge e_n$ where any event $e_i \in$ (BPE $\cup$ RFE $\cup$ CPE) and n is the number of events considered.
- **OR** ($e_1, e_2,..., e_i,..., e_n$): A complex event that indicates $e_1 \vee e_2 \vee ... \vee e_i \vee... \vee e_n$ where any event $e_i \in$ (BPE $\cup$ RFE $\cup$ CPE) and n is the number of events considered.
- **SEQ** ($e_1, e_2,..., e_i,..., e_n$): For all events, $e_i > e_{i+1}$ where any event $e_i \in$ (BPE $\cup$ RFE $\cup$ CPE).
- **NOT** ($e_1, e_2,..., e_i,..., e_n$): A complex event that indicates $\neg e_1 \wedge \neg e_2 \wedge ... \wedge \neg e_i \wedge ... \wedge \neg e_n$ where any event $e_i \in$ (BPE $\cup$ RFE $\cup$ CPE) and n is the number of events considered.

A primary event is considered to be instantaneous, that is, it has no time duration. However, a complex event should have duration, since the time stamps of component events are different. The *start_time* and *end_time* of the complex event e_{CP} can be computed using the following equations:

- $e_{CP}.\text{start\_time} = \min(e_i.timestamp)\ (i = 1, 2,..., n)$
- $e_{CP}.\text{end\_time} = \max(e_i.timestamp)\ (i = 1, 2,..., n)$

Table 2 The ECA rules for inter-workflow patterns

Pattern	ECA (Event-Condition-Action) rules	
CSM	**RULE (R1-1:CSM-I)** **ON:** p_A.Finish **IF:** { $p_B.a_{\text{pre\_condition}}$ } **DO:** $p_B.initiate()$ **ENDRULE**	**RULE (R1-2:CSM-II)** **ON:** $p_A.t_{A8}$.Finish **IF:** { $p_B.a_{\text{pre\_condition}}$ } **DO:** $p_B.initiate()$ **ENDRULE**

Table 3 The RFID-ECA rules for auto-triggering of RFID-based processes

Define	Example 1 : RFE = { e_1 }, BPE = { e_2 }, CPE = { e_3 }
	$e_1 = <\text{tag\_1, reader\_001, 2009.03.14.15:30:45}>$ $e_2 = <P_{A1}, \text{finish, 2009.03.14.15:32:55}>$ $e_3 = <(e_1 \square e_2), t_1, t_2>$
Pattern	**CSM-I**

RULE (R 1-1 : CSM-1)
ON { e_3 }
IF { (e_1.tag_001.*quantity* == '100' **AND** e_1.reader_101.*location* == 'Warehouse #1') **AND** (e_2.*eventType* == 'finish') }
DO { $P_{C1}.initiate()$ }
ENDRULE

Define	Example 2 : RFE = { e_4, e_5 }, BPE = { e_6 }, CPE = { e_7 }
	$e_4 = <\text{tag\_4, reader\_401, 2009.03.14.16:25:15}>$ $e_5 = <\text{tag\_5, reader\_501, 2009.03.14.16:30:45}>$ $e_6 = <P_{B1}.t_{B8}, \text{begin, 2009.03.14.15:32:45}>$ $e_7 = <(e_4 \wedge e_5)> e_6, t_1, t_2>$
Pattern	**CSM- II**

RULE (R 1-2 : CSM- II)
ON { e_7 }
IF { (e_4.tag_4.*productName* == 'PNU_1' **AND** e_5.tag_5.*currentInventory* < '50') **AND** (e_4.tag_4.*productName* == e_5.tag_5.*productName*) **AND** (e_6.*eventType* == 'begin') }
DO { $P_{C2}.initiate()$ }
ENDRULE

5 Controlling Inter-Workflow Patterns

5.1 ECA Rules for Patterns

After the relations between processes are specified, they can be converted to ECA rules for run-time controlling. ECA rules, in our approach, use basic objects such as 'process', 'task', 'attribute', and 'user'. Almost every commercial BPM system predefines and provides events and functions for objects; we use them to specify the event, condition, and action elements in the ECA rules. The principal rules are summarized in Table 2.

5.2 RFID-ECA Rules for Auto-Triggering Using Complex Event

We provide RFID-ECA rule in order to trigger RFID-based processes automatically using complex event with RFE and BPE. This rule is more complex than ECA rule. Table 3 shows the examples of the rules for inter-workflow.

6 Conclusions

While a global logistic process involving multiple organizations is executed, each unit process of a participatory company is managed by its own BPM system, and all of the interactions between the participants are controlled by a rule engine according to the codes of the ECA rules. In ubiquitous environment, RFID technology will be applied whole logistic environment, so BPM systems must have ability to handle RFID event to have the power of competition.

We provide inter-workflow patterns and introduce RFID-ECA rules for auto-triggering of RFID-based logistic process using complex event. Our contribution lies in integrating processes operated by multi-organizations and thereby enabling the connections between processes to be effectively and efficiently executed.

Acknowledgments This work was supported by the Grant of the Korean Ministry of Education, Science and Technology (The Regional Core Research Program/Institute of Logistics Information Technology).

References

Bae, J., Bae, H., Kang, S.–H., & Kim, Y. (2004). Automatic Control of Workflow Process Using ECA Rules. IEEE transactions on knowledge and data engineering, 16(8), 1010–1023.
Basu, A., & Kumar, A. (2002). Research Commentary: Workflow Management Systems in e-Business. Information Systems Research, 13(1), 1–14.

Casati, F., Ceri, S., Pernici, B., & Pozzi, G., (1996). Deriving Active Rules for Workflow Enactment. Proceedings of 17 Int'l Conference on Database and Expert Systems Applications (pp. 94–110).

Chen, L., Li, M., & Cao, J. (2006). ECA Rule-Based Workflow Modeling and Implementation for Service Composition. IEICE Transactions on Information and Systems, E89-D (2), 624–630.

Goh, A., Koh, Y.-K., & Domazet, D. S. (2001). ECA rule-based support for workflows, Artificial Intelligence in Engineering, 15 (1), 37–46.

Gunasekaran, A., & Ngai, E. W. T. (2004). Information systems in supply chain integration and management. European Journal of Operational Research, 159(2), 269–295.

Jung, H., Chen, F. F., & Jung, B. (2008). Decentralized supply chain planning framework for third party logistics partnership, Computers & Industrial Engineering, 55(2), 348–364.

Liu, R., & Kumar, A. (2005). An Analysis and Taxonomy of Unstructured Workflows. Third International Conference on Business Process Management (BPM 2005) Nancy, France, Springer-Verlag, Lecture Notes in Computer Science, 3649, 268–284.

Liu, R., Kumar, A., & Aalst, W. (2007). A Formal Modeling Approach for Supply Chain Event Management. Decision Support Systems, 43 (3), 761–778.

Lucia, A. D., Francese R., & Tortora, G. (2003). Deriving workflow enactment rules from UML activity diagrams: a case study. Proceedings of IEEE Symposium on Human Centric Computing Languages and Environments (pp. 211–218).

Rhee, S.-H., Cho, N., & Bae, H. (2007). A More Comprehensive Approach for Enhancing Business Process Efficiency. Lecture Notes in Computer Science, 4558 (HCI International 2007), 955–964.

Rhee, S.-H., Bae, H., & Choi, Y. (2007). Enhancing the Efficiency of Supply Chain Processes through Web Services. Information Systems Frontier: Special Issue on from Web Services to Services Computing, 9(1), 103–118.

Tan, C. W., & Goh, A. (1999). Implementing ECA rules in an active database, Knowledge-Based Systems, 12(4), 137–144.

van der Aalst W. M. P., ter Hofstede A. H. M., Kiepuszewski B., & Barros A. P. (2000). Advanced workflow patterns, Lecture notes in Computer Science, 1901, 18–19.

WfMC: Workflow Management Coalition the Workflow Reference Model. WfMC Standards, WfMC-TC00-1003 (1995), http://www.wfmc.org.

Selectivity of EPC Data for Continuous Query Processing on RFID Streaming Events

Mikyung Choi, Byungjo Chu, Gihong Kim and Bonghee Hong

1 Introduction

EPCglobal as a standard association for RFID systems established EPCglobal Architecture Framework. EPCIS (EPC Information Services) as a component of EPCglobal architecture framework provides store and search services over EPC related information (EPCglobal Inc. 2005). EPCIS provides two types of query services, onetime query service (poll) getting query results immediately and continuous query service (subscribe) getting query results continuously. For continuous query processing, query index is used in general. However, existing query index causes some problems in EPCIS, a curse of dimensions, a partial infinite data problem, and a partial infinite query problem. Because EPCIS has thirteen domains and all query domains are not mandatory condition and all event domains are not mandatory condition.

To improve performance of EPCIS continuous query processing this paper proposes a scheme of query index with multiple 1-dimensional indexes as a basic approach, and proposes dynamic query execution plan techniques considering selectivity of an each domain condition.

The next section discusses related work. In Sect. 3, we define a target environment and problems, which are inappropriate features of existing query index

M. Choi (✉), B. Chu, G. Kim and B. Hong
Department of Computer Engineering, Pusan National University, Pusan, Republic of Korea
e-mail: choimk48@pusan.ac.kr

B. Chu
e-mail: cbj1004@pusan.ac.kr

G. Kim
e-mail: buglist@pusan.ac.kr

B. Hong
e-mail: bhhong@pusan.ac.kr

H.-J. Kreowski et al. (eds.), *Dynamics in Logistics*,
DOI: 10.1007/978-3-642-11996-5_31, © Springer-Verlag Berlin Heidelberg 2011

for EPCIS continuous query processing. Section 4 proposes a scheme of query index with multiple 1-dimensional indexes and dynamic query execution plan technique as a solution. Section 5 shows experimental results of performance evaluation for proposed techniques. Finally, Sect. 6 discusses conclusion of this paper.

2 Related Work

2.1 Queries of EPCIS

EPCIS is a component of EPCglobal architecture framework, and it is an object of this study. Main roles of EPCIS are store and search for business information related with EPC, therefore, EPCIS stores historical data, and deals with business information which is generated from EPCIS capturing application.

EPCIS has four event types, which are Object Event, Aggregation Event, Quantity Event, and Transaction Event to represent events of a business environment (EPCglobal Inc. 2005). Object Event represents actual observations of EPCs. Aggregation Event represents an event that is happened to one or more entities denoted by EPCs that are physically aggregated together. Quantity Event represents an event concerned with a specific quantity of entities sharing a common EPC class, but where the individual identities of the entities are not specified. Transaction Event represents an event in which one or more entities denoted by EPCs become associated or disassociated with one or more identified business transactions.

EPCIS provides EPCIS Query Control Interface to search EPCIS events. Figure 1 shows EPCIS query control interface. Core methods are poll and subscribe. The poll is a onetime query getting query results immediately. On the other

```
<<interface>>
EPCISQueryControlInterface
---
subscribe(queryName : String, params : QueryParams, dest :
URI, controls : SubscriptionControls, subscriptionID :
String)
unsubscribe(subscriptionID : String)
poll(queryName : String, params : QueryParams) :
QueryResults
getQueryNames() : List  // of names
getSubscriptionIDs(queryName : String) : List // of Strings
getStandardVersion() : string
getVendorVersion() : string
<<extension point>>
```

Fig. 1 EPCIS query control interface

hand, the subscribe get query results continuously according to query conditions and report periods. The subscribe called continuous query is processed continuously for a certain period. This paper addresses a study for efficient processing of the subscribe query which is EPCIS continuous query.

2.2 Query Index

A representative technique to process continuous query is query index. Registered queries are indexed as data, and then streaming data searches the query index to find registered queries which match with conditions of streaming data. In other words, queries become data and data becomes queries in query index. It is appropriate for RFID environment that streaming data is inserted continuously. The query index is aiming to process continuous query efficiently (Kalashnikov et al. 2002a; Wu et al. 2004). So processing of query index is based on main memory to provide high speed of search. In general case, query index guarantees great speed of search, although the number of inserted queries is increased. There are some representative query indexes, VCR (virtual construct rectangle) index and CQI (cell-based query) index.

2.3 Selectivity of DBMS

The aim of query execution procedures is minimizing volume of processing for getting same query results. Selectivity, which represents a rate of desired results in set of processing domains, effects on decision of query execution procedures significantly. It is necessary to estimate selectivity for optimized query execution procedures (Achaya et al. 1999). In DBMS (relation database management system), selectivity is estimated by the number of distinct values in total rows. Cardinality, which is result counts of a criterion or middle result counts for the next step, is estimated multiply selectivity by count of total rows. Selectivity is just a rate about a domain set, so estimation of cardinality is required. 1% of a million and 1% of a hundred are same rate, but an absolute quantity, cardinality may be different.

3 Problem Definition

3.1 Target Environment

EPCIS continuous query can be utilized for various applications such as system automation, automated inventory, and supply-chain management. For example, if administrators of automated inventory system want to monitor inventory information about Samsung LCD products continuously, then they may use EPCIS continuous query. First, accessing application registers a continuous query to EPCIS about inventory information about Samsung LCD products as 10 min of a report period. Then, they can get results of the continuous query every 10 min. EPCIS continuous query provides efficient filtering technique for desired events over streaming data.

3.2 Problem for Adopting Query Index to EPCIS

It is inappropriate for EPCIS continuous query to apply existing general query index, because of three problems. As the first problem, the EPCIS standard specifies thirteen fundamental event domains, so if the all domains are indexed, then search performance falls considerably by curse of dimensions, which means that if dimensions are increased, the rate of duplication is increased (Berchtold et al. 1996).

As the second problem, all query domains are not mandatory condition. So, multi-dimensional query index of EPCIS may cause a partial infinite data problem, by omitted query domains. Figure 2 shows an example of partial infinite data. For example, if constructs 2-dimensional index with two query domains such as EPC and eventTime, in case of query conditions of the two domains are described exactly, then query conditions are indexed as point or finite range types, however, if any of the query conditions of domain is omitted, then the omitted query condition becomes data of index covering partial or all domain area, so this problem raises heavy insertion load and search cost.

As third problem, all domains of EPCIS input event are not mandatory condition. In case of a general query index, input data is represented as a point type. But, the EPCIS standard specification describes most fields of EPCIS events as optional conditions. It may cause a partial infinite query problem. Figure 3 shows

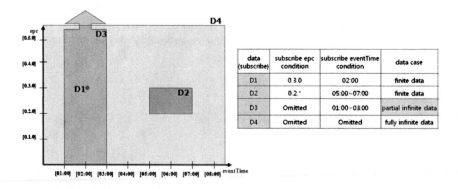

data (subscribe)	subscribe epc condition	subscribe eventTime condition	data case
D1	0.3.0	02:00	finite data
D2	0.2.*	05:00~07:00	finite data
D3	Omitted	01:00~03:00	partial infinite data
D4	Omitted	Omitted	fully infinite data

Fig. 2 Example of partial infinite data

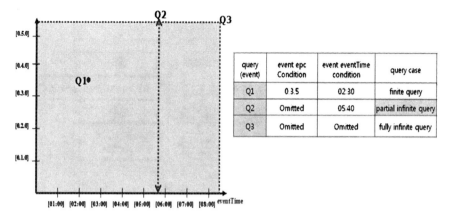

query (event)	event epc Condition	event eventTime condition	query case
Q1	03.5	02.30	finite query
Q2	Omitted	05.40	partial infinite query
Q3	Omitted	Omitted	fully infinite query

Fig. 3 Example of partial infinite query

an example of partial infinite query. Omissions of event domains means that query conditions of the domains become partially infinite. A searching area of omitted domains become partial or all area. According to above three critical problems, it is inappropriate for EPCIS continuous query processing to apply existing general query index.

4 Selectivity Scheme of EPC Data

4.1 Basic Approach

To solve mentioned three problems. This paper proposes a basic approach which is a scheme of query index with multiple 1-dimensional indexes. Figure 4 shows an

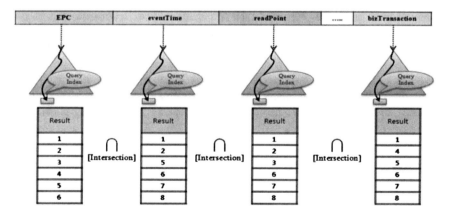

Fig. 4 Example of basic approach

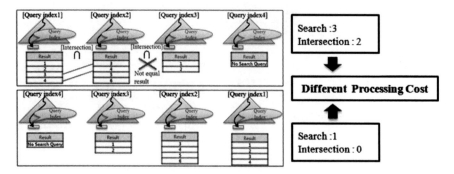

Fig. 5 Problem of basic approach

example of a basic approach. If conditions of input events match with all 1-dimensional indexes, then the events are appended as results. On the other hand, if conditions of input events match with some 1-dimensional indexes, then intersection of each result of indexes are appended. So the basic approach can solve above three problems simply

But the basic approach has a blind point, which is processing in fixed order among indexes. It may cause disproportion of performance. That is, if query processing follows wrong query execution order, then processing cost become large. So optimized query execution procedure is required for efficient continuous query processing. Figure 5 shows a problem of processing in fixed order. This paper defines query execution plan as query execution procedure. It is efficient to process low selectivity indexes first, on other words, should process an index first which makes a few query results.

4.2 Dynamic Query Execution Plan

This section proposes dynamic query execution plan technique considering selectivity. Selectivity of each domain is estimated by query conditions and features of domains. Flow of Processing is following; first, updating initial data need to selectivity estimation when continuous queries are registered. Second, selectivity is estimated based on event conditions when events are inserted. Third, dynamic query execution plans are generated based on estimated selectivity. Finally, continuous queries are processed as the generated execution plans.

4.3 Selectivity Estimation

It is necessary to estimate selectivity to generate optimized query execution plans. This paper proposes various techniques for selectivity estimation based on features of domains, a distribution or features of events. The next section proposes three types of selectivity estimation techniques.

4.3.1 Min/Max Value

A query condition type representing time such as eventTime is processed as range mainly. And there is proximity among values of query domains. To make the characteristic of the domains in selectivity estimation min/max values of minimum bounding value of input events is used. Selectivity is estimated by Eqs. 1 and 2.

$$\text{Cardinality} = \frac{\text{Sum of data size}}{\text{MBV}_{max} - \text{MBV}_{min}} \tag{1}$$

$$\text{Selectivity} = \frac{\text{Cardinality}}{\text{the number of data}} \tag{2}$$

4.3.2 Data Rate in Index

There is no proximity among query conditions in a domain such as readPoint representing URI (uniform resource identifier), so selectivity is estimated based on a rate of data occupation in an index. Selectivity is estimated by Eq. 3.

$$\text{Selectivity} = \left(\frac{\text{Sum of data size}}{\text{Domain size}} + \text{the number of omitted} \right) * \frac{1}{\text{the number of data}} \tag{3}$$

4.3.3 Aggregation Transformation

EPC is an identifier of products or objects in RFID environment, and consist of <company. product. serial number>. Three parts of EPC is represented by numeric, [low–high], or * each (EPCglobal Inc. 2005). Because of representations of [low–high] and *, so many empty spaces are raised in minimum bounding value. Many empty spaces may cause inaccurate selectivity estimation, therefore, to solve this problem, uses an aggregation transformation. This technique was proposed by (Berchtold et al. 1996) and is aim to decrease insertion cost of multiple segment data in indexes. Aggregation transformation technique of this paper estimates exact selectivity based on that, if multiple segments are represented as a single rectangle, then, empty spaces can be reduced.

A flow of processing is following; first, when conditions of EPC domain of continuous queries are registered, then transforms to combination of cell-ID and serial number. Second, generate minimum bounding rectangles containing all transformed data. Third, when an EPC condition is inserted, transforms to combination of cell-ID and serial number. Fourth, decides that a transformed EPC (cell-ID, serial number) is existed or not in a minimum bounding rectangle. Finally, selectivity is estimated by Eq. 4.

$$\text{Selectivity} = \left(\frac{\text{Sum of data size}}{\text{Size of MBR}} + \text{the number of omitted} \right) * \frac{1}{\text{the number of data}} \tag{4}$$

5 Performance Evaluation

5.1 Experimental Setup

We implemented our proposed techniques using Java 2 Platform Standard Edition 6.0. An experimental platform was Windows XP and used a PC with 2 GB main memory, CPU Pentium IV 2.4 GHz. Experiments measured rates of CPU occupancy by each method over same input of events. Performance of each technique was measured according to following steps. First, continuous queries are registered to EPCIS. Second, events are inserted as a specified number. Finally, we measured rates of CPU occupancy of an EPCIS process continuously.

5.2 Experimental Result

In sequential matching processing, cost of continuous query processing grew considerably as increasing of continuous queries. In the two proposals of continuous query processing; a basic approach, dynamic query execution plan technique, there was not considerable decrease of performance, although the number of continuous queries is increased. Dynamic query execution plan based on selectivity shows evaluation of performance; 40% in average, 79% in maximum compared to fixed order processing of a basic approach. So it means that optimized query execution plans can decrease cost of search and intersection considerably. Figure 6 shows changes of CPU occupancy rate by increasing of continuous queries.

Fig. 6 CPU occupancy rate by increase of continuous queries

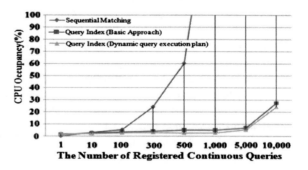

In conclusion, performance of dynamic query execution plan technique was improved 18,274% in average compared to sequential matching processing, 80% in average compared to processing in fixed order.

6 Conclusions

In continuous query processing, existing sequential matching processing causes heavy load and processing delay as increase of registered continuous queries. For the problems, multi-dimensional query index technique was proposed by (Kalashnikov et al. 2002b). But existing multi-dimensional query index technique is not appropriate for EPCIS because of 13-dimensional domain condition, an optional event domain condition, and an optional query domain condition. Above inappropriateness causes three problems: curse of dimensions, a partial infinite query problem and a partial infinite data problem.

About the problems, this paper proposed a scheme of query index with multiple 1-dimensional indexes. And final results are appended as query results after intersection of results of each index. But wrong query execution procedure makes large search and intersection cost. To solve this problem, this paper proposed dynamic query execution plan technique based on selectivity estimation techniques, which are Min/Max value, Data rate in index, Aggregation transformation.

In performance evaluation, we made experiment on rates of CPU occupancy by changing of the number of continuous queries and events. A representative experimental result shows that proposed dynamic query execution plan technique makes performance evaluation of 18,274% in average compared to sequential matching processing, 80% in average compared to processing in fixed order.

Future work will be related to scheme of many multi-dimensional query indexes after solving the omitted query condition problem.

Acknowledgments This work was supported by the Regional Research Universities Program (Institute of Logistics Information Technology, LIT) granted by the Ministry of Education, Science and Technology, Korea.

References

Achaya S., Poosala V., Ramaswamy S. Selectivity Estimation in Spatial Databases (1999) ACM SIGMOD
EPCglobal Inc. EPC Information Services (EPCIS) Specification. Working Draft Version 1.0 June 8, 2005
EPCglobal Inc. EPCglobal Architecture Framework. Final Version July 1, 2005
D. V. Kalashnikov, S. Prabhakar, W. G. Aref and S. E. Hambrusch (2002a) Efficient evaluation of continuous range queries on moving objects, Proc. of 13th Database and Expert Systems Applications 731–740

D. V. Kalashnikov, S. Prabhakar, W. G. Aref and S. E. Hambrusch (2002b) Efficient evaluation of continuous range queries on moving objects, Proc. of 13th Database and Expert Systems Applications 731–740

Kun-Lung Wu, Shyh-Kwei Chen and Philip S. Yu (2004) Interval query indexing for efficient stream processing, CIKM 88–97

Stefan Berchtold, Daniel A. Keim, Hans-Peter Kriegel (1996) The X-tree: An Index Structure for High-Dimensional Data Stefan Berchtold, Proc. 22nd VLDB Conf. 28–39

K. L. Wu, S. K. Chen and P. S. Yu (2004) Processing Continual Range Queries over Moving Objects Using VCR-Based Query Indexes, Proc. of International Conference of Mobile and Ubiquitous Systems 226–235

Criticality Based Decentralised Decision Procedures for Manufacturing Networks Exploiting RFID and Agent Technology

Hermann Küehnle, Arndt Lüeder and Michael Heinze

1 Introduction

Applying network principles in manufacturing replacing hierarchical management gives competitive advantages, as the "certainties" of command and control approaches evidently seem to no longer "hold true". A company may see itself primarily as unit in a network, getting value out of this loosely coupled enterprise (Norri and Lee 2006) by focusing on distinct process segments and by excellence in attracting a maximum of network resources towards its visions and objectives. Analyzing operation networks through the lens of Complex Adaptive Systems (Kauffman 1995) is advantageous for the fact that contemporary operation setups rather resemble dynamic, complex, interdependent and globally distributed webs, than the static well determined systems, which have traditionally dominated our thinking (Slepniov and Waehrens 2008). Within such simple settings of collocated operations, the challenge of managing can still be achieved by conventional planning systems and other intra-organizational decision mechanisms. For networks, the control becomes much more complicated, as the involved units and their roles are not stable, but evolve dynamically. However precisely these properties, activated for incorporating changing external partners as well as varying capabilities and knowledge, enormously increase the companies' adaptabilities and strongly amplify differentiations and uniqueness. This means continuous restructurings and adaptations for manufacturing networks. For the decisions on structuring, re-linking, or breaking up connections in manufacturing networks, models and instruments are introduced that support adequate

H. Küehnle (✉), A. Lüeder and M. Heinze
Otto-von-Guericke-University Magdeburg, Universitaetsplatz 2, D-39196, Magdeburg, Germany
e-mail: hermann.kuehnle@ovgu.de

A. Lüeder
e-mail: arndt.lueder@ovgu.de

H.-J. Kreowski et al. (eds.), *Dynamics in Logistics*,
DOI: 10.1007/978-3-642-11996-5_32, © Springer-Verlag Berlin Heidelberg 2011

interventions. The outline attempts to extend solutions that have been successfully implemented for distributed automation in manufacturing, which are based on fractal structures and self similarity (Kuehnle and Peschke 2006).

2 Criticality and Decisions in Network Improvements and Restructuring

2.1 Theoretical Foundation

By interpreting the network nodes as points, a manufacturing network may be identified as a specific Hausdorff Space. Such topological structures support smooth mappings and appear rich enough to model all configurations occurring in manufacturing networks (Kuehnle 2007). Configurations may be modelled by indicators, attributes, parameters and all other relevant figures of the unit and the views are expressed by "attached" tangent spaces to the nodes. The resulting topology is also referred to as "manifold with boundaries" (Lee 2000). These attachments as well as all projections thereof are assumed to be homeomorphisms.

Following these thoughts we may represent any production network unit by its Space of Activity (SoA, tangent space of the network), representing the feasible configuration values of the units which is, in the case of dimensions and values for the relevant figures, mostly represented by a polyhedron (Fig. 1).

2.2 Spaces of Activity and Criticality

As a common representation of the nodes, the Spaces of Activity may be described by the units' objectives, the resources and constraints. In consequence, the SoA volume may be identified by the unit's decision space or admitted zones for the units' positions. The unit's SoA position, e.g. expressed by corresponding indicators, gives input for decisions on maintaining the self-organization mode or reducing autonomy and calling for external interference. In cases of a unit's inability to cope with the objectives or the changes in the environment, network

Fig. 1 Production Networks envisioned as Hausdorff Spaces, attached Spaces of Activity (Tangent Space) models, including derived projections (homeomorphisms)

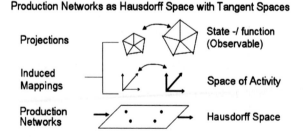

Fig. 2 Units' Space of
Activity (SoA)—viewing
(valid or invalid) positions of
relevant indicators and
observable

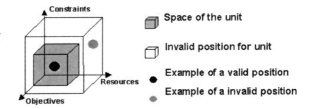

"order parameters" may gain influence on the units' activities [(self) reproduction (self) destruction (self) structuring].

For the decision, if harder efforts or even an adaptation of the network are needed, we may envision the Spaces of Activity as criticality thresholds (Karasavvas et al. 2005). In situations of overriding this "interaction flag"—i.e. an observable moves into an invalid position (Fig. 2), the units' criticality exceeds the assigned threshold and immediate actions are initiated. Usually there is more than one way to interpret a particular critical situation, so the solution options may range from "the status may just be adjusted or adapted" up to "the situation initiates severe interactions". If any units repeatedly fail to supply the promised/necessary capabilities, these units become "critical units", e.g. units that's roles within the network must be checked. In the repeated cases of criticality, the question is to be raised, whether keeping a unit in the network that is unable to avoid criticalities isn't a waste of potential and resources.

For adequate differentiations of criticalities, a comparison integer for characterising repeated "Not admitted" situations may be introduced indicating the least number of comparisons that may be accepted in a particular decision situation: The higher the number of "Not admitted" situations, the more critical that unit becomes. If the number of comparisons, resulting in "Not admitted positions", is higher than the accepted integer, the unit will run into a "more severe" decision cycle. Applied for manufacturing network decisions, such "criticality thinking" will result in a levelled manufacturing network adaptation procedure, similar to findings in complex adaptive systems (Ivanov et al. 2006).

2.3 Interrelated Criticalities for Network Decisions

Good network units' decisions will evolve the networks in economising resources, fulfilling objectives and strengthening/enabling the networks in total. Most promising for Planning and Decision in Manufacturing Networks seem to be approaches, engaging distributed and concurrent procedures that continuously and progressively generate "evolutive" solutions (Bennett and Dekkers 2005). All processes appear as embedded in rich structures of actors, units and connections, which may arbitrarily be compressed/detailed by fold/unfold properties (Fig. 3) applying self similarity properties (Kuehnle 2007). Any critical state on a lower level may have an impact on the criticality of the involved unit as well as on units on more aggregated network levels or even the configuration of the total network.

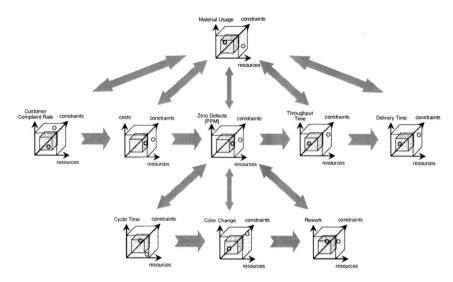

Fig. 3 Self-Similar Breakdown into Levels (of Detail) of SoAs and Criticality Thresholds

Arising criticalities are to be negotiated and harmonized with other units' objectives and resources. In extreme cases, the total networks' objectives have to be refocused in order to eventually obtain a consistent set of criticality thresholds.

If the number of comparisons, resulting in "not admitted positions", is higher than the assigned integer, it will run into a "more severe" decision cycle. Eventually the units' SoA positions result in decisions on maintaining the self-organization mode, reducing or removing the autonomy and calling for PN interference. In criticality terms, each unit may.

1. Decide on appropriate methods, tools, etc. in order to achieve the objectives negotiated and agreed upon. Units remaining within the admitted SoA are allowed to execute autonomous decisions. Prerequisites are resources, e.g. budgets, competencies, technical and personnel availability and constraints (a unit may have to face may be e.g. legal restrictions and capacity limits).
2. Loose its autonomy, if non critical positions within the units' SoA are not achieved by own efforts. Instantly network mechanisms are activated preventing the deviations and providing for the network plans' achievement.
3. Be replaced by new or other network units and be removed from the network if (2) is repeatedly experienced.

Dependent on the unit's (un)ability to cope with changes in the environment, network order parameters may gain influence on the units' activities according to the subsequent scheme:

A. Admitted position: No action;
B. Sporadically no admitted position: Non critical, self organised optimisation by the unit is demanded;

C. Repeatedly no admitted position (within threshold): (critical) autonomous self organisation, where the critical state is overcome by the respective unit;

D. Repeatedly no admitted position (exceeding threshold): (critical) interaction, where the network asks for changes in criticality values (Space of activity Volume, while presenting expected benefits/drawbacks that account for the critical situation) and

E. Repeatedly no admitted position (exceeding threshold by far): (critical) restructuring, where alternative structures (breaking up of links, generation of new interconnections, and introduction of new entities) are checked and the results are compared.

The ability to do quick, precise and reliable parameter settings and monitoring, concerning the objectives as well as the resources' states, is essential for efficient network management. Necessary improvements, reconfigurations, realignments or restructuring actions as well as adaptations should be possible without any delay or reaction times. The proposed criticality framework offers these options and makes the management of manufacturing networks easier and more structured. Plans, assignments, units, responsibilities etc. may continuously be rearranged, processes (re)established or (re)configured, if necessary by making use of additional units or by eliminating certain units or collaborations.

3 Applications for Decentralised Manufacturing Execution Decisions (MES)

For factory automation applications, the objective and resource axes may be downsized and "rescaled" after being broken down onto the manufacturing equipment unit level in a manner that loads and resource consumptions can be mapped. After this specification, the SoA visualizes and evaluates unit states and objectives for process steps and order loads. Details of objectives, resources and constraints are formalized and may be checked, determined and negotiated by the use of agent technology.

One generic concept for distributed order control, based on this approach, has been PABADIS (Plant Automation BAsed on DIstributed Systems (Lüder et al. 2005), and its extension PABADIS'PROMISE (Lüder et al. 2007). They aim at creating an architecture for distributed plant automation as a standard ensuring flexibility, scalability features and plug-and-participate properties for distributed control of PN. The resulting PABADIS'PROMISE control architecture is offering a complete vertical integration solution for distributed plant automation, providing an agent-based system for the MES layer, which collaborates tightly with the other two levels of the automation pyramid: the service-oriented ERP layer and the function block oriented field control layer. For factory automation, as such also within the PABADIS'PROMISE architecture, flexibility is one of the major, and challenges. Reasons may be introduction of new products or product change,

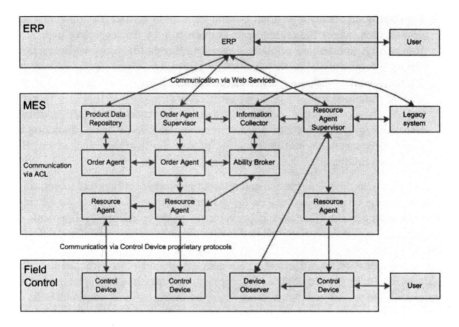

Fig. 4 PABADIS'PROMISE control architecture—the overview

integration of new control devices, integration of new device capabilities, breakdown or recall of capabilities, and change of orders by customers.

The networking structure for the MES level as one integral part of the PABADIS'PROMISE architecture is depicted in Fig. 4.

The top level of the overall system is represented by the ERP. Here the high level order control is executed with respect to resource capability load of the system. Hence, the resulting SoA is characterised by the mapping of orders to manufacturing capabilities over time. Following this range the ERP monitors the availability of manufacturing capabilities in general and controls the manufacturing order release process (self-similar to the process segment navigation on the resources).

In order to execute decision and control in the navigation logic at MES layer, two types of supporting agents can be defined: the Order Agent, and the Resource Agent. The Order Agent is a mobile (communicating) agent, carrying all information necessary for processing orders. Its necessary information including SoA characteristics are provided and maintained from ERP via Order Agent Supervisor and Product Data Repository. The Resource Agents carry unit profiles and information about units' states and provide manufacturing abilities to Order and Resource Agents representing its SoA. These abilities are announced to the overall agent community via Ability Broker. Modifications of products are supported by the Product Data Repository.

Finally, the field control layer implements the interface between the MES and the plant. The Control Device realizes the field control logic, it is physically

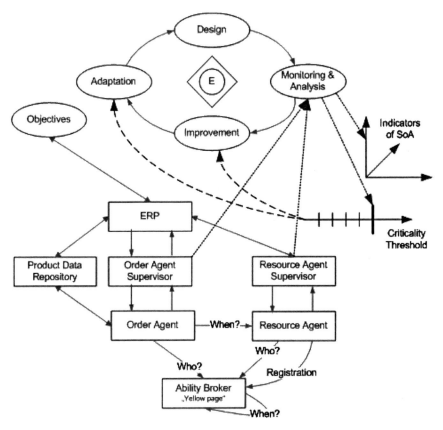

Fig. 5 PABADIS PROMISE control architecture—decentralised decision making in MES by specific implementation of the criticality concept

connected to the plant, and exports suitable functions to the Resource Agent. The Device Observer is a part of a Multi Agent System devoted to discover Control Devices and to register them in the MES via Resource Agent Supervisor.

The resulting procedure, totally executed by the agents, is not yet powerful enough for Decision Support on other manufacturing networks' levels. Adequate extensions will have to support the network in achieving all vital objectives, monitoring and analysing factors and indicators, especially focussing on those putting the objectives' achievements in question (Fig. 5).

4 Implementation of RFID and MAS

The following section will describe some aspects of the implementation of the most important agent types in PABADIS'PROMISE: Order Agent (OA) and Resource Agent (RA).

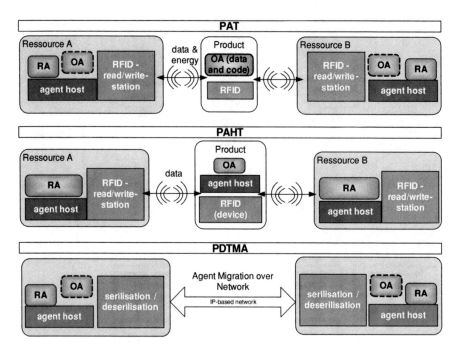

Fig. 6 Migration of agents in combination with RFID

One basic principle for the Order Agent is the fact that this agent has to interact with a number of other agents across the MAS, but closely control the execution of production operations for the product it is responsible for. For the implementation of this behaviour mobility of OA seems to be a useful approach, as this enables a kind of synchronisation between OA and product. As described by (Kuehnle and Peschke 2006) mobility of agents has some challenging requirements for the agent platforms, which might limit the applicability at shop floor level. The utilisation of RFID-technologies for agent storage and migration can help to overcome these limitations. Out of some general possibilities for agent-RFID integration the most important shall be discussed with a focus on applicability within P2.

Basically RFID-technologies can improve the migration of agents by transferring the agent "attached" to a product and thus supporting some natural synchronisation between agent and product (Treytl et al. 2008). In contrast to this approach a conventional migration of agents using serialisation/de-serialisation in combination with a transfer over IP-based networks requires additional mechanisms to synchronise agent and product. It has to be noted that for these synchronisation reliability is a very important issue as any error within this process may result in lost product. As shown in Fig. 6 there are three main approaches:

(1) Product and Agent Tag (PAT) means that the agent execution is implemented on local resource (control device) but there is a physical migration of OA stored within a RFID-Tag and transmitted via RFID-reader. A problem for this

solution is the fact that during migration the agent cannot communicate within the MAS and thus is not reachable for other agents. Nevertheless it can be useful for production systems where the usage of sophisticated active RFIDs is not reasonable for cost reasons.

(2) Product and Agent Host Tag (PAHT) requires the usage of an advanced RFIDs-tag (more a smart device) for the direct execution of an agent on this "RFID-device". In this solution the whole agent communication relies on RF-communication. The requirements for the implementation of this solution are very high as there is the need for RFID-systems capable of executing agents (active RFID-devices) as well as for an agent system with limited resource consumption. Thus, PAHT is only applicable for expensive products with complex processing operations, but it provides a kind of "intelligent product" consisting of the product itself and the appropriate order agent which could also be used after the manufacturing process during the complete lifetime of a product for purposes of maintenance etc.

(3) Product Data Tag with mobile agents (PDTMA) describes the agent execution on local resources in combination with a classical agent-migration via (IP) network. This solution does not require explicitly RFID (identification can also use other systems) and, therefore, present the typical implementations used in former agent based approaches such as PABADIS. As described, it has essential drawbacks in the synchronicity of product and agent.

For the implementation of a PABADIS'PROMISE-system all three approaches can be used but PAT is the typical one. Basically it combines the most advantages of PAHT and PDTMA but is applicable to a wide range of production types and ensures a close connection of agent and product increasing the reliability of an order agent migration.

Nevertheless, also the application of PAHT is possible following (Treytl et al. 2008) the execution of OA is on an advanced RFID device.

In each case PAT or PAHT the OA has to transport and use its SoA for the control of the order execution. Therefore, it is implemented by a set of bordering conditions and tangent space coded within the order related data. They cover beneath the description of the Bill of Operation related to the product given by necessary manufacturing processes also economic details like due dates, quality levels etc. All this information is coded within the so-called process segments, a data structure derived from ISA 95 standard by improving it with control related data within PABADIS'PROMISE (Diep et al. 2007; Georgoudakis et al. 2007). The data, coded in this way, can be stored and applied using PAT or PAHT efficiently.

The second agent type in PABADIS'PROMISE is the Resource Agent representing a resource and the functionality provided towards the MAS. According to the implemented roles the RA has to fulfil three main tasks:

- To provide the abilities of the manufacturing resources within the P2-system
- To maintain a schedule for production steps to be executed on the resource and negotiate with OA requesting an ability
- To control and monitor the execution of production steps on the local resources.

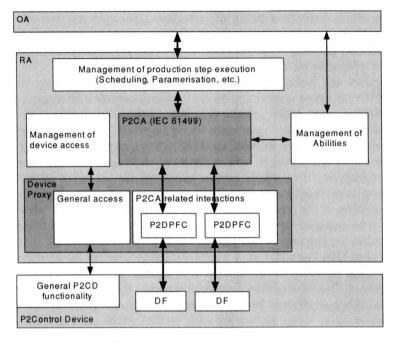

Fig. 7 Interactions between RA and control device

While the first two tasks are typical functionalities within a distributed production control system as regular part of the MES-layer, the third task requires the direct access to the control level and an interaction with the control devices. As nowadays there is still a technological gap between MES and field control layer, specific mechanisms are required to support these interactions. It is a fundamental requirement that legacy control devices have to be integrated in the system with reasonable costs. Many agent-based approaches fail to fulfil this requirement as the prerequisites are to complex (e.g. implementing an agent host on a control device). If this requirement can be fulfilled it realizes (in combination with the MES-layer of a PABADIS'PROMISE-system) a real vertical integration, enabling ERP-systems not only to collect data on the field level but also control some aspects of the order execution. In order to provide this possibility the RA has to enable an abstract view on the resource. For this purpose the RA provides a multistage interface to the control device. Figure 7 shows the PABADIS'PROMISE-concept for the interaction between RA and control device.

Within the MAS, the RA provides an ability describing the capabilities of the resources following the needs of the bill of operation representation of the OA. Thereby, it covers parts of the SoA of the RA related to the manufacturing capabilities of the RA. Additional parts are given by the economic boundaries of the RA like resource scheduling and allocation, costs, quality etc., all integrated within the SoA description again within sets of process segments and additional information.

Within the field layer the control devices are able to execute Device Functions (DF) controlling the physical processes realizing the ability of the resource.

The RA has to map these both views and ensure a flexible usage of DF. To enable the integration of a wide range of devices a P2ControlDevice is represented inside the RA by a Device Proxy. This is a software entity with a defined interface towards the agent world, while the communication to the field layer is device specific and can be tailored to the specific needs of different devices to be integrated. The Device Proxy (DP) ensures a general access to the device (collection general information, requesting a device shutdown etc.) as well as the invocation of DF (via a P2 Device Proxy Function Call—P2DPFC). This invocation is executed in the second stage of the interface the so-called P2ControlApplication (P2CA). This is not a real-time control application as on the field level, but a coordinated access to DF realizing e.g. a sequence of operations. The typical implementation is done as an IEC 61499 application (see Vyatkin 2007), where the DF are represented as Service Interface Function Blocks (SIFB). More details about this type of application are given in IEC61499 (2006). The P2Control-Application can be changed during runtime of the resource, allowing the interaction of new devices and Device Functions as well as a product specific usage of DF. This concept ensures much more flexibility compared to older approaches, where only a parameterisation of running control applications was possible.

Facing this structure the SoA of the RA is determined by the device functions related to the RA and the available control building blocks to be used within the P2ControlApplication. As they are integrated within the system automatically via control device and device observer for device functions as well as via Product Data Repository for the control application building blocks the SoA of the RA will be adapted automatically if the system changes.

5 Conclusions

Decision procedures in networks should be of gradual and evolving nature. The mental models behind evidently go beyond systems thinking and include complexity (Olson et al. 2001), as lots of random interactions may be observed. They induce different decision behaviour, which optimises the networks' structures in total and which smoothly direct networks in ever-changing environments. In order to understand how interdependencies and connectivity evolve over time and what the implications of that in manufacturing networks are, the application of CAS frameworks seems to be appropriate (Bennet and Bennet 2004). Advantages of a network interpretation of manufacturing, based on topology could be demonstrated. The focus was on coordination and control aspects of organizational units, constituting the manufacturing system. Therefore, a selection of models is proposed for better production networks' planning and control problem solving. As optimization of given processes and not dynamic interlinking of units has been emphasized in the past, there is a lack of models and methods for dynamic linking (emergent processes) on all levels (personal, informational, process ...). There have been attempts already to extend the application of complexity theory to the management of supply chains and operations networks (e.g. Surana et al. 2005).

The new PABADIS'PROMISE-architecture introduced within this paper overcomes some essential drawbacks of older architectures for distributed production control systems with respect to practical applicability. This is reached by a modelling approach enabling an implementation independent description of the architecture based on roles and associated activities and protocols.

The next steps in the development of the PABADIS'PROMISE-architecture will be analysis of realised complex simulations and (smaller) industrial demonstrators. The application of the described architecture will enable a most flexible, adaptable, and efficient control of manufacturing systems, covering all requirements of future manufacturing. Although these works lay down some initial ideas for the analysis of global operations networks using the principles of complex adaptive systems, they definitely call for more systematic studies to further refine and examine the ideas developed (Kuehnle and von der Osten 2008). Introducing self-similarity for the derived models and the criticality thinking may be considered substantial steps towards efficient manufacturing network management. There are promising agent approaches in bioinformatics to be considered as well (Merelli et al. 2007). Further research is needed for the development and integration of improvement techniques as well as coupling, uncoupling, breaking up and (re)linking instruments.

References

Bennet, A., Bennet, D. (2004), The Intelligent Complex Adaptive System, KMCI Press Elsevier.

Bennett, D., Dekkers, R. (2005), Industrial Networks of the future—a critical commentary on research and practice, 12th International EurOMA Conference on Operational and Global Competitiveness, Budapest Proceedings pp. 677–686.

Diep, D., Alexakos, Ch., Wagner, Th. (2007), An Ontology-based Interoperability Framework for Distributed Manufacturing Control, 12th IEEE International Conference on Emerging Technologies and Factory Automation, 25–28th September 2007, Patras Greece, Proceedings.

Georgoudakis, M., Alexakos, Ch., Kalogeras, A., Gialelis, J., Koubias, S. (2007), Methodology for the efficient distribution a manufacturing ontology to a multi-agent system, 12th IEEE International Conference on Emerging Technologies and Factory Automation, 25–28th September 2007, Patras Greece, Proceedings.

Ivanov, D., Kaschel, J., Sokolov, B., Arldiipov, A. (2006), A Conceptional Framework for Modeling Complex Adaptation of Collaborative Networks. In: Network-Centric Collaboration and Supporting Frameworks, IFIP International Federation for Information Processing, Volume 224, [Camarinha-Matos, L., Afsarmanesh, H., Ollus, M. (eds.)], Springer. Boston, pp. 15–22.

Karasavvas, K., Burger, A., Baldock, R. (2005), A Criticality-Based Framework for Task Composition in Multi-Agent Bioinformatics Integration Systems, Bioinformatics 21(14), pp 3155–3163.

Kauffman, S. (1995), At Home in the Universe: The Search of Laws of Self-Organization and Complexity, New York.

Kuehnle, H. (2007), A system of models contribution to production network (PN) theory. Journal of Intelligent Manufacturing, 18(5), pp 543–551.

Kuehnle, H., Peschke, J. (2006), Agent technology enhancement by embedded RFID for distributed production control, In: Moving up the value chain. University of Strathclyde, Glasgow, pp 731–739.

Kuehnle, H., von der Osten, D. (2008), Planning and Decision Procedures for Networked (Network Centric) Manufacturing, Proceedings of CAMSIM 2008, Institute for Manufacturing, Cambridge.

Lee, J.M. (2000), Introduction to Topological Manifolds, Graduate Texts in Mathematics, 202, New York.

Lüder, A., Peschke, J., Klostermeyer, A., Bratoukhine, A., Sauter, T. (2005), Distributed Automation: PABADIS vs. HMS, IEEE Transactions on Industrial Informatics, 1,1, pp. 31–38.

Lüder, A., Peschke, J., Bratukhin, A., Treytl, A., Kalogeras, A., Gialelis, J. (2007), The Pabadis'Promise Architecture, Journal Automazione e Strumentazione, Nov. 2007, pp. 93–101.

Merelli, E., Armano, G., Cannata, N., Corradini, F, d'Inverno,.M., Doms, A., Lord, P., Martin, A., Milanesi, L., Moller, (2007), Agents in bioinformatics, computational and systems biology, Brief Bioinform, 8(1), pp. 45–59.

Norri, H., Lee, W.B. (2006), Dispersed network manufacturing: Adapting SMEs to compete on the global scale, Journal of Manufacturing Technology Management, Vol. 17 No. 8, pp. 1022–1041.

Olson, E. E., Eoyang, G. H., Beckhard, R., Vaill, P. (2001), Facilitating Organization Change: Lessons from Complexity Science, Jossey-Basss/Pfeiffer, San Francisco.

Slepniov. D., Waehrens, B.V. (2008), Evolving interdependencies in the context of global operations, Proceedings of the 15th EurOMA Conference proc., Groningen.

Surana, A., Kumara, S., Greaves, M., Raghavan, U. N. (2005), Supply-chain Networks: a Complex Adaptive Systems Perspective, International Journal of Production Research, Vol. 43, no. 20, pp. 4235–4265.

Treytl, A., Spenger, W., Riaz, B. (2008), A Secure Agent Platform for Active RFID, 13th IEEE International Conference on Emerging Technologies and Factory Automation, September 15–18, 2008, Hamburg, Germany, Proceedings.

Vyatkin, V. (Ed.), (2007), IEC 61499 Function Blocks for Embedded and Distributed Control Systems Design, ISA Publisher.

The Application of the EPCglobal Framework Architecture to Autonomous Control in Logistics

Karl A. Hribernik, Carl Hans and Klaus-Dieter Thoben

1 Introduction

Autonomous control in logistic systems is characterized by the ability of logistic objects to process information, to render and to execute decisions on their own (Böse and Windt 2007). The requirements set by complex logistics systems towards the integration of data regarding the individual entities within them prove immensely challenging. In order to implement complex behaviour with regards to autonomous control, dynamism, reactivity and mobility, these entities, including objects such as cargo, transit equipment and transportation systems but also software systems such as disposition, Enterprise Resource Planning (ERP) or Warehouse Management Systems (WMS) require the development of innovative concepts for the description of and access to data (Hans et al. 2007). One recent standard approach to providing data visibility and integration throughout complex logistics processes on an item-level is defined by the EPCglobal Architecture Framework. This paper aims to evaluate whether the framework can be applied to successfully support autonomous control in logistics and if so, where the limitations of the current set of specifications lie in this regard.

K. A. Hribernik (✉), C. Hans and K.-D. Thoben
Bremer Institut für Produktion und Logistik GmbH (BIBA), Bremen, Germany
e-mail: hri@biba.uni-bremen.de

C. Hans
e-mail: han@biba.uni-bremen.de

K.-D. Thoben
e-mail: tho@biba.uni-bremen.de

H.-J. Kreowski et al. (eds.), *Dynamics in Logistics*,
DOI: 10.1007/978-3-642-11996-5_33, © Springer-Verlag Berlin Heidelberg 2011

2 The EPCglobal Architecture Framework

Figure 1 shows the principle components of the EPCglobal Architecture Framework, as defined by EPCglobal (Armenio et al. 2007). The Tag Protocol standards (UHF Class 1 and HF Generation 2) define the physical and logical requirements for Radio Frequency Identification (RFID) systems operating in the ultra-high and high-frequency ranges, respectively. The Low-level Reader Protocol (LLRP) deals with the standardisation of the network interface of RFID readers. The Reader Management standard is the wire protocol used by management software to monitor the operating status and health of RFID readers. The Application Level Event (ALE) standard specifies an interface through which clients may obtain filtered, consolidated RFID data whilst abstracting from the underlying physical infrastructure. The Electronic Product Code Information Services (EPCIS) represents an event-based service interface for the distributed, interorganisational sharing of application-level Electronic Product Code (EPC) data. The Object Name Service (ONS) defines a mechanism by which authoritative metadata and services associated with EPC Identifiers may be located in the network. In the following sections, those components most relevant to data integration are described in more detail—the EPC Identifier itself, followed by the ALE, ONS and EPCIS standards.

2.1 The EPC Identifier

In EPCglobal Inc (2008a, b), the EPC is described as a scheme for universally identifying physical objects. The media targeted foremost by EPC is RFID tags but

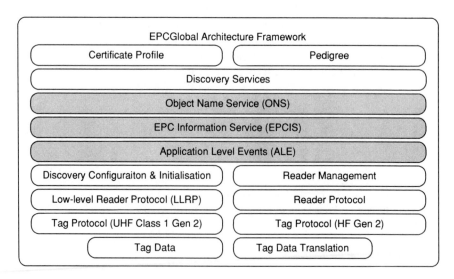

Fig. 1 The EPCGlobal Architecture Framework (Armenio et al. 2007)

other means are not excluded. The standard tag encodings consist of an EPC (or EPC Identifier) that uniquely identifies an individual object, as well as a filter value, if necessary, to enable effective and efficient tag reading. Within the EPC Identifier scheme, "Pure Identity" refers to the abstract name or number used to identify an entity, regardless of the media used to encode it, such as barcode, RFID or a database field. The Pure Identity may be represented as an Identity URI (Uniform Resource Identifier). The General Identifier, e.g., consists of three fields. Encodings of the GID include a fourth field, the header, to guarantee uniqueness within the EPC namespace. The General Manager Number identifies an organizational entity (company, manager or other organization) responsible for managing the subsequent fields. The General Manager Number is assigned by GS1. The Object Class identifies a class or "type" of thing. The Serial Number code, or serial number, is unique within each object class and denotes a specific, individual physical entity.

2.2 Application Level Events

The ALE standard specifies interfaces for the communication between multiple devices and clients which exchange EPC data, and facilitates the abstraction of logical readers from the physical infrastructure. ALE allows a client application to subscribe to a network of readers for read events. The term "readers" is not restricted to RFID readers as such, but may be interpreted as any device capable of capturing EPC data.

2.3 EPC Information Services

EPCIS stands for Electronic Product Code Information Services. It is a standard that defines interfaces for the sharing of data among trading partners. Its aim is primarily to enable supply chain participants in the EPCglobal network to gain real-time visibility into the movement, location and disposition of assets, goods and services throughout the world (Soon and Ishii 2007). EPCIS can be leveraged to track individual physical objects and collect, store and act upon information about them. By providing a standard interface to that information, EPCIS enable cooperation partners to seamlessly query such information throughout supply chains. The EPCIS standard consists of three layers. The first is the Abstract Data Model Layer, which specifies the generic structure of EPCIS data. The second is the Data Definition Layer, in which the syntax and semantics of the data exchange via EPCIS are defined. In this layer, the concept of EPCIS events is defined and the Core Event Types Module specified. The third layer is the Service Layer, which specifies the service interfaces for both querying and capturing EPCIS events. On top of these three layers, recommendations for bindings are given which exemplify how the specifications may be implemented in practice.

2.3.1 EPCIS Events

The individual EPCIS event classes inherit their basic structure from the parent EPCIS event following the object-oriented paradigm. Object events are the simplest type of EPCIS event, and are used to capture and communicate information about objects identified by an EPC. They can, for example, be used to document that a specific object has been observed at a specific position and time, or capture changed to the status of that object. Aggregation events are used to document events in which an object identified by an EPC is brought into relation with a number of other such objects. In the example presented in this paper, it is used to capture which component parts of are installed into which overall product. It can be used to both create and destroy such relations. Quantity events are specified to document events which influence a set of objects which can be identified as belonging to a common class. An example for such an event could be the change of price for an entire assortment of products. Transaction events can be employed whenever an event triggers a business transaction. The mapping of these events is facilitated by means of a business transaction list, which is a core element of the suggested specification of this event.

2.3.2 EPCIS Capture and Query Interfaces

The standard defines two services—EPCIS Capture Interface and the EPCIS Query Interfaces. The latter consists of two services, the Query Callback Interface and the Query Control Interface. The capture interface consists of only one method, "capture". This method is responsible for accepting EPCIS events and storing them in the respective repository. It accepts a list of EPCIS events as its sole argument and returns no results. The Query Interfaces only provide a general framework for the definition of an interface by which applications may query EPCIS data. The interface caters both for on-demand queries via the Query Control Interface and standing queries via the Query Callback Interface.

2.3.3 EPCIS Event Data Storage

The EPCIS standard only specifies event types and the interfaces to arbitrary systems capable of storing such events. It neither makes a statement about the characteristics of the system used to store EPCIS event data, nor about the representation in which the events are stored. ERP systems, tracking and tracing applications, or WMS are examples of systems which might implement one or both of the EPCIS interfaces (EPCglobal Inc 2007a, b).

2.4 Object Name Service

The function of the ONS is to transform the EPC stored, e.g., on RFID tags, via their corresponding Identity URI encodings into URLs, which may respectively point to a Web Service or other information resource. First, the EPC read from the RFID tag is transformed into its standard URI encoding. The next step is to rewrite the given URI into a DNS name, which is accomplished using a simple regular expression. Then, the DNS name is resolved into a valid URL. This is accomplished by using a DNS Type Code 35 (Naming Authority Pointer, NAPTR) request, which offers a number of different options for specifying the type of URL to resolve the DNS name into. Neither the method for carrying out this request nor the specification of the DNS resolver is within the scope of the ONS specification (EPCglobal Inc 2008a, b). The most common use for ONS is the discovery of specific services for an object class. For example, Web Services may be discovered using this mechanism. Similarly, ONS can be used by an application to discover the EPCIS Capture and Query Interfaces responsible for the object class. The application may then bind to the identified EPCIS services.

3 Integrating EPCIS into Autonomous Control in Logistics

According to Böse and Windt (2007), "Autonomous control in logistic systems is characterized by the ability of logistic objects to process information, to render and to execute decisions on their own." They furthermore define logistics objects in this context as "material items (e.g. part, machine and conveyor) or immaterial items (e.g. production order) of a networked logistic system, which have the ability to interact with other logistic objects of the considered system." In Scholz-Reiter et al. (2007), the former are further differentiated as commodities and all types of resources and whilst constraining the immaterial logistics objects to orders. In Windt et al. (2005) a catalogue of criteria for autonomous cooperating processes is suggested. Within this catalogue, three criteria explicitly address the "information system" layer, as illustrated in Fig. 2 below. Specifically, these criteria deal with the properties of data storage, data processing and the system's interaction ability. The first two criteria are directly related to the problem of data integration. The properties of these criteria make apparent that the less central the data storage and processing of an information system is, the higher the level of autonomous control is. The third criterion, interaction ability, relates implicitly to the data integration problem, in that the interaction ability of the information system in question is based upon its ability to access and process data stored according to the initial two criteria.

Consequently, an IT architecture can be said to contribute to the autonomous control of a logistics system if it provides support for information processing and decision-making on the part of logistics objects, both material and immaterial. Furthermore, the information system is required to exhibit the properties data

System layer	Properties				
Information system	Data Storage	Central	Mostly central	Mostly decentral	Decentral
	Data Processing	Central	Mostly central	Mostly decentral	Decentral
	Interaction Ability	None	Data allocation	Communication	Coordination

Increasing level of autonomous control

Fig. 2 Information system layer criteria for autonomous cooperating processes (Windt et al. 2005)

storage, data processing and interaction ability with respect to these types of logistics objects. The following sections examine the EPCglobal Framework Architecture in its support for the criteria shown in Fig. 2 from the perspective of material and immaterial logistics objects.

3.1 Material Logistics Objects

Auto-identification, that is a method for the unique, item-level identification of material objects coupled with a mechanism for the automated retrieval of that identity, forms the basis of any data processing operation with regards to individual material objects. This has been extensively shown for the Intelligent Product, a class of autonomous object similar to the material logistics object discussed here, e.g. in McFarlane et al. (2003), Wong et al. (2002), Kärkkäinen et al. (2003) and Ventä (2007). The discussion with regards to the Intelligent Product encompasses the entire product lifecycle, from the production, distribution, and use to the disposal of the products. This includes autonomously controlled logistics processes, e.g., increased intelligence in supply chains (Ventä 2007), manufacturing control (McFarlane et al. 2003), production, distribution, and warehouse management (Wong et al. 2002). Consequently, the findings with regards to auto-identification in the field of Intelligent Products are applicable to that of the autonomous control of material logistics objects. The EPC identification schema fulfils that requirement by providing a standardized means for the unique identification of individual physical logistics entities. Along with the specifications for the encoding of EPC on media such as RFID, the EPCglobal Architecture Framework provides the foundation for processing information regarding individual physical entities. An EPC may be associated with each individual material logistics object within an autonomously controlled logistics system. Leveraging the standard structure of, for example, the general identity type General Identifier (GID-96) allows for a rough classification of logistics objects at the identifier level with the possibility of the expression of ownership relations (General Manager Number) and type of logistics object (Object Class). The Serial Number can then be used to uniquely identify the logistics object within that structure.

The low-level standard specifications along with the ALE components support basic auto-identification mechanisms which satisfy the needs outlined above. Building on the unique identification and hierarchical structure provided by the EPC itself, ONS offers a mechanism for the automated discovery of arbitrary services on the Object Class level. Within the context the autonomous control in logistics, that means each type of logistics object can be associated with a specific set of services using ONS. The types of services exposed by the logistic objects types are not restricted by the standard, although recommendations are given and all further services are to be considered experimental. The recommended use encompasses html, WSDL and EPCIS. The second of these is relevant to autonomous control due to the fact that by means of WSDL and methods of automatic service discovery and composition, logistics objects may be enabled to expose, discovery and consume information services, facilitating data processing and providing a basis for their ability to interact with each other. The third option allows for logistics objects in Autonomous controlled logistics systems to access standard EPCIS. Furthermore, the design of this service discovery functionality anticipates the ad hoc integration of stakeholders into a trading network as opposed to the traditional, pre-arranged approach (Soon et al. 2007).

The service interface and event model of EPCIS provide a modular, extensible and decentral means of storing and exchanging data pertaining to the uniquely identified material objects. According to EPCglobal Inc (2007a, b) the perspective EPCIS takes is *"visibility of the physical world, incorporating common notions of What Where When and Why."* Each EPCIS event represents data corresponding to a unique spatio-temporal constellation of one or more physical entities associated with an EPC. Hribernik et al. (2007) demonstrates the applicability of EPCIS events to seamlessly modeling, tracking and tracing products and their subcomponents throughout their life-cycles, including logistics processes. Attributes of physical logistics objects such as the time, position, status and the current related business transaction can be modeled using the basic event objects. Further attributes, such as the properties of autonomous controlled material logistics objects like speed, weight, enterprise affiliation, as well as capabilities such as transportation or storage capabilities, or sensors for measuring humidity as described by Langer et al. (2006) may be modeled by extending the basic objects with user vocabulary extensions.

3.2 Immaterial Logistics Objects

Not all events related to logistic objects take place on the level of the material flow of specific physical entities. In Scholz-Reiter et al. (2007), the kinds of logistics objects with relevance to autonomous control are identified as commodities, all types of resources and orders. The former two kinds are readily addressable by EPCIS as they refer to specific, individual physical entities which may be identified using an EPC as discussed above. Commodity and resources types are also addressable using EPCs object class hierarchy level. The latter kind of logistics object, the order, corresponds to the immaterial logistics object according to Windt et al. (2005). Whilst Pure Identity types of EPC Identifier generally deal with the unique identification of physical objects, types such as the Global Document Identification Number GDI, which can refer to both a physical document as well as, for example, a database entry, demonstrate that unique identification in the EPCglobal Architecture Frame is not conceptually restricted to physical objects.

As a standard dealing solely with the handling of RFID capture data, the ALE component has no immediate relevance to immaterial logistics objects. Furthermore, whilst EPCIS does not directly support the modeling of orders, it does facilitate the association of order information to identifiable, physical entities on the material flow level. As soon as an order is explicitly related to physical logistic objects, an EPCIS event may be generated to exchange information about it. This is the aim of EPCIS transaction events. Here, immaterial logistics objects may be identified and described by participating stakeholders using URIs. Purchase orders may be mapped to physical entities via the BusinessTransactionID vocabulary which may point to an URI describing the transaction. A schema for the unique identification of such transactions is, however, not specified by the standard. Furthermore, Master Data attributes are not given by the standard (EPCglobal Inc 2007a, b). Ultimately, even with regards to Transaction events, the event data handled by EPCIS is always based upon physical handing steps (SupplyScape 2008), meaning the relevance of the event object model is to the material flow of individual physical entities.

4 Summary and Conclusions

The EPCglobal Framework Architecture provides explicit technical support for the first type of autonomous logistic objects defined by Böse and Windt (2007)— the individual physical entity. The EPC identification schema provides a means for the unique identification of individual physical logistics entities. Along with the specifications for the encoding of EPC on media such as RFID, it provides the foundation for processing information regarding individual physical entities. ONS supports automated service discovery on the basis of that unique identifier. The discovery of standard EPCIS for each individual physical logistics entity is also facilitated in this way. The service interface and event model of EPCIS provide a modular, extensible and decentral means of storing and exchanging event data pertaining to the identified entities. EPCIS events are proficient at capturing, storing and providing access to event information regarding material logistics objects in a decentral fashion. This distributed, fine-grained event data is based upon physical handing steps (SupplyScape 2008), meaning the relevance of the event object model is primarily to the material flow of individual physical entities. Attributes such as the time, position, status and the current related business transaction can be modeled using the event object model. Whilst the abstract data model can be extended in a modular fashion by user vocabularies to encompass application-specific attributes, the event-driven concept imposes restrictions upon the applicability of EPCIS and does not explicitly support the interaction of autonomously controlled logistics objects. The EPCglobal Architecture Framework focuses on providing support for data exchange centred on identifiable, physical entities. Immaterial logistic objects such as orders are indeed considered by the architecture and its data structures, for example, in EPCIS transaction events and their related properties, or in specific EPC Identifier types. However, the perspective of the framework upon immaterial logistics objects is always from the material object to which they are related.

From the perspective of the information system layer criteria for autonomous cooperating processes, through the EPC Identifier and EPCIS the Framework Architecture explicitly fulfills the requirements towards a high level of autonomous control for decentral data storage pertaining to the visibility of individual physical logistics objects. With regards to the second criterion, decentral data processing, it provides implicit support by specifying standard interfaces by means of that data may be discovered and accessed, whereas the specification of actual data processing facilities for autonomous control are outside the scope of the standards. The third and final criterion, the interaction ability of autonomous controlled logistic objects lies well outside of the scope of the specifications. To conclude, the EPCglobal Architecture Framework's contribution towards the standardization of auto-identification technology and increased visibility of the material flow represent a significant step towards a platform capable of functioning as a foundation of autonomous control in logistics.

References

Armenio, Felice et al. *The EPCglobal Architecture Framework.* Lawrenceville, New Jersey: EPCglobal Inc, 2007.

Böse, Felix, and Katja Windt. "Catalogue of Criteria for Autonomous Control in Logistics." In *Understanding Autonomous Cooperation and Control in Logistics—The Impact on Management, Information and Communication and Material Flow,* by Michael Hülsmann and Katja Windt, 57–72. Berlin: Springer, 2007.

EPCglobal Inc. *EPC Information Services (EPCIS) Version 1.0.1 Specification.* Lawrenceville, New Jersey: EPCglobal Inc, 2007.

EPCglobal Inc. *EPCglobal Object Name Service (ONS) 1.0.1.* Lawrenceville, New Jersey: EPCglobal Inc, 2008.

EPCglobal Inc. *EPCglobal Tag Data Standards Version 1.4.* Lawrenceville, New Jersey: EPCglobal Inc, 2008.

EPCglobal Inc. *EPCIS (Electronic Product Code Information Service) Frequently Asked Questions.* Lawrenceville, New Jersey: EPCglobal Inc, 2007.

Hans, Carl, Karl Hribernik, and Klaus-Dieter Thoben. "An Approach for the Integration of Data within Complex Logistics Systems." *LDIC2007 Dynamics in Logistics: First International Conference Proceedings.* Heidelberg: Springer, 2007. 381–389.

Hribernik, Karl A., Martin Schnatmeyer, Andreas Plettner, and Klaus-Dieter Thoben. "Application of the Electronic Product Code EPC to the Product Lifecycle of Electronic Products." *RFID Convocation.* Brussels, Belgium: European Commission, 2007.

Kärkkäinen, Mikko, Jan Holmström, Kary Främling, and Karlos Artto. "Intelligent Products—A Step Towards a More Effective Project Delivery Chain." edited by J C Wortmann. *Computers in Industry* (Elsevier) 50, no. 3, (February 2003) 41–151.

Langer, Hagen, Jan D. Gehrke, Joachim Hammer, Martin Lorenz, Ingo J. Timm, and Otthein Herzog. "A Framework for Distributed Knowledge Management in Autonomous Logistic Processes." *International Journal of Knowledge-Based & Intelligent Engineering Systems,* 10, 2006:277–290.

McFarlane, Duncan, Sanjay Sarma, Jin Lung Chrin, C Y Wong, and Kevin Aston. "Auto ID Systems and Intelligent Manufacturing Control." *Engineering Applications of Artificial Intelligence,* 2003: 365–376.

Scholz-Reiter, Bernd, Jan Kolditz, and Thorsten Hildebrandt. "Specifying adaptive business processes within the production logistics domain – a new modelling concept and its challenges." In *Understanding Autonomous Cooperation & Control in Logistics—The Impact on Management, Information and Communication and Material Flow,* by Martin Hülsmann and Katja Windt, 275–301. Berlin: Springer, 2007.

Soon, Tan Jin, and Shin-ichii Ishii. "EPCIS and Its Applications." *Synthesis Journal,* 2007: 109–124.

SupplyScape. *Combining EPCIS with the Drug Pedigree Messaging Standard.* Woburn, MA: SupplyScape, 2008.

Ventä, Olli. *Intelligent Products and Systems. Technology Theme—Final Report.* VTT, Espoo: VTT Publications, 2007, 304 s.

Windt, Katja, Felix Böse, and Thorsten Phillipp. "Criteria and Application of Autonomous Cooperating Logistic Processes." *Proceedings of the 3rd International Conference on Manufacturing Research—Advances in Manufacturing Technology and Management.* Cranfield, 2005.

Wong, C Y, Duncan McFarlane, A Ahmad Zaharudin, and V Agarwal. "The Intelligent Product Driven Supply Chain." *Proceedings of IEEE International Conference on Systems, Man and Cybernetics, 2002.* Tunisia: IEEE, 2002.

Design of Middleware Architecture for Active Sensor Tags

Haipeng Zhang, Bonghee Hong and Wooseok Ryu

1 Introduction

The Radio Frequency Identification (RFID) as a frontier technology is an automatic identification technology which is now becoming part of our daily life through a variety of applications. Generally, a basic RFID system is composed of RFID tag, RFID readers, RFID middleware, and RFID applications (Finkenzeller 2003). Tags store the unique ID (EPC code) and related data in their memory to uniquely identify the object. Readers are used for reading the information stored at RFID tags placed in their interrogator zone and sending the streaming data to middleware through wired or wireless interfaces. Middleware systems collect data from readers, process receiving data according to the requests of applications, and generate reports sending to applications. Applications are the software components which issue requests to middleware and provide business services to users, such as supply chain management or warehouses management.

Using RFID techniques, many industries can track and monitor their products in real time or increase products visibility while they are in transit along the supply chain. Besides these, people maybe want to know the surrounding environmental conditions or condition of the product itself like temperature, pressure, humidity, light etc. A sensor-enabled RFID tag can be the key to addressing these requirements. A sensor-enabled RFID tag (later also: "sensor tag") is an RFID tag which

H. Zhang (✉), B. Hong and W. Ryu
Department of Computer Engineering, Pusan National University,
Busan, 609-735, Republic of Korea
e-mail: jsjzhp@pusan.ac.kr

B. Hong
e-mail: bhhong@pusan.ac.kr

W. Ryu
e-mail: wsryu@pusan.ac.kr

H.-J. Kreowski et al. (eds.), *Dynamics in Logistics*,
DOI: 10.1007/978-3-642-11996-5_34, © Springer-Verlag Berlin Heidelberg 2011

contains one or more sensors to monitor some physical environmental parameter but also contains the same identification function as a normal RFID tag does. This kind of sensor tag may fall into class2, class3 or class4 in EPCglobal's tag classification (EPC Tag Class Definitions 2008). As class2 and class3 are passive, these tags with sensors are simply called passive sensor tag, while class4 sensor tag is called active sensor tag. Passive sensor tag is backscatter and short reading distance. The advantages of active sensor tag are increased reading reliability, reading distance and also can initial a communication which can send alert information, such as battery low, memory full, etc.

Using sensor-enabled RFID tags, manufactures can monitor quality of products during transportation. Firstly, tags are attached on product and configured by the manufacturer to start recording a set of specific environmental variables. Then they measure and store the environmental variables inside memory while they are in transit along the supply chain. When passing through a reader's gate, the measured data can be retrieved from the tag so that the evaluation of the product's quality can be immediately assessed according to real measurements.

With the widely deployed of RFID technology, there exists a huge amount of data being generated from heterogeneous sources that needs to be processed to support user applications. However, handling large volumes of sensor tags generated data streams imposes some challenges. First, the amount of data stream which comes from different sources should be efficiently processed in real time. Second, the useful data should be transmitted to the correct user applications. Finally, the processing system has to be able to manage complex components in a scalable manner to answer user queries. The solution to these challenges is to deploy a middleware system.

The RFID middleware is an intermediary between RFID reader and application systems, which is responsible for effectively process RFID tag stream data by using a number of functions and methods. Generally, there are three primary motivations for using RFID middleware: providing connectivity with RFID readers; processing raw RFID data such as filtering, grouping and aggregating; and providing application-level interface to manage readers and capture filtered RFID events. But currently, most RFID middleware systems only support passive tags, but not support active tags.

In this paper, according to the features of active sensor tags, we have analyzed the requirements of designing a middleware for RFID system to support active sensor tags. The main contribution of this paper is designing a middleware architecture which is based on the EPCglobal standard RFID middleware (EPCglobal Standard 2009) for supporting active sensor tags as well as passive tags. Since the standard middleware can't support for active sensor tags.

The rest of this paper is organized as follows. Section 2 describes related works. Section 3 represents the requirements of designing the architecture. Section 4, represents the shortages of ALE for supporting active sensor tags. Section 5 represents how to extend ALE for active sensor tags. Section 6 represents our extended architecture. Finally Sect. 7 concludes the paper with the future work.

2 Related Works

In order to widely deployed RFID system, EPCglobal specifies EPC Network which is built on many fundamental technologies and standards. ALE 1.1 (EPCglobal Inc 2009) is a standard RFID middleware specified by EPCglobal. The ALE is an intermediary between RFID reader and capture application. Its purpose is to process the reduction of the data volume directly coming from various sources such as RFID readers and send the reports to applications. ALE provides the standard interfaces for processing real time RFID data and delivering the reports to correct applications. It defines five standard APIs: Reading API, Writing API, Tag Memory API, Logical Reader API and Access Control API. Although ALE is the de facto international standard, it does not support active tags.

Beside the standard RFID middleware, there are some related RFID middleware systems and solutions that have been developed. LIT (Ashad et al. 2007) middleware is fully based on EPCglobal standard, which can efficiently process RFID tag data with some unique techniques. Siemens RFID middleware (Fusheng and Peiya 2005) is an integrated RFID data management system which is based on an expressive temporal data model for RFID data. It enables semantic RFID data filtering and automatic data transformation based on declarative rules, provides powerful query support of RFID object tracking and monitoring, and can be adapted to different RFID-enabled applications. But these middleware systems cannot handle active sensor tags. UbiCore system (HunSoon and SeungIi 2007), an XML based middleware system, can handle various kinds of sensors, and support continuous queries. Also, the system provides efficient query processing for data stream by reusing filtering and intermediate result, and automatically delivers the results to the related back-end applications. Oracle Sensor Edge Server[1] is a sensor based service integration platform supporting data acquisition, event processing, data dispatching and extracting specific data using pre-defined filters as well as new filter which can be defined by users to process data in detail. Although these systems can process sensor data, systems lack of consideration for the de facto international standard and the methods to support active sensor tags. Open Geospatial Consortium (OCG) (http://www.opengeospatial.org/) specifies Sensor Web Enablement (SWE) standards for developers to make all types of sensor and sensor data repositories discoverable accessible and useable via the Web. But it lack of consider for integrating with RFID tag. There exist some researches on integrating RFID and wireless sensor network. In Loc et al. (2005), it uses WSN and RFID to build an in-home elder healthcare system. Although it can handle both RFID and sensor data, it does not consider standard middleware architecture. ESN architecture, Weixin et al. (2008) and Jongwoo et al. (2007) are integration systems of RFID and WSN which are based on EPCglobal architecture. Both of them can process RFID tag data and sensor data. However the current EPCglobal

---

[1] Enterprise Information Architecture for RFID and Sensor-Based Services, Oracle White Paper, February 2006

middleware does not support active tags, as well as active sensor tags. BRIDGE (http://www.bridge-project.eu/) project uses its own sensor-enabled RFID tag based on EPC Gen 2 tag which is also a passive tag.

3 Middleware Requirements for Supporting Active Sensor Tags

Generally speaking, the RFID Middleware has some characteristics and design requirements (Christian and Matthias 2005). For designing a middleware to support active sensor tags, there also have some requirements that should be considered.

3.1 Sensor Data Processing for Active Sensor Tags

Active sensor tags have both identification information and sensing information. ALE only supports filtering and count operations by using tag ID. For supporting active sensor tags, the middleware system should support more operations for sensor data, such as extends the filtering operation for sensor data that can filter out of non-meaningful values from the series of sensor data and sensor data fusion which sensor data are fused to provide more accurate and reliable data than a single sensor. Active sensor tags have user memory which stores historical sensor data, so the aggregation operations for sensor data are needed to support, such as AVG, SUM, MAX, MIN, etc.

3.2 Alarm Processing for Active Sensor Tags

As active tag can initiate communications by itself, so it has been used in many monitoring applications by sending alarm messages to inform the user that some important events were happened. Those alarms which sent by readers should be processed in time in the middleware system which then informs the user. In order to process the alarm, the middleware should provide interface for applications to define the alarm conditions, such as the alarm threshold.

3.3 Management for Active Sensor Tags

Active Sensor tags have large user memory to store the sensing data, sensor device to sense the environments, battery to provide power to tag. So the middleware should provide interfaces for managing the tags. Through those interfaces, user can define different memory banks for storing sensor data and RFID data, define sensor sampling interval or conditions, for examples, senses the temperature once per hour or stores the temperature when it is higher than 1$\overset{\circ}{0}$, monitor tag battery healthy and memory status, sleep and wake up the tag.

3.4 Reader Extensibility for Supporting Active Sensor Tags

There are heterogeneous active RFID readers and many existing applications. The middleware system should provide independent from types of readers as well as different applications. Various new readers should be easily added and processed without any modifications, which can avoid many complex issues about multi-connection maintenance (Ashad et al. 2007). As the number of RFID readers increasing, the volume of data generated from the readers is sharply increasing. So the middleware system should have the ability to tolerate the huge amount of data as the number of readers is increased.

As view of these requirements for active sensor tag, in next section, we analyze the ALE standard and get the reason that why ALE is not suitable for active sensor tags.

4 Shortage of ALE for Supporting Active Sensor Tags

The ALE standard defined by EPCglobal is used to process the reduction of the data volume that comes directly from heterogeneous EPC data sources such as RFID readers and to generate the events of interest to applications. It provides interfaces for defining the operation and delivery of the filtered and collected tag read data, and then user applications can access to ALE to obtain the interested data.

In order to obtain the data or execute operations on tags, client needs to specify the standard reading query (ECSpec) or writing query(CCSpec) in XML format (refer to Figs. 1, 2). Both of them have to specify a list of readers, the time interval (event cycle or command cycle) for gathering tags and executing user operations. The difference between them is ECSpec need to specify the reporting type

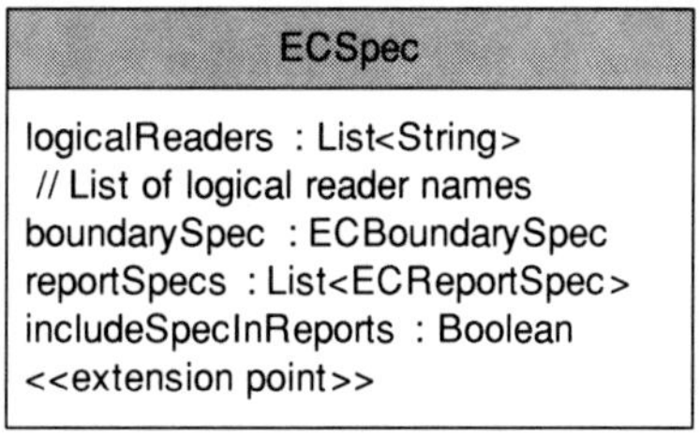

Fig. 1 ECSpec

ECSpec
logicalReaders : List<String> // List of logical reader names boundarySpec : ECBoundarySpec reportSpecs : List<ECReportSpec> includeSpecInReports : Boolean <<extension point>>

Fig. 2 CCSpec

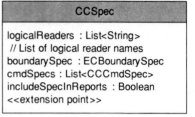

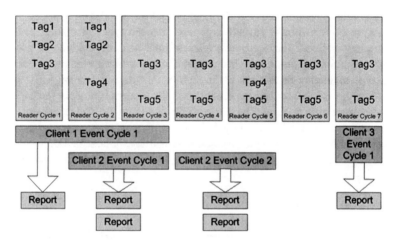

Fig. 3 ECSpec operation description of ALE

(ECReportSpec), but CCSpec need to specify the user commands (CCCmdSpec). After client registers the query to ALE, it can ask for the subscription to the specific query which the client is interested in. After ALE receives the subscription, it starts to receive EPC data from data sources. The data is collected in every read cycle, filtered by removing duplicated or irrelevant data and organized in groups in very event cycle. At the end of the event cycle, a report is generated. Figure 3 shows the operation description of ALE.

For filtering the EPC data, ALE has defined an ECFilterSpec for defining filter conditions, which can be used in both ECSpec and CCSpec. The filtering condition of tag represented as an *EPC tag pattern* (EPCglobal Inc 2009). Only the matched EPC data can be added in the final report. However, the current filtering and aggregation operations are only for EPC data, there are no operations for sensor data. So ALE need define more filtering conditions and aggregations for sensor tags.

As active sensor tags can initiate the communication to the reader, so they can send some emergency alarm messages when the predefined threshold is exceeded or some emergent events occurred. Whenever an alarm message sent to ALE, the report is generated only at the end of an event cycle, which causes emergent alarms cannot be sent to user (refer to Fig. 4). Active sensor tags also have internal

Fig. 4 Alarm processing

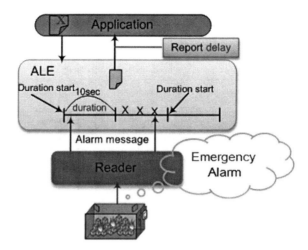

battery and user memory, so it supports wireless programming by using active readers. For example, we can on/off the tag, set sensor sample interval, etc. However, current ALE writing operations do not support these functions for active sensor tags. So ALE need extend more writing operations for active sensor tags.

ALE collects EPC data from a variety of readers which are passive readers. But active sensor tags only can be read by active readers, so for supporting active sensor tags, ALE should have ability to communicate with active readers. The different types of readers can easily attach to the ALE at real time without any modification.

5 Extensions of ALE Interfaces for Active Sensor Tags

So far, the shortages of ALE middleware are described in previous section. In this section, we will introduce our solutions for supporting active sensor tags through extending ALE interfaces as well as defining some new interfaces.

5.1 Extend ALE Reading Interface for Active Sensor Tag

Through the ALE reading interface, clients can define and manage event cycle specification, read tags on-demand by activating ECSpec synchronously, and enter standing requests (subscriptions) for ECSpec to be activated asynchronously. Results from standing requests are specified as ECReports according to the ECSpecs.

For filtering and collecting the data, ALE has specified some filtering conditions and aggregation operation in ECFilterSpec. However the defined filtering conditions and aggregation in ECFilterSpec are only for EPC data. ECFilterSpec does

Table 1 Filtering conditions and aggregations

	Filtering condition	Aggregation
ALE	Value, * [Lo–Hi], Mask	Count
Extended	MIX, MAX, ≥, ≤, <, >	AVG, SUM, MIN, MAX

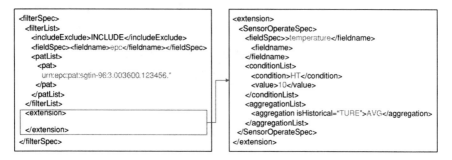

Fig. 5 Example of FilterSpec and Reports

not define any operations for sensor data. For supporting sensor tag, we can define a new Spec for processing sensor tag data and a new report Spec for representing the result. But the new Spec maybe has several same operations which cause redundant for system designing. As the standard Specs have the extension point, we can extend the Spec and define some new operations for processing sensor data. Table 1 shows the filtering conditions and aggregations defined in ALE 1.1 and the extended filtering conditions and aggregations for sensor data. Filter conditions for sensor data are executed on single tag, while aggregation operations are executed on a group of tags collected within a event cycle. Figure 5 shows an extension example of manipulating sensor data.

5.2 Extend ALE Writing Interface for Active Sensor Tag

By using ALE writing interface, user can define his commands operated on tags and get the result. In ALE specification, it supports nine operations such as READ, WRITE, PASSWORD, KILL, LOCK, etc. Although ALE supports those operations, it doesn't support for setting sensing conditions. Active sensor tags can continuous sense the environment no matter whether it in the reader zone or not. The sample interval may be different according to different applications. So the middleware should have the ability to set the sensing conditions. For saving the battery power, the operations for controlling the tag are also needed, such as on/off the tag, sensor data logging on/off, tag active/sleep. These operations can be extended in ALE writing Spec.

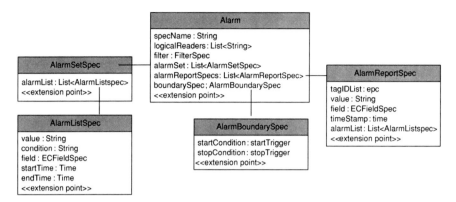

Fig. 6 AlarmSpec

Active sensor tags have internal battery and user memory, they should let user know when their battery power is low or memory is full by sending an alarm message. This is the original goal of alarm manage. Since active sensor tag can initiate communication by itself, it is usually used for monitoring abnormal events which cause alarm and send to user. From previous section, we know ALE is not suitable for processing alarm events. So in our middleware, we use alarm manager for managing the alarms. In order to compliant with ALE, we design an Alarm-Spec (Fig. 6) for defining an alarm and generating the alarm report. Alarm-BoundSpec is responsible for defining alarms, AlarmSetSpec contains a list of AlarmListSpec which is responsible for defining a single alarm and AlarmRe-sportSpec is responsible for defining the alarm report.

5.3 Reader Abstraction Interface

The primal role of these interfaces is used for providing seamless link between the middleware and various types of readers as well as providing reader independence to the middleware. This component abstracts the reader protocols, supports various types of reader devices which can be easily added at run time.

According to the protocols used for readers, there are two types of readers: one is Standard Reader Protocol (RP) compatible readers and other is non-RP compatible readers. It is easy to communicate with RP compatible readers using Standard Reader Protocol. However, non-RP compatible readers use their own vendor protocols. So we need create a new adapter to a non-RP reader according to its own properties files when it come into the middleware. The detail structure of reader abstraction is shown in Fig. 7. The adapter is designed for non-RP compatible readers. Using the adapter, different types of non-RP compatible readers can be easily added at run time. The adapter manager is designed for creating/

Fig. 7 Reader Abstraction
Framework

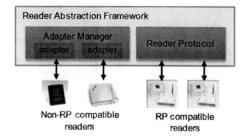

deleting adapters when add/remove a reader for middleware and coordinating
these adapters and providing transparency for the various types of readers.

6 Design of Middleware Architecture

So far, the extensions of ALE for supporting active sensor tag are described above.
In this section, we will introduce our designed architecture for supporting active
sensor tags, which is based on LIT ALE middleware (Ashad et al. 2007). Figure 8
shows the overview of the designed architecture. It is mainly divided into three

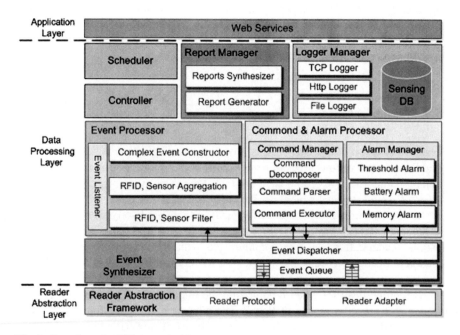

Fig. 8 Overview of middleware Architecture

layers. In the bottom-up turn they are: Reader abstraction layer, Data processing layer and Application layer.

Reader Abstraction Layer is responsible for gathering data from different sources. It abstracts the standard reader protocols, supports heterogeneous reader devices and allows users to configure and control all these devices. It consists of two modules: Reader Protocol and Adapters. The Reader Protocol module is used to communicate between RP compatible readers and the middleware, while for each non-RP compatible readers individual adapter is generated and used for communication between reader and middleware. Therefore, this layer can provide seamless link between various types of readers and middleware.

Data Processing Layer is the core layer which performs filtering of redundant tag reads, aggregating the filtered data, executing writing operations on tag, managing the sensing alarms and delivering the reports to applications or different destinations. Event Synthesizer collects data from readers, generates tag events and dispatches the event to other modules. Event Processor filters and aggregates the streaming sensor events according to the operator conditions specified by user. Command and Alarm Processor executes writing operations on user required tags, defines different types of alarms for specified active sensor tags and processes the alarms generated by tags. Finally, Reports Manager generates reports and middleware uses different types of loggers to send the reports in response to user's queries. Controller and Scheduler initiate different modules, manage whole system and schedule of the middleware.

Application Layer contains web service and provides clients access to the middleware via common application interfaces which provided by EPCglobal. Web applications can use standard protocol to send user queries to the middleware and get the report.

7 Conclusions

In this paper, we have discussed about the features of active sensor tag, analyzed the requirements for designing a middleware for supporting active sensor tag, and discussed the shortages of ALE for supporting active sensor tags. According to those requirements and shortages, we extended the ALE interfaces and defined new interfaces for active sensor tags. At last, we designed a middleware architecture which is based on the ALE middleware specification. Our middleware system has the functions of filtering, grouping and aggregating sensor data in real-time as well as configuring sensor tag and managing emergency alarms. By using our architecture, both active sensor tag and passive tag can be supported. In the future, our work will focuses on system implementation and system performance optimization.

Acknowledgments This work was supported by the Grant of the Korean Ministry of Education, Science and Technology (The Regional Core Research Program/Institute of Logistics Information Technology).

References

Ashad K, Bonghee H, Wooseok R (2007) LIT Middleware: Design and Implementation of RFID Middleware Based on the EPC Network Architecture. The 1st International Conference on Dynamics in Logistics, pp 221–229

Christian F, Matthias L (2005) RFID middleware design: addressing application requirements and RFID constraints. Proceedings of the 2005 joint conference on Smart objects and ambient intelligence, pp 219–224

EPC Tag Class Definitions, http://www.epcglobalinc.org/standards/TagClassDefinitions_1_0-whitepaper-20071101.pdf Accessed 23 December 2008

EPCglobal Inc. The Application Level Events (ALE) Standard Specification Version 1.1, http://www.epcglobalinc.org/standards/ale/ale_1_1-standard-core-20080227.pdf. Accessed 10 January 2009

EPCglobal Standard, http://www.epcglobalinc.org/home. Accessed 10 January 2009

Finkenzeller. K. RFID handbook. Wiley Hoboken, NJ, 2003.

Fusheng W, Peiya L (2005) Temporal Management of RFID Data. Proceedings of the 31st International Conference on Very Large Data Bases, pp 1128–1139

HunSoon L, Seungli J (2007) An Effective XML-Based Sensor Data Stream Processing Middleware for Ubiquitous Service. Computational Science and Its Applications, vol. 4707, Springer, pp 844–857

Jongwoo S, Tomas L, Daeyoung K (2007) The EPC Sensor Network for RFID and WSN Integration Infrastructure. Proceedings of the Fifth Annual IEEE International Conference on on Pervasive Computing and Communications Workshops, pp 618–621

Loc H, Melody M, Zachary W, etc. (2005) A prototype on RFID and sensor networks for elder healthcare: progress report. SIGCOMM'05 Workshops, 2005, pp 70–75

Weixin W, Jongwoo S, Daeyoung K (2008) Complex Event Processing in EPC Sensor Network Middleware for Both RFID and WSN. 11th IEEE Symposium in Object Oriented Real-Time Distributed Computing, pp 165–169

Part V
Production Logistics

Investigation of the Influence of Capacities and Layout on a Job-Shop-System's Dynamics

Bernd Scholz-Reiter, Christian Toonen and Jan Topi Tervo

1 Dynamic Behaviour of Job-Shop-Systems

Job-shop-systems are making high demands on its underlying production and logistic system due to their usually complex structure of material flow. This situation is intensified by dynamical effects within the system's behaviour, e.g. the so-called Bullwhip effect (Lee et al. 1997; Chen et al. 2000), that describes the amplification of oscillatory amplitudes of delivery rates along the supply chain. Generally, it can be distinguished between two elementary forms of dynamics: system internal and system external dynamics. Both commonly affect the logistic performance of the system in a negative way. Varying stock levels, increasing lead and cycle times or decreasing workload can be the results (Helbing and Lämmer 2005).

External dynamics can occur because of the formation of production networks. Since companies concentrate on their core competences and cooperate with suppliers and distributors they generate strong linkages and connections between one another. The resulting networks are characterised by permanently increasing complexity and dynamics, forcing them to adapt to today's rapidly changing markets (Daganzo 2003).

But also the interdependence of processes within companies was intensified in the last years. This is due to several reasons. Increasing competitive pressure in global dynamically changing markets and the demand for individualised products

B. Scholz-Reiter, C. Toonen (✉) and J. T. Tervo
BIBA-Bremer Institut für Produktion und Logistik GmbH, University of Bremen,
Hochschulring 20, 28359 Bremen, Germany
e-mail: too@biba.uni-bremen.de

B. Scholz-Reiter
e-mail: bsr@biba.uni-bremen.de

J. T. Tervo
e-mail: ter@biba.uni-bremen.de

H.-J. Kreowski et al. (eds.), *Dynamics in Logistics*,
DOI: 10.1007/978-3-642-11996-5_35, © Springer-Verlag Berlin Heidelberg 2011

and services lead many enterprises to customise their products for as many customers as possible. Thus, diversity of product options increases.

Otherwise, the interaction of competitive pressure and dynamically changing markets causes a reduction of the product life cycle time. The product design is strongly enhanced by modern computer technology, so new products or product updates can be developed in shorter time periods. This results in the fact, that new product models or updates are introduced earlier into the market. Thus, the time period shortened in that products can be sold. Since the competitive pressure will increase further in future, only companies will survive which have modern products that match the spirit of the time (Meyer 2007).

Another driver of product innovations are new emerging technologies. The following product generation is developed in constantly decreasing intervals. Thus, products obsolesce more quickly, series-production readiness has to be achieved faster and development times are enormously shortened.

All these facts will lead to an increasing complexity of the products themselves. The complexity of the market, the environment and the demand will be transferred to the complexity of the products (Aurich et al. 2007) which induces a complex material flow (for example re-entrant systems). The effort of production planning and control will rise to handle these emerging challenging tasks. This especially affects production processes and production logistics.

In different models of production systems evidence was found that internal complexity induces internal dynamics. For example, non-linear behaviour has been discovered (Radons and Neugebauer 2004) and different models were found to exhibit complex, oscillatory and even chaotic behaviour (Ushio et al. 1995; Larsen et al. 1999; Katzorke and Pikovski 2000; Helbing et al. 2004a). Thus, stability analyses of different models and topologies were performed (Helbing et al. 2004b; Nagatani and Helbing 2004) to detect critical sets of parameters and stabilising influences. Scholz-Reiter et al. (2002a, b) identified in a re-entrant model of a production system four main groups of variables, which mainly influence the internal dynamic behaviour of these kinds of systems: structure, capacity, priority rules and order release time.

There are various methods and approaches to cope with dynamics, either internal or external dynamics. Mostly all address the groups "priority rules" and "order release time". For example, an integrated and holistic supply chain management or autonomous control can be useful to handle external dynamics. On the other side, powerful methods and tools of production planning and control are needed to manage internal dynamics, for example multiagent systems (Monostori et al. 2006), neural networks (Scholz-Reiter et al. 2004, 2006) or control theory (Pritschow and Wiendahl 1995; Duffie 1996). These methods mainly address the short- and middle-term planning and control process (see Fig. 1). But the influence of long-term strategies that reach into the area of factory planning may not be neglected. Here, planning methods are mostly statical and therefore cannot cope with dynamics sufficiently.

This leads to a new approach which will be presented in this paper. The aim is to analyse the possibilities of influencing the parameter groups "structure" and

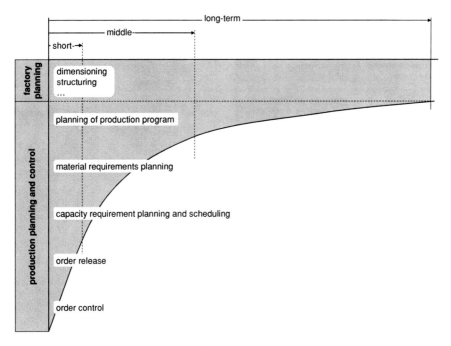

Fig. 1 Determinants of a system's dynamics and logistic performance and their range in time (following Scherer 1991)

"capacity" of production systems at the early stage of factory planning to avoid the negative impacts of internal dynamics. This means to manipulate positively the dynamic behaviour already in the phase of layout, design and configuration of factories. The next section will introduce the research topic and the course of action, followed by descriptions of the model and the simulation studies.

2 Aims and Course of Action

The dynamic behaviour of a production system in operational condition reflects its logistic performance. To achieve a satisfactory performance those factors which influence internal dynamics must be designed in a beneficial way. This paper focuses on job-shop-productions. Due to complex material flows these systems are particular sensitive for internal dynamics reducing logistic performance. In the following an approach will be described, which strives to investigate the influence of capacities and structure of a job-shop-system on its dynamic behaviour.

To obtain representative and general results the project accesses in its first step production programmes which are typical for job-shop-productions. The long-term production programme is basis for an initial design which is generated employing standard planning methods of dimensioning and structuring. The initial design is subject to simulation and data analysis studying its dynamics and logistic

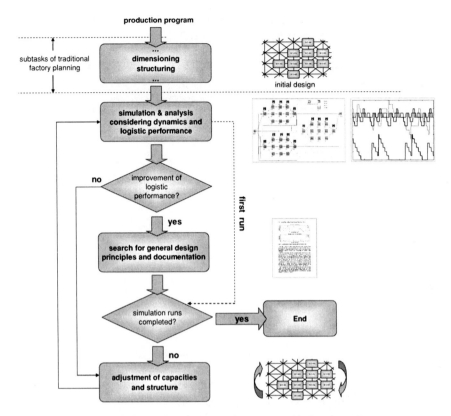

Fig. 2 Course of action to find general design principles considering dynamics

performance. The simulation runs follow a specific simulation plan based on structured parameter variation which will be complemented throughout the simulation cycles. According to the simulation plan the job-shop-system's design is altered in its capacities and layout to achieve improvements. While adjusting the design and processing simulation runs the effects on dynamics and logistic performance are examined. Goal is to indentify dependences between capacity and structure settings on the one hand and the resulting dynamic behaviour and effects which reduce performance on the other hand. Measures which improve the logistic performance of the system are documented and analysed with respect to their similarities and general principles. The course of action is depicted in Fig. 2 and will be detailed in the following sections.

3 Model Description

A job-shop-production consists in its elementary form of workstations and buffers. Characteristic feature of a job-shop-system is the arrangement of workstations with

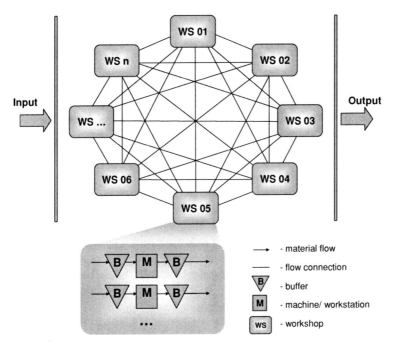

Fig. 3 Basic elements of a job-shop-production

similar features in groups forming so-called *workshops* (Wiendahl 2007). Figure 3 shows a set of workshops. Due to their characteristics all workshops are connected to one another underlining the possibilities of material flow. To address the area of highest sensitivity to dynamics the approach concentrates on production and machining opposed to assembly. Each workshop comprises one or more machines or workstations with an up-stream and a down-stream buffer. To realise short lead times the model does not include any warehousing except for the inventories resulting in the buffers. Machines allow to be attributed with a specified capacity; their locations are defined by the workshop they belong to and are described by a position on a grid. The location of workshops and the capacity of machines are subject to the adjustments within the simulation runs later on. While the capacity of buffers is not limited their location is linked to the machine they enclose.

The *input* into the job-shop-system consists of production orders with a specific succession and influx-rate given by the master production schedule. The orders vary in their lot sizes and machining requirements depending on their work plan. The work plan comprises information about the succession of machining steps and the necessary amount of machining time for each step.

Within the simulation runs machines and buffers can be described by their input, inventory and output over time. For all workshops this information reflects the dynamics of the job-shop-system. The structure of the material flow results from the information within the work plans. Here, re-entrant structures may cause complex dynamics. To exclude strong effects of production planning and control on the system's dynamics priority rules at workstations are limited to the first-in-first-out-principle. Rule for allocating incoming orders to a workshop is to address the machine with the lowest inventory waiting in front of it. The *output* of the production system is determined by all upstream inputs and processes. By considering input and output dates of an order when it enters and leaves buffers and machines the total lead time can be calculated. This enables to distinguish waiting and machining times and gives information about bottlenecks.

4 Production Program and Initial Design

A typical production programme for job-shop-production is characterised by a high variety of products manufactured in low quantities up to medium lot-sizes. As described above the production programme is linked to information about the work plan giving data about modality and order of specific machining requirements and processing times. This information is the basis of an initial design regarding dimensions and structure of the job-shop-system.

The *dimensioning*, meaning the determination of kind and extend of capacities, is based on the work plan and total processing times for the expected quantities given by the master production schedule. While the necessary types of machines are directly defined by the work plan, their quantity for a workshop can be calculated roughly by comparing demand and supply for a set of given standard-machines within a period of time.

The *structure* of the job-shop-system is specified by the material flows between all workstations. Based on a matrix of material flow, giving information about the number of transports between the workshops, the structuring applies a standard method to arrange the workshops within a simple grid, for example Schmigalla (1969). Primary objective here is to realise a small total transportation distance and thus to achieve short transport times. Figure 4 shows a set of workshops which have been arranged in an initial design. As the machines of one workshop are close

Fig. 4 Initial design

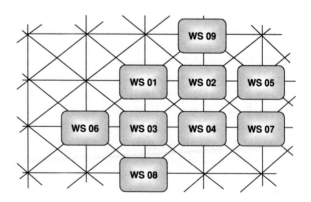

to one another, it can be assumed that transportation times of one workshop to another are equal for all incorporated machines.

5 Simulation Runs, Data Analysis and Adjustments in Design

Once an initial design for the job-shop-system has been established, it is analysed with respect to its dynamic behaviour and logistic performance. The *simulation model* is implemented within an environment of event-discrete programming. Figure 5 shows a simple simulation model with three workshops, each of which contains three machines and linked buffers. As described above the input into the model is defined by the master production schedule.

The simulation data is input for the following *analysis*. To analyse the *logistic performance* data about inventories, lead times and capacity utilisation for each workshop and the total system will be gathered and interpreted (Wiendahl 2007). The use of capacity will be limited in a way that a minimum level cannot be underrun. That way the unlimited growth of capacities as an approach to reduce lead times within the design is eliminated. The data on logistics will help to indentify the existence and causes of bottlenecks and thus lead to further design adjustments.

Instruments applied for the *dynamic analysis* comprise for example time series analysis and methods of nonlinear dynamics. With linear methods regular

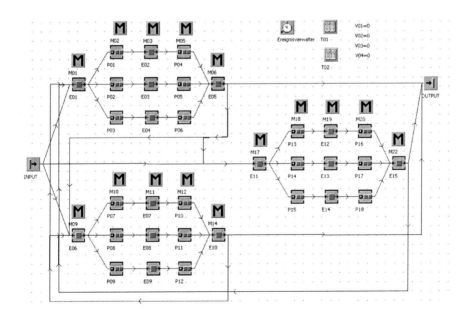

Fig. 5 Simulation model

structures in a data set can be interpreted. Examples are dominant frequencies with Fourier analysis or correlations with auto- or cross-correlation analysis. Also standard statistic tools like average and variance can give first impressions of the dynamical properties of the system. However, irregular behaviour can hardly be addressed with these methods. Irregularities are basically caused by stochastic or non-linear deterministic processes. In the regarded case stochastic influences are excluded so that non-linear analysis methods can be applied to characterise the system dynamics.

A first impression of the system can be achieved by reconstructing its dynamics in phase-space. This means to embed the observed data in a coordinate system which suits best the given constraints. The Lyapunov Exponent of a dynamical system is a measurement that characterises the rate of separation of infinitesimally close trajectories. Thus, it is a qualitative and quantitative measurement for the convergence or divergence of nearby starting trajectories. It can be used to examine the stability of dynamical systems and can show if a systems has a sensitive dependence on initial conditions, which is an indication for possible chaotic behaviour (Kantz and Schreiber 1997).

The following *adjustments* are subject to an initial simulation plan which is complemented throughout the cycles. Here, adaptations in capacities and layout will focus on the identified bottlenecks to ensure relevance and reduced complexity. All new relevant design configurations will be noted in the simulation plan and run in succession. While structured parameter variations will cover a wide range of different design configurations, a statistical design of experiments will guarantee validity of the results. The cycle of simulation, analysis and adjustment ends when the simulation runs are completed according to the simulation plan.

Each time an adjustment proves to be beneficial it is documented and subject to further analysis with respect to similarities. Main idea is to identify design constellations and therefore *design principles* which help to reduce internal dynamics. To obtain general results varying production programmes will serve as input, Fig. 6. Furthermore, succession and influx-rate of production orders will be varied to simulate different workloads.

6 Summary and Outlook

The logistic performance of production systems is directly linked to their dynamic behaviour. Especially, job-shop-systems are characterised by complex dynamics which occur due to highly branched structures within the material flow. This is the reason for dynamic effects which reduce the logistic efficiency.

Conditions for a system's internal dynamics are set by factory planning and production planning and control. Here, dimensioning and structuring determine long-term conditions for dynamics. This paper described an approach which develops and simulates varying job-shop-systems assuming typical production programmes and master production schedules. By analysing their dynamic

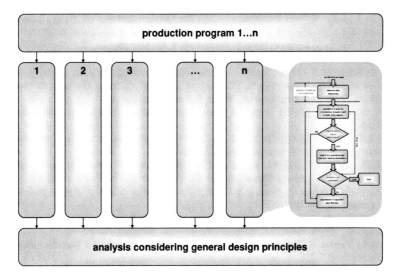

Fig. 6 Varying production programmes and superior analysis

behaviour and logistic performance it strives to discover dependences of specific design configurations and their effects on the system's dynamics. Thus, the design process can be improved in a way that allows considering dynamic behaviour at a very early state.

While the described approach and its applied model are still elementary, further research work will extend the scope gradually. For instance, availabilities of workstations will be reduced by breakdown and maintenance related downtimes. Furthermore, means of transportation will be implemented which realise the material flow between the workshops and which will be limited in their quantity and availability.

Acknowledgments This work was funded by German Research Foundation (DFG) under the reference number SCHO 540/15-1 "Application of Methods of Nonlinear Dynamics for the Structuring and Dimensioning of the Logistic System in Job-Shop-Systems".

References

Aurich J, Grzegorski A, Lehmann F (2007) Management vielfaltsinduzierter Prozesskomplexität in globalen Netzwerken. Industrie Management 6:13-16
Chen F, Drezner Z, Ryan J, Simchi-Levi D (2000) Quantifying the Bullwhip Effect in a Simple Supply Chain: The Impact of Forecasting, Lead Times, and Information. Management Science 46(3):436 - 443
Daganzo C A (2003) Theory of Supply Chains. Springer, New York
Duffie N A (1996) Heterachical control of highly distributed manufacturing systems. International Journal of Computer Integrated Manufacturing, 9/4:270-281

Helbing D, Lämmer S, Witt U, Brenner T (2004a) Network-induced oscillatory behaviour in material flow networks and irregular business cycles. Physical Review E 70:056118

Helbing D, Lämmer S, Seidel T, Seba P, Platkowski T (2004b) Physics, Stability and Dynamics of Supply Networks. Physical Review E 70:066116

Helbing D, Lämmer S (2005) Supply and production networks: From the bullwhip effect to business cycles. In: Armbruster D, Mikhailov A S, Kaneko K (eds.) Networks of Interacting Machines: Production Organization. World Scientific, Singapore

Kantz H, Schreiber T (1997) Nonlinear Time Series Analysis. Cambridge University Press, Cambridge

Katzorke I, Pikovski A (2000) Chaos and complexity in a simple model of production dynamics. Discrete Dynamics in Nature and Society 5:179-187

Larsen E R, Morecroft J D W, Thomsen J S (1999) Complex behaviour in a production-distribution model. Europ J Op Res 119:61-74

Lee H L, Padmanabhan V, Whang S (1997) The bullwhip effect in supply chains. Sloan Management Review 38:93-102

Meyer C (2007) Integration des Komplexitätsmanagements in den strategischen Führungsprozess der Logistik. Haupt, Bern

Monostori L, Vancza K, Kumara S R T (2006) Agent-Based Systems for Manufacturing. CIRP Annals 55(2):697–720

Nagatani T, Helbing D (2004) Stability Analysis and Stabilization Strategies for Linear Supply Chains. Physica A 335:644–660

Pritschow G, Wiendahl H-P (1995) Application of control theory for production logistics - results of a joint project. CIRP Annals - Manufacturing Technology, 44/1:421–424

Radons G, Neugebauer R (eds.) (2004) Nonlinear Dynamics of Production Systems. Wiley, New York

Scherer E (1991) The Reality of Shop Floor Control – Approaches to Systems Innovation. In: Scherer E (ed.) Shop Floor Control – A Systems Perspective. Springer, Berlin

Schmigalla H (1969) Methoden zur optimalen Maschinenanordnung. VEB Verlag Technik, Berlin

Scholz-Reiter B, Freitag M, Schmieder A (2002a) Modelling and Control of Production Systems Based on Nonlinear Dynamics Theory. Cirp Annals 51/1:375-378

Scholz-Reiter B, Freitag M, Schmieder A (2002b) A dynamical approach for modelling and control of production systems. In: Boccaletti S et al. (eds.) Proc. 6th Experimental Chaos Conference. AIP Conference Proceedings 622(1):199–210

Scholz-Reiter B, Hamann T, Höhns H, Middelberg G (2004) Model Predictive Control Of Production Systems Using Partially Recurrent Neural Networks. Proceedings of the 4th CIRP - International Seminar on Intelligent Computation in Manufacturing Engineering, 93–97

Scholz-Reiter B, Hamann T, Zschintzsch M (2006) Cased-Based Reasoning for production control with neural networks. Proceedings of the 39th CIRP International Seminar on Manufacturing Systems, 233–240

Ushio T, Ueda H, Hirai K (1995) Controlling chaos in a switched arrival system. System Control Letters 26: 335–339

Wiendahl H-P (2007) Betriebsorganisation für Ingenieure. Hanser, München

Modelling Packaging Systems in the Furniture Industry

Dennis Reinking and Hans Brandt-Pook

1 Introduction

The large product variety, the decreasing product life cycles and sinking batch sizes raise the number of packing variants and the complexity of packing systems of the furniture industry for knocked down furniture enormously. At the same time, the competitive pressure becomes higher and higher (BBE 2005). In addition to that, packing is not regarded as core competency of the furniture companies. These are the main arguments to deal with the problem of packaging within the project *lean packaging.*[1]

There is no model or method for analyzing packaging processes, finding weak points or possibilities for improvements. This paper introduces the first model of packaging systems in the furniture industry. It is structured as follows: after this introduction, Sect. 2 discusses the basics of the value stream analysis and explains the application to a packaging system. Section 3 presents two additional and more detailed analyses: in a second level, the so called *packaging system level* is introduced, where the workflow of a packaging process has to be modelled and the third level, the *packaging item level,* regards the packaging items itself. A conclusion completes this paper in Sect. 4.

D. Reinking (✉)
Hochschule Ostwestfalen Lippe, Projekt "lean packaging" im Fachbereich Produktion und Wirtschaft, Liebigstr. 87, D-32657, Lemgo, Germany
e-mail: Dennis.Reinking@lean-packaging.de

H. Brandt-Pook
Hochschule Ostwestfalen Lippe, Fachbereich Produktion und Wirtschaft, Liebigstr. 87, D-32657, Lemgo, Germany
e-mail: Hans.Brandt-Pook@HS-OWL.de

[1] See http://www.lean-packaging.de.

H.-J. Kreowski et al. (eds.), *Dynamics in Logistics,*
DOI: 10.1007/978-3-642-11996-5_36, © Springer-Verlag Berlin Heidelberg 2011

This paper focuses on packaging systems for knocked down furniture. We want to point out, that other types of furniture are packed in a completely different way. They are not in the scope of this paper.

2 The Value Stream Analysis

2.1 Basics

The value stream analysis is a central method of the "Lean Thinking – Concept". It enables the user to analyze the production process and to document the production flow by using certain symbols for the supplier, the production process, the logistical linkage and the customer (Rother and Shook 2006). The most characteristic feature of the value stream analysis is a comprehensive view on the production flow, from the supplier of the raw material up to the customer. An outstanding issue of this method is to work from the customer's point of view. Another important feature of this method is to simplify the analysis by using experienced data and valuations. Erlach (2007): "Due to the simplification, the necessary expenditure sinks enormously and rises the clarity and the force of expression at the same time." An example of a value stream analysis is shown in Fig. 1. In addition to the value stream analysis, two more methods are useable, specially for the packaging process level. ARIS as well as UML are adequate methods for modelling the packaging workflow on the packaging process level (Brandt-Pook and Kollmeier 2008; Lehmann 2008; Oestereich 2006; Rupp et al. 2007; Scheer 1996).

2.2 Application of the Value Stream Analysis to the Packaging System of the Furniture Industry

The application of the value stream analysis to the packaging system of the furniture industry follows the general proposal of Erlach (2007) and is divided into four steps:

Fig. 1 Elements of the value stream analysis

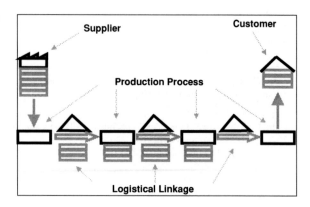

(1) creating product families,
(2) analyze customer's demands,
(3) mapping the packaging flow, and
(4) search for improvements.

2.3 Creating Product Families

For creating product families, the entire product range has to be divided into individual product families. A different workflow or different requirements to the packaging process helps to distinguish them from each other. The different dimensions of e.g. bedroom furniture (e.g. walk-in wardrobes vs. small bed shelves) make it necessary to pack them in a different way.

2.4 Analyze the Customer's Demands

The customer's demand is the most outstanding issue for the value stream analysis. Therefore, the whole packaging system has to be adapted to the customer's needs. The customer's needs are measured on the basis of sales figures. Regarding the time which is required for the packaging process, the tact rate can be calculated. The tact rate is the time of packing divided by the quantity of the items. Example: every 28 s a packaging is finished. This tact rate is one of the most important key figures for a packaging system.

2.5 Mapping the Packaging System Flow

By mapping the packaging system flow, the actual situation of the individual value stream should be analysed and outlined directly in the work shop. A clear schematic representation can be drawn by using the simple symbolism. It is also an instrument of communication for the entire analysis. In a first step the value stream analysis has to be worked out as shown in Fig. 2. Most case studies of packaging systems can be idealized in the steps supplier, incoming goods, incoming store, temporary store, packing process, finished good store, outgoing goods and customer. The tact rate has to be registered in the customer symbol. Personnel requirements and operating time are collected in the step of the packaging process. The quantity of the packaging goods on stock, the inventory time, the warehousing costs have to be written in the symbols for incoming store, temporary store, finished goods store and outgoing goods store. The costs for the in-house transport have to be written in the arrow of the logistical linkage. These are the most

Fig. 2 Idealized value stream analysis in the furniture industry

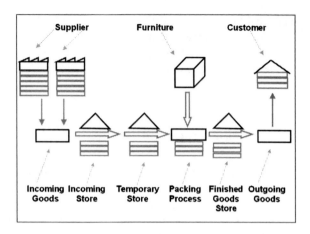

important key figures of the logistical linkage. There have to be two more symbols for the suppliers. One for the usual packaging and one for the packaging aids. Here the quantity and the quality of the goods have to be written in. The number of different types is also a very significant information of a packaging system.

2.6 Search for Improvements

With the help of the drawn value stream and the determined key figures, a detailed and complete overview of the quality of the packaging system is possible. This is the basis to identify weak points and possibilities for the improvement of the value stream. This can be done by the method of value stream design (Erlach 2007).

3 Additional Analysis for Packing Systems in the Furniture Industry

For the analysis of packing systems in the furniture industry two more levels of aspects have to be considered: the process—level, where the packing process itself has to be analysed and the packing item—level, where the packing items has to be analysed (see Fig. 3).

3.1 The Packing Process Level

The first additional level is called the process level. On this level the packing process itself has to be analysed. The packing process in the furniture industry can be divided into seven typical steps (see Fig. 4).

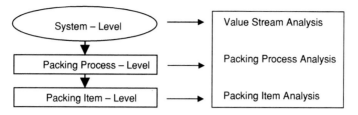

Fig. 3 Levels of a packaging system (on the basis of Erlach 2007, p. 36)

Fig. 4 Packaging process
(on the basis of Rupp et al.
2007, p. 259 ff)

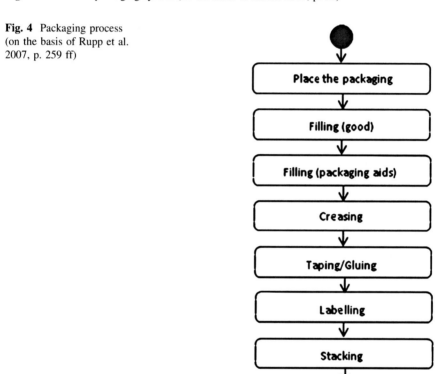

In the first step, the packaging is placed on the packing table or on the conveyor. Than the furniture parts have to be put on the packaging sheet. In a third step the void has to be filled with void fillers. After that, the packaging has to be creased, taped or glued, labelled and stacked. By analysing this workflow, it is also possible to identify weak points and possibilities for improvements. This workflow can be organized in two different modalities. Depending on the batch size, the packing process can be done in line mode and in cell mode. In the line mode, as shown in Fig. 5, the packaging is packed by six or more workers at a driven conveyor belt. Each of the workers has to place several parts of the furniture and the packaging aids on the sheet and has to move it from one worker to the other. The creasing,

Fig. 5 Line mode and cell mode

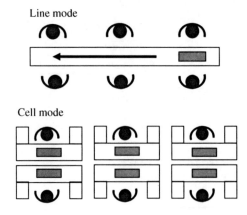

aping, labelling and stacking is done automatically. Is the batch size lower than roughly 250 pieces, the cell mode should be used (Achatz 2008). In the cell mode, the packaging is packed completely only by one worker. In this mode, the worker places the packaging on a simple table, fills it with the good and the packaging aids, creases, tapes, labels and stacks it. None of the steps is done automatically.

3.2 The Packing Item Level

The second additional level, the package level, describes the packing item itself—regarding the packaging design and its associated characteristics such as size, weight or volume utilization. For analysing the packing item, one should have a closer look on how the packaging design is done. The packaging design is one of the most critical points in a packaging system. With the packaging design all of the following costs of the process chain will be fixed. Mistakes can only be debugged very cost-intensive. To design the packaging of furniture, the following aspects have to be taken into account (see Fig. 6).

3.2.1 Splitting the Components

The splitting is the first step in the packaging design. In order to have a convenient handling for the customer, all furniture parts are divided into several packing

Fig. 6 Packaging design

1	Splitting
2	Arrangement
3	Void Fillers
4	Fefco-Type
5	Corrugated Cardboard

items. Sides of a bed or sides of a wardrobe are packed separately, because they are very long and heavy, for example. Freight carriers also give some rules in terms of dimensions and weight to optimize their logistical affairs.

3.2.2 Arrange the Components

In the next step the arrangement of the furniture components has to be done. With the definition of the arrangement, the dimensions of the packing item are defined. An outstanding goal of the arrangement is, to optimize the utilization degree and to minimize the cavity. Otherwise it has to be filled with void fillers and that will increase the material costs, the storage costs and the logistic costs.

3.2.3 Void Fillers

In order to prevent a slipping inside the packing item, the cavity has to be filled with void fillers. In the same step the protection of the front ends will be realized. The quality demands of this protection depend on the channel of distribution. The distribution with own people has lower demands on the product protection as the distribution with other fright carriers such as parcel cervices.

3.2.4 Define the Fefco-Type

Fefco means: "Fédération Européenne of the Fabricants de Carton Ondulé" and is an organisation of the European manufacturers of corrugated cardboard. The organization was founded in 1952 and has its head office in Brussels and is concerned—among other things—with the harmonization of standards. The Fefco-standard is an international standard for corrugated boxes. It was developed, in order to standardize packaging constructions and to make it understandable for all involved parties The definition of the Fefco-Type belongs to the further packaging design process (FEFCO 2009). Figure 7 shows a Fefco 412, often used for knocked down furniture.

Fig. 7 Fefco 412 (on the basis of Kappa 2004, p. 43)

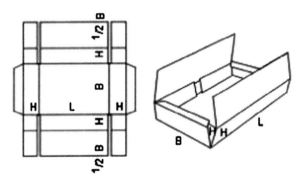

The main goal of the definition of the Fefco-Type is to minimize the folding and taping activities and to ensure the product protection. The quality of the product protection is given by the numbers of layers at the front end.

3.2.5 Define the Corrugated Cardboard

The main goal of the definition of the corrugated cardboard is to ensure the product protection. In order to achieve the quality demands of customers, it is open for the manufacturer of the cardboard to combine certain qualities paper. The quality of the corrugated cardboard depends on the converted paper and the produced kind of the wave. The quality of the cardboard is measured in bursting strength, in edge crush resistance and in puncture resistance. To determine the bursting strength, a sample is fixed on two sides and exposed to a constant pressure up to bursting. The edge crush resistance is the resistance, which a sample can apply towards the wave up to the collapse. To determine the puncture resistance, a sample is fixed on two sides and is exposed to a punctual impact up to the breakthrough (Kappa 2004). A further possibility to achieve different qualities is the use of different flutes. They are determined in wave height and wave thickness. In this way rough waves, medium waves, fine waves and purifying waves can be manufactured. Also the combination of these different profiles of the cardboard is possible.

4 Conclusion

This paper has introduced a method for modelling and analyzing packaging systems for knocked down furniture. The goal of modelling packaging systems is to create a basis for communication, optimization, controlling, knowledge management and cost planning.

This model contains the well known value stream analysis on its first general level—the *packaging system level*. On this level, the value stream from the supplier, through logistical linkage elements, packaging process or the customer is analysed. This paper also has introduced two more detailed levels: on the *packaging process level*, the workflow of the packaging process has been analysed. Idealized steps are: placing the packaging, fill it, crease it, tape ore glue it, label it and stack it. The second additional level is the *packaging item level*, where the packaging item itself, concerning the splitting, the arrangement of the components and the definition of void fillers, the Fefco-Types and the corrugated cardboard is regarded. Object of further studies are the other sectors of the furniture industry and other products like spare parts.

References

Achatz R (2008): Verpackung von RTA-Möbel, Diplomarbeit an der Hochschule OWL, Lemgo

BBE Unternehmensberatung (2005): Branchenreport Mitnahmemöbel 2005, Köln: Koschel Verlag

Brandt-Pook H, Kollmeier R (2008): Softwareentwicklung kompakt und verständlich. 1. Auflage, Wiesbaden: Viehweg und Teubner

Erlach K (2007): Wertstromdesign. 1. Auflage, Berlin, Heidelberg, New York: Springer Verlag

FEFCO Organisation/Fédération Européenne of the Fabricants de Carton Ondulé (2009): http://www.fefco.org [04.01.2009], Brüssel

Kappa Packaging Corporate Communication (2004): Wellppappe. Zahlen, Daten, Fakten, Perspektiven. 1. Auflage

Lehmann F (2008): Integrierte Prozessmodellierung mit ARIS.1. Auflage, Heidelberg: dpunkt Verlag

Oestereich B (2006): Analyse und Design mit UML 2.1.8., aktualisierte Auflage München, Wien: Oldenbourg Verlag

Rother M, Shook J (2006): Learning to see. Version 1.2, Aachen: Lean Management Institute

Rupp C, Queins S, Zengler B (2007): UML 2 GLASKLAR. 3. Auflage, München, Wien: Hanser Verlag

Scheer A W (1996): Heft 133: ARIS House of Business Engineering

Monitoring Methodology for Productive Arrangements (Supply Chain)

Breno Barros Telles do Carmo, Marcos Ronaldo Albertin, Dmontier Pinheiro Aragão Jr. and Nadja G. S. Dutra Montenegro

1 Introduction

It is observed a constant productive arrangement's (PA) competitiveness, where enterprises look for productive and efficient performance. Speed, quality and flexibility are being emphasized as means of responding to the unique needs of customers and markets (Yusuf et al. 2004).

Donath et al. (2002) explains that the biggest problem in the supply chain is that the enterprises do not see themselves as actors in the process. The concurrence forces the enterprises to be, each time.

The concurrence forces the enterprises be better structured and organized in a supply chain (Carmo et al. 2008). This is more accentuated in the global market, where there are many opportunities with increasing demands in lower price and better quality. Chase et al. (2006) states that these opportunities can be more profitable if the enterprises work together to get this market share. To measure the supply chain development, Donath et al. (2002) defines some principles:

- Understanding costumers' needs;
- Understanding enterprises' costs and logistic services and how they are related;

B. B. T. do Carmo (✉), M. R. Albertin, D. P. Aragão Jr. and N. G. S. Dutra Montenegro
Master Program in Transportation Engineering-PETRAN, Federal University of Ceará-UFC, Avenida Norte, 2155-apto 1203, Luciano Cavalcante, Fortaleza, Ceará, Brazil
e-mail: brenotelles@det.ufc.br

M. R. Albertin
e-mail: albertin@ufc.br

D. P. Aragão Jr.
e-mail: dmontier@ot.ufc.br

N. G. S. Dutra Montenegro
e-mail: nadja@det.ufc.br

H.-J. Kreowski et al. (eds.), *Dynamics in Logistics*,
DOI: 10.1007/978-3-642-11996-5_37, © Springer-Verlag Berlin Heidelberg 2011

- Seeing the supply chain as a process;
- Identifying and stopping the inefficient activities.

Among the different ways of productive systems organization, PA is emphasized where enterprises of a same sector interact with local actors and search for competitive advantages unattainable in an isolated way through cooperation (Albertin 2003). PAs have many competitiveness approaches, some of them are considered new:

- Clusters, where the competition is between regions (Porter 1990; Nadvi and Schmitz 1999);
- Value chains, where there are new relationships between enterprises (Gereffi 2001).

In these cases there is competition between enterprises, but the relationship is privileged to get the PA objective.

But there is a question to be responded: How do we evaluate the PA to propose actions to benefit not only a singular enterprise, but all of them?

This question will be answered by this paper. The objective of this work is to propose a methodology to evaluate the PAs based on management tools. These tools are used not just to look inside the enterprise, but to evaluate the relation with other organizations or suppliers. To identify the main constraints and opportunities of a PA, some information was collected in a dynamic and cooperative way.

The paper structure begins with the PA concepts and management tools presentation. The second step brings the system idea presentation, with its benefits and its restrictions. To finish, conclusions about the system will be pointed, presenting some application done in Ceará's PAs.

2 Supply Chains' Concepts

Slack et al. (2002) understands that the productive operation does not exist alone, they are all included in a bigger interconnected mesh, including suppliers and customers. Farina and Zylbersztajn (1992) identified that this process creates and aggregates value to the customer. Yusuf et al. (2004) understands that customers and technological requirements force manufacturers to develop agile supply chain capabilities in order to be competitive. So, these enterprises need to work as they were only one.

The enterprises are less independent, but connected as a part of a system inside the PA. Chase et al. (2006) defines that the PAs are essential to understand all the productive cycle, beginning with the raw material enterprises and ending in the customers' hands.

Chopra and Meindl (2003) see the PAs like a mesh, where one enterprise can have many suppliers and can provide to different organizations in the subsequent steps of the process. So, Pires (2004) defines these relationships like the structure of the most PAs. To resume this concept, Slack et al. (2002) defines that the

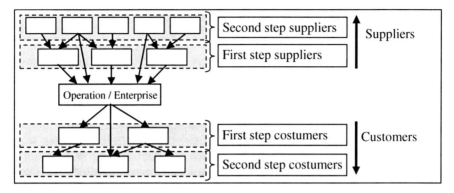

Fig. 1 Productive arrangement representation

enterprises must be all interconnected, like productive units to provide the products and services to the chains' customers.

The Fig. 1 represents a generic PA.

By this picture, it can be observed the multiple relationships between enterprises, suppliers and customers.

Looking at the picture, it can be seen that there are many levels, links and actors in the PA. In order to get its objective it is necessary that the enterprises work in determinate standard to attend customers' requirements. In the other hand, to get a good enterprise management it is necessary the application of some management tools to propose a better logistic and production performance. The next topic shows some tools used in the system to evaluate the PAs.

3 Management Tools

There are many tools that can be used to improve the enterprises performance. Some of them were chosen like the most important for the PA development. Murray (2009) understands that the companies must identify the reasons why their processes are not as efficient as the competitors'. This process is called benchmarking. The same author defines that "for companies that have performed internal benchmarking and want to investigate new ways in which they can improve performance of their internal processes, external benchmarking can produce significant improvements".

Murray (2009) identifies some benchmarking components:

- Financial benchmarking;
- Performance benchmarking;
- Strategic benchmarking;
- Functional benchmarking.

These components can be measurable by management tools and these tools impact in different ways in the enterprise. In each supply chain some tools are more important than others. Because of that, they must have a ranking of importance from 1 to 5 (unimportant to very important) and are grouped into management subsystems.

To identify these tools, a research with some enterprises was done with managers, executives, researchers and consultants to find out which management characteristics the enterprises must have to be considered an excellent leader. This process was based on Delphi methodology. With these data, a cognitive map (Axelrod 1976) was done to organize these concepts and feature the excellence.

It was structured seven subsystems, where the correlated attributes were joined. The enterprise which achieves a good performance in these areas is considered an excellent enterprise. To define what is good and where the enterprises should be, it was identified a benchmarking. This information is organized like a tree.

These subsystems were established using the benchmarking components' concepts. The first one, financial benchmarking is called Financial Subsystem. The second benchmarking, Performance, is formed by Logistics Subsystem, Human Resources Subsystem and Integrate Management Subsystem. The third one, the strategic benchmarking evaluates tools in Strategy Subsystem. The last one, functional benchmarking inspired Production and Products Subsystem. The Fig. 2 shows these areas and the tools correlated to each one.

Each one of these tools will be explained according to the management subsystem where they are located in the next step.

3.1 Integrated Management System

In general terms, Hegering et al. (1998) understands that management systems comprise all measures necessary to ensure the effectiveness and efficiency according to organizations' goals. This first group of tools consists of the best practice of management. It comprises the quality, ambient, social responsibility, and occupational health and safety norms. They were put together because they work in the same way.

3.2 Production Management

Chase et al. (2006) defines production management as the "design, operation and improvement of the systems that create and deliver the enterprise's primary products and services". So, this second subsystem works with the tools applied in production systems. The impacts of these tools are observed, many times, in the production line, proposing actions to reduce production costs, to improve the products quality and reduce de losses and scraps in the line.

It also involves the suppliers' development, aiming the PA development.

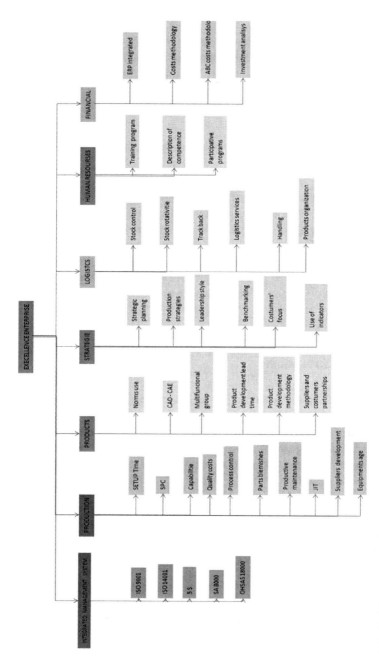

Fig. 2 Management tools

3.3 Products Development

Griffin (1997) understands product management as a methodology to change and improve the quality of the products according to the customer's necessity. There are many tools used for the development of new products. These techniques aid to develop new products and improve them, reducing the time to get to the market and ensuring quality.

3.4 Strategy Subsystem

Ansoff and McDonnell (1993) understand strategy as the rules collection to aid the enterprises' decision. This subsystem is formed by tools that work with the future plans. It gets strategies to enterprise and production optimization. It works to establish indicators to meet clients and market requirements. These indicators are based on benchmarking.

It is also formed by strategies based on customers' focus, leadership style and other tools that cause impact in the enterprise organization.

3.5 Logistics Subsystem

Donath et al. (2002) understands that logistics manager must "find innovative ways of gaining and maintaining customers and keeping them loyal". In this subsystem, it is included the logistic actions evaluation into customers and suppliers. This includes the stock control, the track back tools, if it uses or not logistic operators, how handling is being done and how the products are organized and delivered to the customers.

3.6 Human Recourses' Subsystem

Huselid et al. (1997) defines that the human resource management involves the policies and practices implementation that ensure a firm's human capital. In this enterprise evaluation part, it is considered how the employees are considered inside the enterprise.

It helps to identify if the enterprise has career plans and programs to develop competence and if the workers take part in training programs and in the enterprise's profits.

Application level/ Tools	0%	25%	50%	75%	100%
			Integrated Subsystem		
ISO 9001 - ISO 14001 - 5S - SA 8000 - OSHAS 18000	Informal procedures	Documented procedures	Formal implantation program	internal audits	Enterprise certificated
			Production Subsystem		
Setup time	Informal procedures	Documented procedures	Setup time < 60 min.	Setup time < 40 min.	<10 (SMED)
PCP	Informal	Spreadsheets (excel)	Software	MRP	MRP II
Capability studies	Informal	Unstable processes	Stable processes	Statistics process control	Cpk > 2
Quality costs	Unknown	Monitoring	1-10% from sales	< 1 % from sales	< 0.5 from sales
Parts blemish	Unknown	Monitoring	1-10 %	< 1000 ppm	< 500 ppm
Productive maintenance	Corrective	Maintenance corrective plan	Preventive maintenance	Predictive maintenance	TPM
JIT	Don't use these kind of tools	Use one JIT tool	Use two JIT tools	Use tree JIT tools	Use many tools
Process Control	Informal parameters	Formal parameters	Controlled parameters	Adjusted equipments	Capability studies
Suppliers development	Informal	Formal	Performance monitoring	Suppliers training programs	Establish partnerships
Equipment age	Unknown	More than 20 years	10 < age < 20	5 < age < 10	Less than 5 years

Fig. 3 System tools evaluation

Products Subsystem					
Use of technical norms	Unknown	Know and uses some of them	Use the principal ones	Always use	Use and update the norms
CAD – CAE –CIM	Unknown	Competence	Use CAD	Use CAD e CAE	Use CAD-CAE-CIM
Multifunctional groups	Don't use	Informal use	Documented procedures	Are implementing	Always use
Development lead-time	Don't control	Informal control	Monitoring	Very short	Use benchmark
Development methodology	Don't know	Informal	Documented procedures	Improving	Use the concepts: *lessons learn*
Suppliers and costumers partnerships	Don't do	Informal	Formal	Suppliers	Suppliers and costumers
Strategic Subsystem					
Strategies	Informal	Formal	Monitoring	All employs are informed	Use indicators like BSC
Production Strategies	Informal	Defined	Monitoring	All employs are informed	Action plans
Leadership style	Controller	Centralized	Decentralized	Participativiness	Ambient promote innovation
Benchmarking use	Don't use	Local Benchmarking	Regional Benchmarking	Nacional Benchmarking	International Benchmarking
Costumers focus	Informal	Monitoring consumers dissatisfaction	Satisfaction consumer research	Monitoring satisfaction	Consumers focus
Use of indicators	Informal	Financial indicators	Quality indicators	Process indicators	PDCA – defined methods

Fig. 3 (Continued)

Logistics Subsystem					
Stock control	Low control, without system	Documented control, with system	Product and stocks documented control	Stock control systems	Integrated suppliers and customers systems
Stock rotativity	Low rotativity	Monitoring	1 to 12 a year	12 to 24 a year	24 or more a year
Track back	Don't consider important	Use in some vehicles	Use in many vehicles	Looks for use in all vehicles	Use in all vehicles
Logistics Services	Don't consider important e has its own vehicles	Use transportation services	Use more than transportation service	Use logistics operators	Use operators with system vision
Handling	Don't use machines	Use some machines with many employs	Use standard machines and some specific ones, with many human control	Automatic systems, without human interference, tools customized	Specific machines, system all automatic.
Product organization	Don't use equipments to aggregate products	Use all type of pallets	Use specific pallets	Standard pallet, use container	Use many types of containers looking the final consumer transportation
Human Resources Subsystem					
Training plan	Informal	Documented procedures	Monitoring the employs training time	< 20 hours	> 20hours
Competence and careers description	Informal	Responsibility description	Competence description	Multifuncional	Competence evaluation
Participative programs	Informal	Formal	More than one program	Many programs	Employs result participation
Financial Subsystem					
ERP: Cost methodology; ABC cost; investment analysis methodology	Don't do	Implanting	Doing in part	Finishing the implementation	Use to get decision

Fig. 3 (Continued)

3.7 Financial Subsystem

Santos and Panplona (2005) define that the enterprise objective is to create an investment system to maximize the company value. This value is the enterprise health. This is the last subsystem evaluated in the enterprises. It identifies the enterprises' financial health and if they use the tools in taking decisions like investment analysis, cost systems, and software that integrates these kinds of tools.

All of these tools explained in this topic are used in the system to evaluate the enterprises and identify in which subsystem, actions must be taken to develop the PA.

The next topic shows how this information is used to evaluate the enterprises, the link and the productive arrangements.

4 The Productive Arrangements System

This part of the paper shows how the PA evaluation is done. The first aspect to be commented is abou the questionnaire. It was built up based on the tools presented in the last topic. These tools were defined by the Delphi technique, applied with many managers, enterprises and professors.

Each subsystem was obtained using interviews to identify the most important tools. To evaluate the subsystems, each tool has a different level of importance, and according to this, the enterprise has an evaluation of this tool inside the subsystem. The Fig. 3 shows how the tool application is evaluated.

All the tools have a methodology to evaluate the level application.

Each enterprise has an evaluation level according to the tool level application. There is also the importance of the level of these tools and subsystems.

Each tool has relative importances that were defined by specialist and enterprises executives. In this case, this importance was done using the MACBATH methodology.

All enterprises are evaluated in these tools and with all the results, it can be observed how the PA is developing or not, over the time. Based on the importance of the requisite and application tool level, each enterprise has an evaluation, which can be compared with the other ones and the benchmarking.

Having this information make it possible to propose actions to the PA development, because when the enterprises' chain are being developed, the entire SC gets an improvement.

The next topic shows the system's benefits and its restrictions.

5 System's Benefits and Restrictions and an Application Example

The system has the objective of monitoring the PAs and its biggest benefits are the possibility to ease the proposition of actions to support the development of the links that show deficiency.

When all the enterprises get the results, they are put together in graphics, where the relative position between the enterprises can be seen and its possible to compare them with the benchmarking established.

An example that shows how the system works is showed in Fig. 4. It can be observed the metal-mechanics enterprises located in Ceará, a state in Brazil.

There are three possibilities in the graphics. The first one is when the subsystem gets a positive evaluation. This means that in specific subsystem the enterprise is better than the market requirement. The value zero means that the enterprise does what is recommended for the PA success. The last situation is when the evaluation has a negative value, which means the enterprise works and harms the PA performance. The Fig. 4 shows this result.

In Fig. 4, the enterprise is represented by the bars and the lines represent the clients and market necessary requirements. It can be observed the subsystems which are very critical, like the production subsystem.

The analysis permits to see each enterprise by each tool. The Fig. 5 shows this kind of analysis. It is illustrated the integrated management subsystem, where the enterprise is not good in any tool and has a negative evaluation.

The analysis also permits us to see each enterprise through each subsystem. The Fig. 5 shows this kind of analysis. It is illustrated the integrated subsystem, where the enterprises are bad and do not use many of the tools presented, representing a problem in this PA.

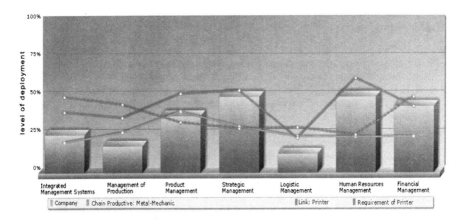

Fig. 4 Chain evaluation

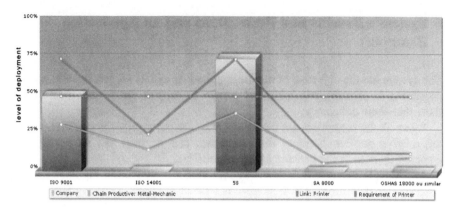

Fig. 5 Integrated subsystem example

6 Conclusion

This paper presented a system that was developed to promote the PAs monitoring. It is based in subsystems that evaluate the management tools application in enterprises that belong to a SC.

It is accessible through the site: http://www.ot.ufc.br, where the enterprises can participate and get their strategic position comparing with the other enterprises. The biggest benefit of this system is giving an overview to promote actions that impact not only in one enterprise but impact in the whole value chain.

This software reveals itself as essential to promote regional development because it identifies the opportunities and constraints of the supply chains and propose solutions for them.

References

Albertin, M.R. O processo de governança em arranjos produtivos: o caso da cadeia automotiva do RGS. Tese de Doutorado do PPGEP da UFRGS. Porto Alegre, 2003.

Ansoff, H. G., McDonnell, E. J. Implantando a administração estratégica. Atlas, São Paulo, 1993.

Axelrod, R. (Ed) Structure of Decision – The Cognitive Maps of Political Elites. Princeton: Princeton Univ. Press, 1976.

Carmo, B. B. T., Albertin, M. R. Moreira, M. E. P. Metodologia para avaliar a aplicação das ferramentas de gestão nas empresas da cadeia produtiva metal-mecânica no estado do ceará. In. XV Congresso Panamericano de Engenharia de Trânsito e Transportes. Cartagena de Indias, Colômbia, 2008.

Chase, R. B., Jacobs, R. F., Aquilano, N.J. Operations management for competitive advantage with global cases. 11 edição. Editora New York McGRAW – Hill – International Edition, 2006.

Chopra, S. & Meindl, P. Gerenciamento da cadeia de suprimentos: estratégia, planejamento e operação. Tradução de Cláudia Freire. São Paulo: Prentice Hall, 2003.

Donath, B., Mazel, J., Dubim, C., Patterson, P. Handbook of logistics and inventory management. New York: John Wiley & Sons, 2002.

Farina, E.M. & Zylbersztajn, D. Organização das cadeias agroindustriais de alimento. ENCONTRO NACIONAL DE ECONOMIA,1992, Campos de Jordão. São Paulo: 1992, pp. 189–207.

Gereffi, G. Schifting governance structures in global commodity chains with special reference to internet. American Behavioral Scientist. Duke University, v. 44, n. 10. 2001.

Griffin, A. PDMA research on new product development practices: updating trends and benchmarking best practices. Journal of Productive Innovation Management. New York: 14: 429–458, 1997.

Hegering, H. G., Abeck, S., Neumair, B. Integrated management of networked systems: concepts, architectures and their operational application. San Francisco: Morgan Kaufmann Publishers, 1998.

Huselid, M. A., Jackson, S. E., Schuler, R. S. Technical and strategy human resource management effectiveness as determinants of firm performance. Academy of Management Journal. United Kingdom: v 40, 171–188, 1997.

Murray, M. Benchmarking in the Supply Chain. Disponible in: http://logistics.about.com/od/qualityinthesupplychain/a/benchmarking.htm. 2009.

Nadvi, K.; Schmitz, H. Industrial in developing countries. Word Development, IDS, v.27, n.9, 1999. Special Issue.

Pires, S.R.I. Gestão da cadeia de suprimentos: conceitos, estratégias, práticas e casos. São Paulo: Atlas, 2004.

Porter, M. E. Clusters and competition: New agenda for companies, governments and instutions. In On Competition. Boston: Harvard Business School Press, 1990.

Santos, E. M., Panplona, E. O. Teoria das Opções: uma atraente opção no processo de análise de investimento. Revista de Administração da USP - RAUSP. São Paulo: V. 40, n. 3, 2005.

Slack, N., Chambers, S. & Johnston, R. Administração da Produção. 2ª Ed. São Paulo: Atlas, 2002.

Yusuf, Y. Y., Gunasekaran, A., Adeleye, E. O., SivayoganathanI, K. Agile supply chain capabilities: Determinants of competitive objectives. European Journal of Operational Research 159 (2004) 379–392.

Complexity-Based Evaluation of Production Strategies Using Discrete-Event Simulation

Reik Donner, Uwe Hinrichs, Christin Schicht and Bernd Scholz-Reiter

1 Introduction

Globalization of markets is a multiple challenge for modern economies, which requires particularly the ability of flexible and fast adaptation to changing economic conditions (i.e. a temporally varying market demand, shortening product life-cycles, etc.). As a result, typical manufacturing networks are characterized by a large structural as well as dynamic complexity and a diversity of goods and services. During the last decades, many enterprises have reacted to the corresponding requirements by a concentration on their specific core competences, incorporating concepts like outsourcing of special tasks of production and logistics. As a result, economic networks have successively developed, which may span a variety of different businesses. The mutual interactions between different companies involved in these networks often lead to a complex dynamics incorporating both cooperative and competitive behavior.

The increasing complexity of production dynamics is however known to have severe negative effects on the individual manufacturing processes. Prominent examples include production breakdowns due to a lack of material as well as the Bullwhip effect (Forrester 1958), i.e. large (and usually irregular) oscillations of stocks which amplify along a supply chain due to a convective instability of the commodity flows (Helbing et al. 2004; Donner et al. 2010). The classical approach to solve this problem is decoupling the individual business units by large buffers

R. Donner (✉) and C. Schicht
Institute of Transport and Economics, Dresden University of Technology, Würzburger
Str. 35, 01187, Dresden, Germany
e-mail: donner@vwi.tu-dresden.de

U. Hinrichs and B. Scholz-Reiter
BIBA-Bremer Institut für Produktion und Logistik GmbH,
University of Bremen,
Hochschulring 20, 28359, Bremen, Germany

H.-J. Kreowski et al. (eds.), *Dynamics in Logistics*,
DOI: 10.1007/978-3-642-11996-5_38, © Springer-Verlag Berlin Heidelberg 2011

for material storage, which allows compensating demand and supply variations. However, a severe disadvantage of this strategy is that it requires large initial investments, and that high inventory levels mean storing large amounts of capital in terms of material inside the buffers. Moreover, this decoupling leads to a systematic increase of the time interval between the delivery of commodities or semi-finished material and the production of the final goods. As the cycles of development and production are gradually accelerating in many businesses, these features mean a high economic risk, especially for smaller manufacturers. In reality, instead of decoupling, companies therefore often tend to closely link to each other by applying strategies such as just-in-time logistics (Hopp and Spearman 2000). As a consequence, the resulting business networks may get even larger and more complex, which may result in a very irregular and unpredictable dynamics (Helbing et al. 2004).

The optimization of production processes in terms of their logistic performance requires the joint minimization of inventory levels and throughput times under a simultaneous maximization of the reliability of the networks (Bloomberg et al. 2002). In order to perform such an optimization (or at least find some approximate solution for this problem), the considered production system has to be mapped onto a suitable mathematical model at first. For this purpose, continuous-flow or discrete-event models are the most convenient approaches that can be used to prescribe the essential structure of interactions and study the main features of the dynamics on such networks. However, for investigating the dynamic effects due to the specific structure of a real-world manufacturing system, using established software packages has important advantages and allows the identification of economic potentials and a successive testing of various optimization and control strategies. In the literature, a vast amount of examples for applications of the different modeling approaches can be found (for a list of topically related references, see Donner et al. (2008a)).

With respect to the mathematical analysis, the complex dynamics and high dimensionality of manufacturing networks require the careful application of existing, and the development of new methods for modeling, analysis, and control that are adapted to the specific features of logistic observables like inventory levels or throughput times. As a particular approach to gain a fundamental understanding of the dynamics at least in small-scale systems, relevant information may be deduced by applying nonlinear methods for analyzing the corresponding time series (Donner et al. 2008a, b).

In this study, an event-discrete simulation model is used to classify different sources of instability of material flows and study the robustness of small-scale manufacturing networks. In Sect. 2, the basic features and variants of this model are described. Sect. 3 includes a review of recent results on the dynamic complexity of manufacturing systems, which motivates a detailed qualitative as well as quantitative analysis of the dynamical patterns that may arise in the considered model. Based on this analysis, the potential sources of instability are identified and investigated in some detail.

2 Model Description

For the development and implementation of a discrete-event simulation model, the commercial software eM-Plant (version: eM-Plant 7.5; Tecnomatix Technologies Ltd.) has been used. In particular, this software is well suited for a realistic modeling of small-scale manufacturing systems, i.e. networks consisting only of a few participating companies. The consideration of such networks allows efficiently investigating the emergence of instabilities, e.g. due to feedback loops.

As a specific system, an idealization of the real-world production network of a manufacturer of pumping systems with five main factories in different Western European countries (Windt 2002) is considered. The corresponding model consists of four factories (nodes) that mutually produce and deliver semi-finished goods to the other network participants, and finished goods to an external market (see Fig. 1). Since full details of the dynamic interrelationships in this model have already been described elsewhere (Donner et al. 2008a, b), only the basic features will be briefly summarized in the following:

Like in other discrete-event simulations (in contrast to continuous-flow models), all quantities as well as all production and transportation processes are discrete. This allows a more detailed representation of individual goods circulating in the network, treating processing and transportation as discrete events that are determined by a basic clock. The internal dynamics of the different manufacturers is represented by a sub-model including all necessary procedures (see Fig. 1). For simplicity, the basic structure of these sub-models is identical for all manufacturers. Each manufacturer consists of four production lines which simultaneously manufacture four different products, and four sort-pure buffers (that is, buffers that

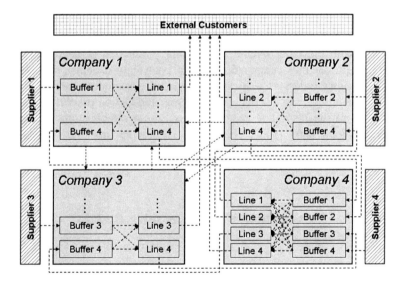

Fig. 1 Schematic representation of the network studied in this work (after Donner et al. 2008a)

contain only one specific commodity) for four distinct types of semi-finished material that have an unlimited size and can be assessed by all production lines. One of these buffers is equipped by a network-external supplier, whereas the remaining ones are supplied by one specific production line of each of the other nodes in the network via predefined connections. While the corresponding goods stay inside the network and are used as semi-finished material by the other collaborators, the remaining product of every manufacturer is supplied to a network-external costumer. To keep this supply scheme as simple as possible, each supplier inside or outside the network delivers only one sort of semi-finished material to the respective customer.

To start the fabrication of a product, certain kinds and amounts of semi-finished materials are required. The corresponding combination is defined in a designated production matrix. For the realistic illustration of the production process, operating times and working-on intervals have been implemented into the model. This setup allows an explicit computation of the time required for completing a product and to permit or prohibit a multiple treatment of the production lines. By prescribing different lot sizes in the production matrix, it is possible to consider different scenarios that correspond to different capacities of transport containers. Deviating transportation routes with different lengths can be modeled by prescribing the transportation times.

In order to consider the effects of distributed processing and transportation times (for example, due to machine failures, missing workers or delayed transportation devices), the discrete basic clock of the system can be modified by additive stochastic contributions with a pre-defined distribution (Scholz-Reiter et al. 2006). However, for understanding the deterministic part of the dynamics, in this work, all events are assumed to follow a timing given by integer multiples of 1 h.

For understanding the influences of real-world effects on the production, several parameters of the model have been systematically varied, including processing and transportation times (also for external suppliers and customers), the initial inventory levels for all commodities in the network, and the order volumes. In connection with the latter quantities, different production strategies have been applied and thoroughly compared: (i) push control (production starts without explicit customer orders as soon as the required commodities are available in sufficient quantities), and (ii) pull control (every production process is initialized by an order of the respective customer). As further modifications of the latter strategy, different order policies have been considered (provision policy, order point policy, periodic order policy), which are distinct in terms of the volumes and timings of orders (Donner et al. 2008a, b).

3 Heuristic Classification of the Dynamics Based on Inventory Levels

It is a well established fact that present-day manufacturing networks are usually characterized by large structural and dynamic complexity (Hülsmann and

Windt 2007). Here, structural complexity is related to the fact that there may be many contributors ("nodes") within the networks corresponding to different levels of suppliers, manufacturers, and customers with multiple vertical as well as horizontal interrelationships. In order to mathematically describe the production process of even a single product completely, all these different nodes have to be taken into account. From the perspective of complex systems sciences, it is known that even with careful planning and control, structural complexity may result in severe instabilities of the material flows, which may make a prediction of the production process impossible (Helbing et al. 2004).

In this work, however, the main question is how dynamical instabilities of material flows may arise in networks with even a low structural complexity. The sources of dynamic complexity are non-linear, possibly time-delayed interactions (in terms of both information and material flows) between the individual nodes. Under ideal conditions, one may assume that a network with a simple structure may be operated in a fully regular and predictable way, so that the temporal variations of inventory levels, production rates, or throughput times follow a certain simple deterministic function (constant values, periodic behavior, etc.). However, in the presence of the mentioned factors, one may frequently observe a very irregular dynamics even in the case of a fully deterministic system.

In the following, some key results on the stability of manufacturing networks and, more general, discrete-event systems are briefly reviewed. Subsequently, it is discussed in some detail which conditions influence the predictability of the production process in the considered model system in a negative way. For this purpose, extensive model simulations for a large variety of parameter combinations as well as production and order strategies have been systematically evaluated.

3.1 Chaos and Complexity in Manufacturing Systems

In many complex systems, a large structural and/or dynamic complexity may lead to the emergence of chaotic behavior. In mathematical terms, chaos means a non-periodic deterministic (i.e. non-random) dynamics with a high sensitivity to the choice of initial conditions. In continuous systems, such deterministic chaos is characterized by an exponential divergence of infinitesimally separated initial conditions. Many nonlinear systems are known to exhibit this kind of behavior, including a variety of continuous-flow models of manufacturing systems (e.g. Chase et al. 1993; Katzorke and Pikovsky 2000; Rem and Armbruster 2003).

In discrete-event systems, the notion of infinitesimally separated initial conditions cannot be adopted. In contrast, one has to note that if there are no complex time-varying factors like supply or demand fluctuations, under certain conditions deterministic chaos cannot exist in the sense of the above mentioned definition. In particular, in the case of manufacturing networks, the crucial point is that storage capacities and production rates are usually limited. Considering the inventory levels of all buffers and the production rates of all machines as the state-space of the system, this space includes only a finite number of elements. This argument still applies if the important role of time-delays is considered, i.e. if a joint embedding of all state-space variables is performed into an extended state-space, with an embedding dimension determined by the maximum cycle time of a product within a network.

In contrast to continuous systems, a finite state-space implies that at least a subset of all possible states must be recurring (Binder and Jensen 1986). This means that after one of these states has been reached, the dynamics must necessarily repeat in a periodic way. However, the associated period is related to the dimension of the (extended) state-space, i.e. both structural and dynamic complexity of the system, and may thus be extremely large. As a consequence, for the time intervals usually considered, the dynamics looks fully irregular. In the following, we will refer to this type of behavior as pseudo-chaos or finite-time chaos without explicitly proving whether the "short-term" evolution of the relevant system variables actually fulfills the standard mathematical definition of chaos.

It has to be underlined that the notion of chaos as well as pseudo-chaos must be clearly distinguished from randomness. Unlike in the idealized models of manufacturing systems, in real-world networks factors like machine failures and external variations of demand and supply are however usually stochastic. As a stochastically forced system must necessarily behave stochastically itself, any stochastic forcing directly implies a stochastic reaction, which may interact with— or even counter-act—the emergence of deterministic instabilities due to structural or dynamic complexity (Donner et al. 2008c).

3.2 Qualitative Characterization of Dynamics

In the presence of a push control, the dynamics of all inventory levels is characterized by a strong regularity. In most cases studied, the system (at least asymptotically) reached a stationary state. Only for certain combinations of processing and/or transportation times, a periodic behavior can be observed. As a stationary system can however be considered as periodic with a period of one time step, one has to conclude that this type of behavior seems to be generic for a push control. The fact that stationarity is not present for all parameter combinations is most probably related to the multiple loops of material flows in the considered systems, which may support the emergence of convective instabilities (Helbing et al. 2004; Donner et al. 2010).

The variety of possible patterns of inventory variations becomes much larger as soon as the production strategy is replaced by an order-based control (pull principle). For a standard pull strategy, in most cases the inventory levels seem to fluctuate irregularly on time-scales of up to some thousands of time steps. Only some parameter combinations result in simple short-periodic variations of the inventory levels. As it was already discussed in (Donner et al. 2008c) in some detail for a setting with only two mutually interacting manufacturers, the main driver of the associated instability is the improper choice of lot sizes, which corresponds to a bad logistic synchronization of the network. In such case, the presence of feedback loops in the material flows within the network triggers oscillations (Helbing et al. 2004). For certain combinations of lot sizes on the one hand, and transportation and processing times on the other hand, even different oscillatory modes can be simultaneously excited. If these modes have long periodicities or no low-order $m{:}n$ frequency ratio, the dynamics of the inventory levels becomes rather complex (in particular, pseudo-chaotic states in the sense discussed in Sect. 3.1 may arise).

If applying more sophisticated order strategies, the behavior of the system becomes even more diverse. In particular, the application of order point (and, to some extent, also provision) policies, i.e. strategies with fixed pre-defined order volumes, seems to have a regularizing effect on the systems' dynamics. In general, the order volume seems to be a particularly important quantity: there is a tendency that large order volumes promote periodic dynamics (but may require larger buffer sizes as one may see in the example shown in Fig. 2), while lower order volumes rather lead to pseudo-chaotic behavior. Especially in case of an order point policy,

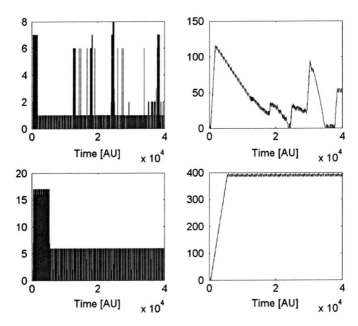

Fig. 2 Variations of the inventory levels of the first (*left panels*) and second (*right panels*) buffers of the first manufacturer (corresponding to materials supplied by the external supplier and the second manufacturer, respectively, see Fig. 1) using a provision policy with constant order volumes of 5 (*upper panels*) and 10 (*lower panels*) pieces for all customers. Note that larger order volumes yield periodic inventory levels, but on the cost of larger buffers. The specific setting of the remaining model parameters used in the considered simulations corresponds to the one described in (Donner et al. 2008a)

the associated time-scales for processing and transportation appear to have an additional crucial influence on the emergence of short-term periodic dynamics. However, it has to be noted that in most cases, pseudo-chaotic behavior is found with a variety of different patterns, ranging from irregular dynamics on a strongly restricted interval of buffer levels to burst-like dynamics. The detailed results demonstrate the effects of the complex interplay of the different relevant variables of the considered systems, which does not allow drawing general conclusions about the occurrence of regular short-term dynamics apart from the above mentioned cases.

3.3 Quantitative Characterization of Complexity

In order to characterize the dynamics especially in the pseudo-chaotic domain in a more quantitative way (in particular, regarding its predictability), simple statistical measures like mean values or moments do not provide sufficient information. As an alternative, it is recommended to complement the corresponding numbers by the values of nonlinear measures that describe the complexity of the observed system. Among others, the consideration of entropies and so-called recurrence quantification analysis has been found to yield a reasonable basis for such quantitative characterization (Donner et al. 2008a, b).

Entropies describe the degree of disorder of observed sequences of logistic observables like inventory levels, and are thus in principle well suited for comparing the dynamics of a model system under various parameter settings. Consequently, entropies and related quantities are natural candidates for dynamic performance measures of logistics systems. As a drawback, estimating entropies from time series however requires a proper discretization of the data, for which there are different possible approaches that may yield different quantitative results (Donner et al. 2008a). In particular, the estimated entropy values are often strongly influenced by the probability distribution of the considered variables, which may strongly differ between model runs with even only slightly different parameter combinations).

As an alternative to the consideration of entropies, recurrence quantification analysis (Marwan et al. 2007) is based on the probabilities that certain sequences of observations closely resemble those found at other time periods. Different aspects of these recurrence probabilities can be quantified in terms of a variety of heuristic nonlinear statistical characteristics, including the degrees of determinism and laminarity in the dynamics (Donner et al. 2008a, b). As an advantage, the values of these measures are usually more robust with respect to the particular choice of the parameters of the method than entropy estimates. However, this benefit comes on the cost of rather high computational demands, since recurrence quantification analysis requires comparing every possible pair of observations from the considered time series.

Finally, in order to filter dynamical features that are related to either short- or long-term variations of inventory levels, a new combination of recurrence quantification analysis with symbolic discretization has been recently introduced (Donner et al. 2008b). In particular, it has been demonstrated that the values of the resulting measures may be more characteristic for specific situations in manufacturing networks than those obtained from other nonlinear approaches. A more systematic application of this approach is outlined to future research.

4 Conclusions

The variety of dynamic patterns shown by small-scale manufacturing networks has recently been addressed elsewhere (Donner et al. 2008a, b), but however not yet been systematically studied in terms of the influence of all relevant control parameters of such a logistic system. In this paper, a general framework for qualitative and quantitative analysis based on the theory of nonlinear dynamical systems has been sketched that may help distinguishing and characterizing the different types of dynamics in logistics systems. A systematic application to a large set of model runs with a large number of different parameter combinations, which is currently performed, is expected to allow new deep insights into the mechanisms of instability acting in manufacturing networks.

The rather general results reported in this contribution underline that the recently reported main mechanisms of instability in manufacturing networks (i.e. imbalance of material flows and incomplete logistic synchronization) (Donner et al. 2008a) may be a severe danger for real-world production systems. In particular, the nonlinear interplay between the different dynamically relevant parameters makes it almost impossible to predict the resulting type of dynamics a priori, which is however mainly due to our incomplete present-day understanding of these complex interactions. However, there are some systematic effects that are at least qualitatively understood, for example, the influence of the lot size matrix and the associated order volumes. Related to this, a possibly relevant observation is the regularization of the dynamics by large order volumes. This finding is actually not surprising, since the corresponding strategy leads to a decoupling of the production process from temporal variations in the availability of commodities and semi-finished products by using rather large buffers. It will be of practical importance to study whether the corresponding strategies are able to not only regularize the variations of inventory levels, but also allow a higher reliability of satisfying customers' orders and a decrease in throughput times, which are other important goals of logistics that may in principle be in conflict with the ideal of low, predictable inventories.

Acknowledgments This work has been financially supported by the German Research Foundation (project no. He 2789/8-1,8-2 and Scho 540/15-1), the Daimler-Benz foundation, and the Volkswagen foundation. Discussions with K. Padberg, K. Peters and D. Karrasch are gratefully acknowledged.

References

Binder PM, Jensen RV (1986) Simulating chaotic behavior with finite-state machines. Phys Rev A 34: 4460–4463

Bloomberg DJ, LeMay SB, Hanna JB (2002) Logistics. Prentice Hall, New York

Chase C, Serrano J, Ramadge PJ (1993) Periodicity and chaos from switched flow systems: Contrasting examples of discretely controlled continuous systems. IEEE Trans Automat Contr 38: 70–83

Donner R, Scholz-Reiter B, Hinrichs U (2008a) Nonlinear characterization of the performance of production and logistics networks. J Manufact Syst 27: 84–99

Donner R, Hinrichs U, Scholz-Reiter B (2008b) Symbolic recurrence plots: a new quantitative framework for performance analysis of manufacturing networks. Eur Phys J Spec Top 164: 85–104

Donner R, Hinrichs U, Scholz-Reiter B (2008c) Mechanisms of instability in small-scale manufacturing networks. In: Haasis HD, Kreowski HJ, Scholz-Reiter B (eds) Dynamics in Logistics—First International Conference, LDIC 2007—Bremen, Germany, August 2007—Proceedings. Springer, Berlin, pp 161–168

Donner R, Padberg K, Höfener J, Helbing D (2010) Dynamics of supply chains under mixed production strategies. In: Fitt AD et al (eds) Progress in Industrial Mathematics at ECMI 2008. Mathematics in Industry 15. Springer, Berlin, pp 527-533

Forrester JW (1958) Industrial dynamics—a major breakthrough for decision makers. Harvard Business Rev 36: 37–66

Helbing D, Lämmer S, Witt U, Brenner T (2004) Network-induced oscillatory behavior in material flow networks and irregular business cycles. Phys Rev E 70: 056118

Hopp WJ, Spearman ML (2000) Factory Physics. McGraw-Hill, Boston

Hülsmann M, Windt K (2007) Understanding autonomous cooperation and control in logistics. Springer, Berlin

Katzorke I, Pikovsky A (2000) Chaos and complexity in a simple model of production dynamics. Discr Dyn Nat Soc 5: 179–187

Marwan N, Romano MC, Thiel M, Kurths J (2007) Recurrence plots for the analysis of complex systems. Phys Rep 438: 237–329

Rem B, Armbruster D (2003) Control and synchronization in switched arrival systems. Chaos 13: 128–137

Scholz-Reiter B, Hinrichs U, Donner R, Witt A (2006) Modelling of networks of production and logistics and analysis of their nonlinear dynamics. In: Wamkeue R (ed) Modelling and Simulation. IASTED, Montréal, pp 178–183

Windt K (2002) Optimierung von Lager- und Distributionsstrukturen in Logistiknetzen am Beispiel eines weltweit agierenden Maschinenbauers. In: Tagungsband zum Wissenschaftssymposium Logistik der BVL. Huss-Verlag, Munich, pp 235–251

Converting Knowledge into Performance Within Global Production and Logistic Systems

Enzo Morosini Frazzon and Bernd Scholz-Reiter

1 Introduction

Global supply chains comprise material and information flows across process boundaries (e.g. production, inventory and transportation), firm boundaries (e.g. manufactures, carriers, logistic services firms and distributors) and context boundaries (e.g. continents, nations and regions). The connection between markets and sources, demand and supply has increased the strategic relevance of global production and logistic systems.

Any economic organisation faces two basic challenges: first, it must survive the challenges of today, and second, it must adapt to tomorrow's challenges. This research explores the latter, addressing the transformation of context-related knowledge into performance within global production and logistic systems. More specifically: which aspects would have to be considered in a contextualisation processes so that it supports the international venturing of production and logistics?

Suitable prior knowledge—mainly prior practical experience—leads to better managerial interpretation and thus enhanced awareness to opportunities. Our a priori knowledge bounds our capacity to interpret the reality (Popper 1999) and to leverage our decision making. In agreement with a holistic perspective (context, organisation and individual levels), relevant aspects impacting on complexity and

E. M. Frazzon (✉) and B. Scholz-Reiter
BIBA-Bremer Institut für Produktion und Logistik GmbH, Hochschulring 20, 28359, Bremen, Germany
e-mail: enzo@deps.ufsc.br

B. Scholz-Reiter
e-mail: bsr@biba.uni-bremen.de

E. M. Frazzon
DEPS-Department of Industrial and Systems Engineering,
UFSC-Federal University of Santa Catarina, Florianópolis, SC, Brazil

H.-J. Kreowski et al. (eds.), *Dynamics in Logistics*,
DOI: 10.1007/978-3-642-11996-5_39, © Springer-Verlag Berlin Heidelberg 2011

dynamics in production and logistics will be considered. Furthermore, issues regarding knowledge management, organisational learning and resulting performance will be handled. Finally, a process for assembling internal knowledge and expertise in order to enhance the performance of global production and logistic systems will be embodied on an application-oriented contextualisation cycle.

2 Global Production and Logistic Systems

Businesses, markets and national economies are becoming interdependent as a result of diminishing barriers and evolving technologies. Communication and transportation evolution has underpinned the internationalisation of trade and services, allowing business networks to expand out of their original national borders, exploit new markets and locate operations in different countries (Frazzon 2009). Global marketing, production and sourcing have shaped material, information and financial streams in more distant, dynamic and complex manners.

The pursuit of attractive markets for products and services, as well as high quality and low cost sources are some of the most prominent facets of globalisation. Production and logistics systems bridge the mentioned sources and markets and can be interpreted as connective networks of material and information flows, of infrastructures and assets, and of organisations. By comparison, global supply chains comprise material and information flows across process, firm and context boundaries (Frazzon 2009).

International manufacturing sources—whether company-owned or external suppliers—have in recent years been sought out by managers because of reduced cost, increased revenues, and improved reliability (Meixell and Gargeya 2005). Manufacturers typically set up foreign factories to benefit from tariff and trade concessions, low cost direct labor, capital subsidies, and reduced logistics costs in foreign markets (Ferdows 1997). Likewise, benefits accrue due to access to overseas markets, organisational learning through close proximity to customers, and improved reliability because of close proximity to suppliers (MacCormack et al. 1994).

The lure of cost savings, largely due to fewer regulatory controls and significantly lower wages has prompted the mass-migration of manufacturing from the developed world to emergent economies in other regions (Christopher et al. 2006). Therefore, organisations today have to deal with multiple interrelations between actors and stakeholders within distinctive environments (Hülsmann et al. 2006).

The global economy has imposed worldwide business and political challenges. In this hypercompetitive landscape, events, competitors, environments, and industries change constantly and, sometimes, unpredictably.

By its very nature, the hypercompetitive landscape has become precipitously more uncertain, dynamic, intense, aggressive, and at the same time deregulated, technology intensive, and global in scope (Harvey and Novicevic 2001). Due partially to this brave new and diversified world, companies need to possess differentiated capabilities to serve different markets. Their customers around the world have specific needs, and high-performance businesses will accommodate and create value from those differences (WSJ 2007).

In fact, customer expectations, the pressure of competition on turbulent global markets and virtualisation of logistics companies result in complex and dynamic logistics systems, structures and networks (Scholz-Reiter et al. 2004).

In a holistic and integrated approach, it is necessary to consider whole global production and logistic systems and business processes. The global supply chain is influenced first by corporate strategy, shaped by products, technology and markets and also by the moves of competing supply chains (Schary and Skjott-Larsen 2001). It is further influenced by: government policies, factor markets, competition, supporting industries and demand in both home and host country markets (Porter 1990).

Furthermore, substantial geographical distances not only increase transportation costs, but complicate decisions because of tradeoffs between cost and service level due to increased lead-time in the supply chain. Furthermore, different local cultures, languages, and practices diminish the effectiveness of business processes, such as demand forecasting and material planning (Meixell and Gargeya 2005).

Global production and logistic systems carry unique risks that influence performance, including variability and uncertainty in currency exchange rates, economic and political instability, and changes in the regulatory environment (Dornier et al. 1998). Similarly, infrastructural deficiencies in transportation and telecommunications, as well as supplier availability, supplier quality, equipment and technology provide additional. These difficulties inhibit the degree to which a global supply chain provides a competitive advantage (Meixell and Gargeya 2005).

3 Adaptation to Uncertain Situations

The connection and further integration of demand and supply must be congruent with strategic choices that determine which activities and processes an organisation will perform and how they will be designed and coordinated. In fact, business strategy is about how to combine and fit activities and processes to obtain and sustain competitive advantage (Porter 1996).

To further preserve this advantage the organisation has to be unique; and, to generate and sustain uniqueness, its resources must be (Barney 1991): (1) valuable, in the sense that they exploit opportunities and/or neutralise threats in the firm environment; (2) rare among the current and potential competitors of a firm; (3) imperfectly imitable, either through unique historical conditions, causal ambiguity, or social complexity; and (4) singular, without strategically equivalent substitutes.

Successful supply chain integration depends on the supply chain partners' ability to synchronise and share information. The establishment of a collaborative relationship among supply-chain partners is a pre-requisite to information sharing. The swift flow of information and materials has a vital function in such interwoven chains/networks, directly impacting the competitive advantage creation. In fact, the increasing dynamic and structural complexity within logistic networks—caused by changing conditions in markets and the increasing importance of customer orientation and individualisation—challenge logistics in a fundamental manner (Arndt and Müller-Christ 2005).

Successful companies will not simply be those who leverage sourcing opportunities worldwide but those who plan and manage their production systems best. But what are key evolving risks in a global environment, and the strategies for resilience in the face of them?

Strategy formulation—typically a concern for economics and management—has recently received more attention in engineering disciplines due to the introduction of new bottom up approaches and methods. For instance, emergent synthetic (Ueda 2000; Ueda et al. 2001) approaches can deal efficiently with the interaction of organisations and their competitive environment. They include evolutionary computation, self-organisation, reinforcement learning, multi-agent systems, and game theory. Traditional approaches such as analytic or deterministic ones might not be adequate to solve business problems where the environment is not completely characterised (Ueda et al. 2008).

In current and forthcoming challenging environments, production and logistic systems should continuously evolve toward fitted strategies, structures and policies. At the same time, they also led to the development and further improvement of processes and networks designated to fit and connect demand and supply fronts (Frazzon 2009).

For that to materialise, it requires an iterative long-term learning process in which a reductionist, narrow, short-run, static view of the world should be replaced by a holistic, broad, long-term and dynamic view (Sterman 2006), helping to understand how complex problems are generated and which factors influence them over time (Senge 1990) and space.

As the landscape becomes more complex, so does the technology that underpins it. In production and logistic systems the focus on technological evolution is fostering the implementation of solutions that ease the sharing and exchanging of information right across individuals and organisations. Not just physical technology but also social technologies (Beinhocker 2006) have to be employed for supporting business results.

Even though the impact of technology is global and turns supply chain management into a rapidly changing but increasingly common understanding of concepts and practices (Schary and Skjott-Larsen 2001), there is a huge locus for acting on the necessary adaptation of socio-technical solutions to specific contexts, where different cultural, administrative, geographic and economic characteristics present themselves as systemic and uncertain constraints (Frazzon 2009).

How successful a company is at exploiting emerging opportunities and tackling accompanying challenges depends crucially on how intelligent it is at observing and interpreting the dynamic world in which it operates. Creating a global mindset is one of the central ingredients required for building such intelligence (Gupta et al. 2008) in international business strategy.

4 Knowledge into Performance

Suitable prior knowledge—mainly prior practical experience—leads to better managerial interpretation and thus enhanced awareness to opportunities. Our a priori knowledge bounds our capacity to interpret the reality (Popper 1999) and to leverage our decision making.

In agreement with a holistic perspective (context, organisation and individual levels), relevant aspects impacting on complexity and dynamics in logistics will be considered. Furthermore, issues regarding integration, long-term learning process, knowledge management will be handled.

Mindsets represent knowledge structures (that is, cognitive templates) (Gupta et al. 2008). As a result, the development of mindsets follows the same generic path as the development of all types of knowledge. As known from research in a variety of areas, including evolution of species, cognitive psychology and technological innovation, developments occurs through a sequence of evolutions and revolutions.

Effective methods based on knowledge integration and sharing should be able to support the core tasks of business strategic management, namely that of decision making and planning (Yim et al. 2004). International strategy problems are characterised by dynamic complexity, tacit knowledge factors, and feedback effects over time, and unstructuredness (Sterman 2001; Yim et al. 2004). It turns out to be even more challenging considering that "in complex systems and dynamic environments human decision making performance is poor relative to normative standards" (Sterman 1989).

Therefore, much research in the field of business strategy feature distinctive capabilities as the basis of competitive advantage and "dynamic capabilities" as the key to lasting success in a rapidly changing economy (Teece et al. 1997).

Strategic decision making deals with capabilities underpinned on interpreted information, i.e. knowledge. Decision making is dependent on existing implicit and explicit knowledge. Furthermore, differences in the socialisation of managers and the business environment that they face affect both their decision-making processes and the choices that they make (Martinsons and Davinson 2006).

In business interactions and decision making across borders special attention should be given not just to the absorption of explicit knowledge but also to the preliminary gathering of implicit knowledge aspects (e.g. existing social rules and cultural aspects). One core challenge in a globalising business world is to facilitate and direct interaction and learning at interfaces, where knowledge, values and experience would be transferred into cross-cultural domains of implementation (Holden 2001). Profiting from the matching between external opportunities and internal capabilities requires organisational boundaries to be conductive. It is important to cultivate awareness (perception and cognitive reaction to a condition or event) and long-term learning capabilities.

In fact, valuable knowledge identification requires the receiver to possess suitable knowledge and skills to assimilate it (Cohen and Levinthal 1990; Zahra and George 2002). Venturing activities enable gaining knowledge and capabilities that allow the exploitation of opportunities in foreign markets. In fact, the effects of international venturing activities depend on the absorptive capacity of a firm (Zahra and Hayton 2007). The construct is particularly important in businesses that rely on interrelations among people (Frazzon 2009).

A process for integrating internal knowledge and expertise in order to enhance the performance of global production and logistic systems including an application-oriented contextualisation cycle will be now proposed.

The main input for the contextualisation cycle is the preliminary, crude, "uncontextualised" business plan. The main output is the contextualised, ready-for-implementation business plan. Business plan are artifacts, schemes for applying organisational resources, knowledge, technologies and processes and competences in order to achieve business objective based on explicit or implicit interests (Frazzon 2009).

Furthermore, the contextualisation cycle application process itself might be considered as an output. In fact, the communication between actors and stakeholders, under a common perspective, shedding light to complex linkages and echelons provide insights, enhance knowledge integration and potentialise long-term and hence strategic learning.

The questions that could come into view in the proposed cycle are diverse, but there is a clear demand for a "hands on" approach on this field, keeping a substantive systems thinking background in order to better cope with an evolving and hence challenging international business scenario.

Corporate strategy, vision, mission, values and goals are the backbone and brain of any business endeavour; they serve as directions to keep an organisation moving (Miller 2005). The objective of business plans is to take the corporate strategy on and translate it into operational terms, easy and worthy as a business guide. Corporate strategy starts the process by establishing the aim for expansion in one specific country.

However, the business plan implementation process relies on both formal and informal processes, explicit and implicit knowledge, whether in relation to the strategic and structural or environmental aspects of organisations (Mooraj et al. 1999). There are written and unwritten rules and both must be considered.

A contextualisation cycle charters part of the business planning process, from the corporate strategy input to the iterative and interactive loops and outcomes. Analogies could be made with individuals' learning processes. The obstacles and barriers along the way present a dynamic behaviour. It could also be compared to a nautical map where some information is well known (e.g. depth) and several others not (e.g. aggregating uncertainties associated with climate, tides, water streams as well as the conditions of unexplored areas).

Therefore, necessarily a business planning cycle should be dynamic and embrace the following objectives: capture business knowledge (internally and externally), support innovation enhancement, subsidise the finding of new technologies, solutions and services, facilitate communication and bridge gaps/opportunities.

These perceptions are consistent with the findings of O'Grady and Lane (1996) that the lack of foreign market knowledge can create obstacles in doing an international assignment. As O'Grady and Lane H (1996) show, "a firm might believe it has knowledge on a specific market but will soon realise it lacks this knowledge". The reason is that market knowledge often occurs through trial and error and cannot always be learnt in advance. Given the information available at a given time, a contextualisation cycle (Fig. 1) tries to describe a dynamic path to achieve the chartered destination.

The contextualisation cycle for business plans enables a conceptual and evolutionary—iterative, interactive—visualisation of how business plans should be planned and implemented (Fig. 1). The proposed contextualisation cycle depicts the cyclical (iterative) loops as the foundation for the interactive business planning and implementation. The following elements have to be considered:

- *Corporate strategy*: long-term objectives, as well as targeted markets.
- *Preliminary business plan*: partially result of the rebated corporate strategy. It may follow different business models but generally includes details about medium to short-term objectives and means for approaching a specific market.
- *First step*: how distant is the targeted market?

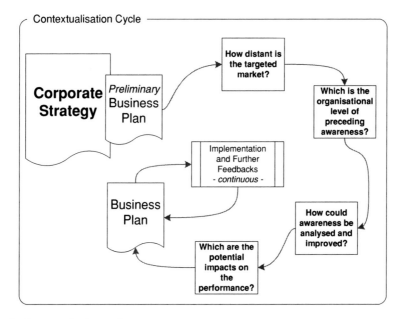

Fig. 1 Contextualisation cycle

- *Second step*: which is the organisational level of preceding awareness?
- *Third step*: how could awareness be analysed and improved?
- *Forth step*: how to measure business plan performance and the causal interrelation with performance indicators?

The main target is to create value for users, i.e. decision makers within the organisation. The final contribution from the contextualisation cycle is bringing some simplicity to very complex processes and systems. The variety, diversity, uncertainty, structure, integration, co-evolution, socio-technical variables impacting the business planning demands "hand on" tools for easing the materialisation and operationalisation of strategic objectives. Possible implementations have to consider that some degree of customisation might be necessary to accommodate real world applications.

5 Conclusions

Globalisation has encouraged companies to expand internationally and create new revenue streams. International venturing may give firms new knowledge and skills that fuel their innovation and business creation (Hamel 1991). Such strategic initiatives impact on both path- and context-dependencies and should take advantage of collaboration to best use available knowledge, resources and competences (Freiling 2004). These initiatives have to reflect a world where knowledge creation and absorption are key determinants for prosperity (Denis et al. 2006). One important element in any entry strategy is strong local knowledge: developing, contracting human capital with a suitable background and experience aiming to shorten learning lead-time and rapidly obtain contextual awareness.

By handling the question on which aspects would have to be considered in a contextualisation processes so that it supports the international venturing of production and logistics this paper addressed a relevant decision-making process within supply chains in an international context. This research focused on the multifaceted understanding of factors and processes that determine the effective and sustainable performance of international venturing initiatives.

The co-evolution of profits in farther markets and sources with concurrent adaptations to these distant contexts will permanently challenge the effectiveness and sustainability of business plans. There are huge opportunities for the ones that challenge wisdom with fit knowledge.

In global production and logistic systems, crossing boundaries involves different domains of decision, practices, values and technologies. Knowledge integration and synchronisation across this kind of boundaries is a major challenge (Schary and Skjott-Larsen 2001). In fact, greater benefits could be realised through the effective integration of new knowledge and capabilities (Zahra and Hayton 2007). International venturing activities enable gaining knowledge and capabilities that allow the exploitation of opportunities in foreign markets. The task of business leaders must be to overcome the paralysis that dooms any organisation and to begin shaping the future. One starting point is to take stock of what they do know about their industries and the surrounding environment (McKinsey 2008).

On this direction, further research should focus on the detailed description of the contextualisation cycles, the proposition of demonstrative scenarios as well as its application for supporting the transformation of knowledge into performance within production and logistic systems.

References

Arndt L, Müller-Christ G (2005) Robustness in the Context of Autonomous Cooperating Logistic Processes: A Sustainability Perspective. In: Haasis H D, Kopfer H, Schönberger J (eds.) Operations Research Proceedings: Selected Papers of the Annual International Conference of the German Operations Research Society (GOR), Springer, Berlin, 67–72

Barney J (1991) Firm Resources and Sustained Competitive Advantage. Journal of Management 17: 99–120

Beinhocker E D (2006) The origin of wealth evolution complexity and the radical remaking of economics. Harvard Business School Press, Boston

Christopher M, Peck H, Towill D (2006) A taxonomy for selecting global supply chain strategies. The International Journal of Logistics Management 17/2: 277–287

Cohen W M, Levinthal D A (1990) Absorptive capacity: A new perspective on learning and innovation. Administrative Science Quarterly 35: 128–152

Denis C, Mc Morrow K, Röger W (2006) Globalisation Trends Issues and Macro Implications for the EU. Economic Paper 254 Directorate–General for Economic and Financial Affairs European Commission, July 2006

Dornier P P, Ernst R, Fender M, Kouvelis P (1998) Global Operations and Logistics Text and Cases. John Wiley & Sons Inc., New York

Ferdows K (1997) Making the most of foreign factories. Harvard Business Review 75: 73–88

Frazzon E M (2009) Sustainability and Effectiveness in Global Logistic Systems. GITO-Verlag, Berlin. ISBN 978-3-940019-71-4

Freiling J A (2004) Competence–based Theory of the Firm. Management Revue—The International Review of Management Studies 15: 27–52

Gupta A, Govindarajan V, Wang H (2008) The Quest for Global Dominance. Jossey–Bass

Hamel G (1991) Competition for competence and inter–partner learning within international strategic alliances. Strategic Management Journal 12: 83–103

Harvey M, Novicevic M M (2001) The impact of hypercompetitive "timescapes" on the development of a global mindset. Management Decision 39/6: 448–460

Holden N (2001) Towards redefining cross–cultural management as knowledge management. American Academy of Management Meeting Washington, August 2001

Hülsmann M, Grapp J, Li Y (2006) Strategic Flexibility in Global Supply Chains – Competitive Advantage by Autonomous Cooperation. In: Pawar K S et al. (eds.) Conference Proceedings of 11th International Symposium on Logistics, Loughborough United Kingdom, 494–502

MacCormack A D, Newmann L J I, Rosenfield D B (1994) The new dynamics of global manufacturing site location. Sloan Management Review 35: 69–84

Martinsons M G, Davison R M (2006) Strategic decision making and support systems Comparing American Japanese and Chinese management. Decision Support Systems, doi:101016/jdss200610005

McKinsey (2008) Leading through uncertainty. The McKinsey Quarterly, December 2008

Meixell M J, GargeyaV B (2005) Global supply chain design: A literature review and critique. Transportation Research Part E 41: 531–550

Miller J A (2005) Practical Guide to Performance Measurement. The Journal of Corporate Accounting & Finance, doi:101002/jcaf20121

Mooraj S, Oyon D, Hostettler D (1999) The Balanced Scorecard a Necessary Good or an Unnecessary Evil?. European Management Journal 17: 481–491

O'Grady S, Lane H W (1996) The psychic distance paradox. Journal of International Business Studies, 2nd quarter 1996: 309–333

Popper K (1999) All life if problem solving. Routledge, UK

Porter M E (1990) The Competitive Advantage of Nations. The Macmillan Press, London

Porter M E (1996) What is Strategy? Harvard Business Review 74/6: 18p

Schary P B, Skjott-Larsen T (2001) Managing the Global Supply Chain. Copenhagen Business Scholl Press, Copenhagen

Scholz-Reiter B, Windt K, Freitag M (2004) Autonomous logistic processes new demands and first approaches. In: L Monostri (Ed.) Proceedings of the 37th CIRP International Seminar on Manufacturing Systems, Budapest, Hungaria 357–362

Senge P (1990) The fifth discipline the art & practice of the learning organisation. Currency Doubleday, New York

Sterman J D (1989) Misperceptions of feedback in dynamic decision making In Organisational Behaviour and Human Decision Process 43, pp 271–287

Sterman J D (2001) System dynamics modelling tools for learning in a complex world. California Management Review 43: 8–25

Sterman J D (2006) Learning from Evidence in a Complex World. American Journal of Public Health 96: 505–514

Teece D, Pisano G, Shuen M (1997) Dynamic Capabilities and Strategic Management. Strategic Management Journal 18/7: 509–533

Ueda K (2000) Emergent Synthesis. Artificial Intelligence in Engineering 15/4: 319–320

Ueda K, Markus A, Monostori L, Kals H J J (2001) Arai T Emergent Synthesis Methodologies for Manufacturing. Annals of the CIRP 50/2: 535–551

Ueda K, Kito T, Takenaka T (2008) Modelling of value creation based on Emergent Synthesis. CIRP Annals - Manufacturing Technology 57: 473–476

Wall Street Journal (2007) Corporations Need a Global Mindset to Succeed in Today's Multipolar Business World. The Wall Street Journal, 18th June 2007

Yim N H, Kim S H, Kim H W, Kwahkc K Y (2004) Knowledge based decision making on higher level strategic concerns system dynamics approach. Expert Systems with Applications 27: 143–158

Zahra S A, George G (2002) Absorptive Capacity: A Review Reconceptualization and Extension. Academy of Management Review 27/2: 185–203

Zahra S A, Hayton J C (2007) The effect of international venturing on firm performance: The moderating influence of absorptive capacity. Journal of Business Venturing 23: 195–220

Dynamic Scheduling of Production and Inter-Facilities Logistic Systems

Bernd Scholz-Reiter, Antônio G. N. Novaes, Thomas Makuschewitz and Enzo Morosini Frazzon

1 Introduction

Supply chains often embrace production around the globe in order to benefit from country-specific advantages. Moreover, organisations are focusing on core competences and thus collaborating with strategic partners. These developments lead to increasing structural complexity and dynamics within global supply chains. Cost and lead-time savings obtained with global manufacturing strategies might be impaired due to the unbalanced, unsynchronized and unstable integration of production and logistic flows. Therefore the production and transportation scheduling should be carried out in a more integrated way.

Planning and control systems underpin manufacturing performance and have been broadly adopted. Advanced planning systems (APS) are often used to synchronise the material flow along supply chains. Their architecture can be illustrated by the Supply Chain Planning Matrix (Rohde et al. 2000). It comprises several modules for the planning tasks, that are characterised by time horizon (strategic, tactical, operational) and involved business functions (procurement, production, distribution and demand forecasting). The degree of detail increases and the planning horizon decreases by shifting from the strategic to the operational level. Due to the large amount of data and the large number of decisions on the operational level, the planning is usually not done in a centralised manner. Thus material requirements planning, production planning and scheduling as well as

B. Scholz-Reiter (✉) and T. Makuschewitz
Planning and Control of Production Systems (PSPS), BIBA-Bremer Institut für Produktion und Logistik GmbH at the University of Bremen, Bremen, Germany
e-mail: bsr@biba.uni-bremen.de

A. G. N. Novaes and E. M. Frazzon
Department of Industrial Engineering, UFSC, Federal University of Santa Catarina, Florianópolis, Santa Catarina, Brazil
e-mail: enzo@deps.ufsc.br

H.-J. Kreowski et al. (eds.), *Dynamics in Logistics*,
DOI: 10.1007/978-3-642-11996-5_40, © Springer-Verlag Berlin Heidelberg 2011

distribution and transportation planning are carried out decentralised in a sequential way (Fleischmann et al. 2004).

Since production and transportation scheduling is carried out in a sequential way and without consideration of internal and external dynamics, an integrated alignment of these two schedules at the operational level holds a great potential for strengthening supply chain's competitiveness. First studies show that a significant improvement can be achieved by an integrated scheduling compared to a sequential scheduling approach (Chen and Vairaktarakis 2005). The integrated production and transportation scheduling problem (PTSP) with capacity constraints is well known in the literature. An optimal solution for it requires solving the production scheduling and transportation routing simultaneous. The characters of the individual problems lead to mathematical programs, which are NP-hard. Thus, even for small scenarios an excessive computational power is needed.

The classic production and transportation scheduling problem focuses rather on the production capacities as constraints than on the transportation times and costs. These approaches often assume the transportation to be instantaneous and do not address the routing of the transportation vehicles. Several concepts for the integration of production and transportation have been developed in the recent years (Sarmiento and Nagi 1999; Cohen and Lee 1988; Chandra and Fisher 1994; Haham and Yano 1995; Fumero and Vercellis 1999). However, most of these concepts focus on the strategic or tactical planning and scheduling (Chen 2004). Research dealing with detailed schedules for the transportation can be classified according to the objectives of to applied mathematical programs and heuristics. The first group focuses on the lead time of orders' production and transportation (Woeginger 1994; Lee and Chen 2001; Geismar et al. 2008). The second group takes associated costs and lead times into account (Chen and Vairaktarakis 2005; De Matta and Miller 2004; Pundoor and Chen 2005; Stecke and Zhao 2007). Although the detailed scheduling of production and transportation presents already a valuable achievement, the transportation system also comprises the routing of the utilised transportation vehicles. This challenge has only been addressed by a few authors (Geismar et al. 2008; Li et al. 2005).

This paper is structured as follows. Section 2 introduces our approach for a decentralised PTSP along global supply chains. In Sect. 3 a mathematical program is presented that integrates production scheduling and routing of inter-facility transports. A test case is studied in Sect. 4. Section 5 presents some conclusions and suggests directions for future research.

2 Integrating Production and Transportation Scheduling

Current approaches focus on the integration of production scheduling of an OEM and the distribution of orders to the customers. In order to adapt the PTSP to a global supply chain with several production facilities in sequence it is proposed

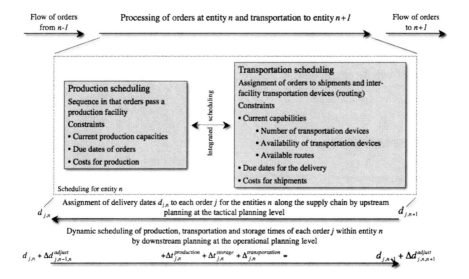

Fig. 1 Characteristics of the PSTP of a given entity along the supply chain

the introduction of scheduling entities along the supply chain. These entities carry out the integrated scheduling for one production facility and the associated transportation to the subsequent production facility. This includes the routing of the transportation devices. Tactical planning aligns these entities by specifying the order delivery dates for consecutive entities. In order to enable the entities to resolve orders' poor schedules that cannot match their delivery date, the entities need to have the flexibility to shift delivery dates according to certain mechanisms. Consequentially each entity has not only to set up a production and transportation schedule that is suitable for its own specifications of delivery dates but also for the specifications of directly connected entities. Figure 1 sketches the characteristics of the proposed PTSP for the integrated production and transportation scheduling.

The scheduling of the orders is based on the order delivery dates $d_{j,n}$, which are specified by the tactical level. The aim of each entity is to meet a certain service level in regard to the delivery of orders in time and to minimise the costs for production and transportation. Therefore a schedule for all orders is set up and the dates for production, transportation and if necessary for storage are specified. This schedule is subject to the constraints given by the date of provision of orders from the preceding production facility, current capabilities of the production and transportation system, delivery dates to the subsequent production facility and associated costs. In addition to this transportation devices become dynamically available as soon as they arrive from the preceding facility. In the case that an order cannot be delivered in time the delivery date $d_{j,n}$ can be shifted by $\Delta d_{j,n,n+1}^{\text{adjust}}$.

3 Proposed Formulation of the PTSP

Section 3 introduces a mathematical formulation of the PTSP for a given entity within the supply chain, considering the described characteristics of Sect. 2.

3.1 Nomenclature

Sets

I	Nodes of the transportation network
I^D	Considered depot/production facility of the transportation network ($I^D \subseteq I$), here only one depot exists
I^{SF}	Subsequent production facilities within the considered transportation network ($I^{SF} \subseteq I$), here only one subsequent production facility exists
I_i^s	Locations that are directly connected to location i; with ($I_i^s \subseteq I$)
J	Orders from subsequent facilities i
N	Production levels of the production facility
M	Machines
M_n^e	Available machines at production level n; with ($M_n^e \subseteq M$)
V	Tours for the delivery of orders to successive facility

Parameters

c^d	Costs for a delayed delivery of order j
c^{dv}	Costs per time unit for a tour v
c^{fv}	Fixed costs for conducting a tour v
c^h	Storage costs of an order at the depot
$c_{j,n,m}^p$	Processing costs of order j at level n on m
$d_{i,i'}$	Travel time between node i and node i'
M	BigM; large scalar
$pt_{j,n,m}$	Processing time of order j at level n on m
r_j	Required transportation capacity by order j
$\bar{r}_v$	Maximum transportation capacity of tour v
$t_{j,n}^a$	Supply date of order j by the tour from the preceding production facility
t_v^{av}	Earliest date of departure of tour v from the production facility
$t_{j,i}^{dd}$	Desired delivery date of order j at the subsequent facility i

Positive variables

$T_{j,n}^c$	Completion time of order j at machine m at level production level n
T_j^d	Delivery delay of order j to the subsequent production facility

T_j^h Storage time of j between the supply date and the start of production at production level n.

$T_{j,n,n'}^{hp}$ Storage time of j between consecutive production levels n and n'

T_j^{hv} Storage time of j between the last production level and the start of the assigned tour

T_v^s Start time of tour v from the production facility

$T_{v,i}^a$ Arrival time of tour v at location i

T_v^{dv} Duration of tour v

Binary variables

$X_{j,n,m}$ Binary variable denoting that order j is processed at machine m at level n

$Y_{j,j',n}$ Binary variable denoting that order j is processed before order j' at level n

$Z_{v,i,i'}$ Binary variable denoting that node i' is visited after node i by tour v

$A_{j,v}$ Binary variable denoting that order j is assigned to tour v

O_v Binary variable denoting that tour v is conducted

3.2 Model Assumptions

The mathematical program combines the production planning of one production facility and the vehicle routing concerning to the orders that have to be delivered to the subsequent production facility. As soon as a tour from a preceding production facility arrives at the considered production facility the processing of the associated orders can start. In order to schedule the production of these orders the mathematical program takes the desired delivery date at the subsequent production facility as well the availability of tours to the designated facility into account. The capabilities of the transportation system are modelled by the time depending availability of the transportation devices. This means that a transportation device for a tour becomes available as soon as it arrives at the considered production facility from a preceding facility.

The applied production scheduling is based on a multilevel production. Hence, orders (jobs) are processed at several consecutive production levels before the production is finished. Each production level can consist of several machines, which are characterised by different properties. In this case each machine features an order specific processing time and processing cost. All orders have to be processed at one machine at each production level. The orders can be stored before the start of production at the first production level, between the production levels and before the assigned tour to the subsequent production facility departs. The storage of orders causes as well additional costs.

The transportation of orders to the subsequent production facility is conducted by assigned tours. Each tour starts at the considered production facility and ends at the designated subsequent production facility. A new tour becomes available as soon as a tour from a preceding production facility arrives. Within a certain tour each location of the transportation network can be visited only once. However, a location can be visited by several tours, in order to satisfy the demand of the subsequent production facility. A partial delivery of an order is not allowed. Each tour has a given transportation capacity, which cannot be exceeded by the assigned orders. In the case that at least one order is assigned to a tour, fixed and variable costs occur. The variable costs depend on the duration of the tour. The actual delivery date of an order should be in line with the desired delivery date. In order to keep the model feasible an order can be delivered late but not early to the subsequent production location. In addition to this the minimal transportation time between two consecutive locations of a tour can be extended. This allows the program to stock orders during their transportation and to save inventory costs.

3.3 Mathematical Model

The production of an order is assigned to one machine at each production level.

$$\sum_{m \in M_n^e} X_{j,n,m} = 1 \quad (j \in J; n \in N) \tag{1}$$

The completion time of an order at a given production level is greater than the sum of the completion time at the previous production level and the required processing time of the assigned machine.

$$T_{j,n-1}^c + t_{j,n}^a + \sum_{m \in M_n^e} pt_{j,n,m} X_{j,n,m} \leq T_{j,n}^c \quad (j \in J; n \in N) \tag{2}$$

Equations 3 and 4 schedule the processing of orders and ensure that at each point in time only one order is processed at a certain machine. For one production level the sequence of all orders is the same at all available machines.

$$Y_{j,j',n} + Y_{j',j,n} = 1 \quad (j,j' \in J : j < j'; n \in N) \tag{3}$$

$$T_{j,n}^c + pt_{j',n,m} \leq T_{j',n}^c + M(2 - X_{j,n,m} - X_{j',n,m}) + M(1 - Y_{j,j',n}) \tag{4}$$
$$(j,j' \in J : j \neq j'; n \in N; m \in M_n^e)$$

Orders can be stored before and between the production levels. The storage times are obtained by Eqs. 5 and 6.

$$T_j^h \geq T_{j,n}^c - \sum_{m \in M_n^e} pt_{j,n,m} X_{j,n,m} - t_{j,n}^a \quad (j \in J; n = 1) \tag{5}$$

$$T_{j,n-1}^{hp} \geq T_{j,n}^c - T_{j,n-1}^c - \sum_{m \in M_n^e} pt_{j,n,m} X_{j,n,m} \quad (j \in J; n = 2, \ldots, N) \tag{6}$$

Before the assigned tour of an order to the subsequent production facility starts the order can be stored after it passed the last production level.

$$T_j^{hv} \geq T_v^s - T_{j,n}^c - M\left(1 - A_{j,v}\right) \quad (j \in J; n = N; v \in V) \tag{7}$$

In the case that a tour is conducted the tour starts at the considered production facility and terminates at the subsequent production facility.

$$\sum_{i' \in I_i^S} Z_{v,i,i'} = O_v \quad (i \in I^D; v \in V) \tag{8}$$

$$\sum_{i \in I: i' \in I_i^S} Z_{v,i,i'} = O_v \quad (i' \in I^{SF}; v \in V) \tag{9}$$

The continuity of a tour between the considered production facility and the assigned subsequent production facility is given by Eq. 10.

$$\sum_{i \in I: h \in I_i^S} Z_{v,i,h} - \sum_{i' \in I_h^S} Z_{v,h,i'} = 0 \quad \left(h \in I/I^D \wedge I^{SF}; v \in V\right) \tag{10}$$

Each order is assigned to one tour; partial deliveries are not allowed.

$$\sum_{v \in V} A_{j,v} = 1 \quad (j \in J) \tag{11}$$

A tour can depart from the considered production facility as soon as all assigned orders are manufactured and the transportation device is available.

$$T_v^s \geq T_{j,n}^c - M\left(1 - A_{j,v}\right) \quad (j \in J; n = N; v \in V) \tag{12}$$

$$T_v^s \geq t_v^{av} - M\left(1 - A_{j,v}\right) \quad (j \in J; v \in V) \tag{13}$$

A lower bound for the arrival time of a tour at the first location is given by the departure time from the depot and the minimal required travel time between the locations.

$$T_v^s + d_{i,i'} - M\left(1 - Z_{v,i,i'}\right) \leq T_{v,i'}^a \quad (i \in I^D; i' \in I : i' \in I_i^s; v \in V) \tag{14}$$

Equation 15 ensures that the arrival time at a consecutive location of a tour is greater than the sum of preceding arrival time and the minimal required travel time.

$$T_{v,i}^a + d_{i,i'} - M\left(1 - Z_{v,i,i'}\right) \leq T_{v,i'}^a \quad (i, i' \in I/I^D : i' \in I_i^s; v \in V) \tag{15}$$

In the case that a location is not part of the tour the arrival time equals zero.

$$\sum_{i' \in I: i \in I_{i'}^s} Z_{v,i',i} M \geq T_{v,i}^a \quad (i \in I; v \in V) \tag{16}$$

Each tour has a limited transportation capacity, which cannot be exceeded by the assigned orders.

$$\sum_j A_{j,v} r_j \leq \bar{r}_v \quad (v \in V) \tag{17}$$

In the case that at least one order is assigned to a tour the tour is conducted.

$$\sum_{j \in J} A_{j,v} \leq O_v M \quad (v \in V) \tag{18}$$

The duration of a tour is greater than zero in the case that the tour is conducted.

$$T_v^{dv} \geq T_{v,i}^a - T_v^s - M(1 - O_v) \quad \left(i \in I^{SF}; v \in V\right) \tag{19}$$

Each order has a desired delivery date. The delivery of an order cannot be early but late.

$$T_{v,i}^a \geq t_{i,j}^{dd} - M\left(2 - A_{j,v} - O_v\right) \quad \left(i \in I^{SF}; j \in J; v \in V\right) \tag{20}$$

$$T_j^d \geq T_{v,i}^a - t_{i,j}^{dd} - M\left(2 - A_{j,v} - O_v\right) \quad \left(i \in I^{SF}; j \in J; v \in V\right) \tag{21}$$

The objective function minimises the costs for delayed deliveries of the orders to the subsequent production facility, the processing and storage costs of orders and as well the fixed and variable costs of each conducted tour.

$$\text{Min.} \sum_{j \in J} T_j^d c_j^d + \sum_{j \in J} \sum_{n \in N} \sum_{m \in M_n^e} X_{j,n,m} c_{j,n,m}^p$$

$$+ \sum_{j \in J} \left(T_j^h + T_j^{hv} + \sum_{n \in N} T_{j,n-1,n}^{hp} \right) c^h + \sum_{v \in V} \left(O_v c^{fv} + T_v^{dv} c^{dv} \right) \tag{22}$$

4 Computational Analysis of the Adapted PTSP

Section 4 aims to identify the limitations of the formulated mathematical program for the integrated production and transportation scheduling problem. Therefore a test scenario is set up in Sect. 4.1 and the obtained results are shown in Sect. 4.2.

4.1 Test Case

The test case consists of two production facilities in northern Germany. The considered facility is located in Bremen and ships orders of intermediate products to the subsequent production facility in Berlin. A two level production process is

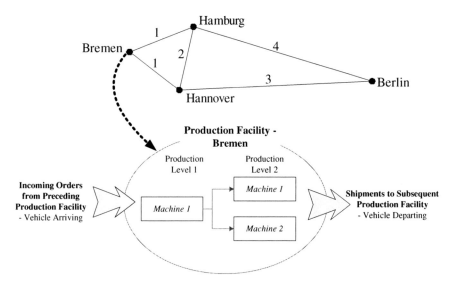

Fig. 2 Structure of test scenario

carried out in Bremen. The structure of the material flow within the production facility and the structure of the transportation network are shown in Fig. 2.

The edges of the transportation network are weighted with the required travelling time of the transportation devices between the locations of the network.

4.2 Computational Results

The proposed mathematical formulation of the integrated production and transportation scheduling problem has been implemented in GAMS 22.8. For simplicity all costs are set to 1, except the cost for machine one that double the cost of machine two at the second production level. Furthermore the cost for an unpunctual delivery is set to 10 for each time unit. The processing at machine 1 at production level 1 requires one time unit. At the second production level machine 1 requires two time units and machine 2 one time unit. The required transportation capacity is assumed to be one for all orders. Each transportation device has a maximal transportation capacity of 5 units. The considered test instances comprise five transportation devices that arrive at the following points in time 2, 10, 18, 26 and 34. At the same point in time new orders become available for the processing. The due dates for the delivery of orders to the subsequent production facility depend on the date of provision of the orders at the planning entity and are given by the following points in time 10, 15, 30, 45 and 60.

Since the mathematical formulation is a mixed integer problem (MIP) the instances could be solved by CPLEX 11. The computation was carried out on a

Table 1 Gap to the optimal solution after 600 s

Transportation devices	Orders	Gap to optimal solution (%)
1	3	0.00
1	5	0.00
2	7	0.00
2	10	21.19
3	13	25.27
3	15	33.23
4	17	34.41
4	20	45.59
5	23	49.27
5	25	49.20

2.67 GHz quad-core computer with 4 GB of RAM in a concurrent mode of CPLEX with four threads. The results for the test instances are given by Table 1.

The table shows the relative gap between the best integer solution and the best node remaining after 600 s. For very small instances the optimal solution can be obtained within this time. Hence, by increasing the number of orders and transportation devices the need for a heuristic that is able to solve larger instances with suitable results is shown.

5 Conclusions and Future Research

In this paper we introduced an approach for the implementation of an integrated production and transportation scheduling along global supply chains. Therefore a mathematical program for the planning tasks of a single entity has been formulated. This formulation takes dynamic changing capabilities of the transportations system into account. The computational analysis has shown that instead of a mathematical program a powerful heuristic is required.

Acknowledgment This research was supported by the German Research Foundation (DFG) as part of the Brazilian-German Collaborative Research Initiative on Manufacturing Technology (BRAGECRIM).

References

Chandra, P.; Fisher, M.L. (1994): Coordination of production and distribution planning. European Journal of Operational Research 72, pp. 503–517.
Chen, Z.L. (2004): Integrated production and distribution operations: Taxonomy, models, and review. Handbook of Quantitative Supply Chain Analysis: Modelling in the E-Business era. Simchi-Levi, D., Wu, S.D., Shen, Z.J. (eds.). Kluwer Academic Publishers, New York.

Chen, Z.-L.; Vairaktarakis, G.L. (2005): Integrated scheduling of production and distribution operations. Management Science, 51, pp. 614–628.

Cohen, M.A.; Lee, H.L. (1988): Strategic analysis of integrated production-distribution systems: Models and methods. Operations Research 36, pp. 212–228.

De Matta, R.; Miller, T. (2004): Production and inter-facility transportation scheduling for a process industry. European Journal of Operational Research, 158, pp. 72–88.

Fleischmann, B.; Meyr, H.; Wagner, M. (2004): Advanced Planning. Supply Chain Management and Advanced Planning. Springer, Berlin.

Fumero, F.; Vercellis, C. (1999): Synchronized development of production, inventory, and distribution schedules. Transportation Science 33, pp. 330–340.

Haham, J.; Yano, C.A. (1995): The economic lot and delivery scheduling problem: Powers of two policies. Transportation Science, 29, pp. 222–241.

Geismar, H.N.; Laporte, G.; Lei, L.; Sriskandarajah, C. (2008): The Integrated Production and Transportation Scheduling Problem for a Product with Short Lifespan. INFORMS Journal on Computing, Vol. 20/1, pp. 21–33.

Lee, C.Y.; Chen Z.L. (2001): Machine scheduling with transportation considerations. Journal of Scheduling 4, pp. 3–24.

Li, C.-L.; Vairaktarakis, G.; Lee, C.-Y. (2005): Machine scheduling with deliveries to multiple customer locations. European Journal of Operational Research, 164, pp. 39–51.

Pundoor, G.; Chen, Z.-L. (2005): Scheduling a production-distribution system to optimize the tradeoff between delivery tardiness and distribution cost. Navel Research Logistics, 52, pp. 571–589.

Rohde, J.; Meyr, H.; Wagner, M (2000): Die Supply Chain Planning Matrix. PPS-Management 1, pp. 10–15.

Sarmiento, A.; Nagi M.R. (1999): A review of integrated analysis of production-distribution systems. IIE Transportation 31, pp. 1061–1074.

Stecke, K. E.; Zhao, X. (2007): Production and Transportation Integration for a Make-to-Order Manufacturing Company with a Commit-to-Delivery Business Mode. Manufacturing & Service Operations Management 9(2), pp. 206–224.

Woeginger, G.J. (1994): Heuristics for parallel machine scheduling with delivery times. Acta Informatica, 31, pp. 503–512.

Part VI
Ports, Container Terminals, Regions and Services

How can Electronic Seals Contribute to the Efficiency of Global Container System?

Kateryna Daschkovska and Bernd Scholz-Reiter

1 Introduction

With about 90% of the volume of world trade moving by ocean transportation (Talley 2004), maritime supply chains play an important role in the world economy. Containerized trade has grown dramatically, from 36 million twenty-foot-equivalent units (TEUs) in 1980 to 303.1 million TEUs in 2003, with up to 468 million TEUs forecast for 2010 (Notteboom 2004; United Nations Conference on Trade and Development (UNCTAD) 2005). And as maritime containers are very popular transport units in the system of global trade more and more international organizations and different private companies concerned about the safety and security of container flows. After 9/11 US Customs and Boarder Protection (CBP) created the Container Security Initiative (CSI) and Customs Trade Partnership Against Terrorism (C-TPAT) which poses to protect the global container trading system and the trade lanes between CSI ports and the U.S. Under C-TPAT, members sign an agreement to cooperate with Customs and Border Protection (CBP) to protect the supply chain, identify security gaps, and implement specific security measures and best practices. Certainly, the prior aim of the government's acts is the national and international security for trans-national container flows. Hence the attention of each partner in container supply chain is focused now on the problem of finding a solution how to balance the security of container transportation without overloading the international transport systems with additional operational costs. Such trade-off can be found in transformation of a dumb box

K. Daschkovska (✉)
International Graduate School for Dynamics in Logistics, University of Bremen, Hochschulring 20, 28359 Bremen, Germany
e-mail: katerynadaschkovska@gmail.com

B. Scholz-Reiter
Bremen Institute for Production and Logistics (BIBA) GmbH, IPS, University of Bremen, Hochschulring 20, 28359 Bremen, Germany
e-mail: bsr@biba.uni-bremen.de

H.-J. Kreowski et al. (eds.), *Dynamics in Logistics*,
DOI: 10.1007/978-3-642-11996-5_41, © Springer-Verlag Berlin Heidelberg 2011

into a smart one by equipping a container with an e-seal or installing an electronic container security device (smart seal). A smart box will communicate evidence of tampering and register every legitimate or unauthorized opening of the container.

According to the current definition an e-seal is a "Read-only, non-reusable freight container seal conforming to the high security seal defined in ISO/PAS 17712 (ISO 2003) and conforming to ISO 18185 (ISO 2005) or revision thereof that electronically evidences tampering or intrusion through the container door". Since October 15, 2008 all maritime containers in transit to the United States are required to be sealed with a seal meeting the ISO 17712 standard. Generally, ISO/PAS 17712 requires that container freight seals meet or exceed certain standards for strength and durability so as to prevent accidental breakage, early deterioration (due to weather conditions, chemical action, etc.) or undetectable tampering under normal usage. ISO/PAS 17712 also requires that each seal be clearly and legibly marked with a unique identification number (Customs and Border Protection 2008). E-seals tend to combine physical seals components with RFID technology. This simplest type of electronic seals contains only a seal ID number. More advanced reusable or permanent active RFID e-seal also includes a seal ID number, a container ID number; can initiate alarm calls and record time/date of container tampering. Smart e-seal or "Container Security Devise" contains additional sensors to indicate environmental status of container content, alarm function to inform in real-time by means of satellite communication systems via GPS/INMARSAT (Daschkovska and Scholz-Reiter 2008).

The functions of e-seals have direct security benefits for container shipments. Nevertheless there are a number of alternative concomitant advantages from investing and using such multifunctional devices (smart seals) for cargo importers, container carriers, service providers and for container port operators. In this paper we expand the list of substantial benefits by proposing to look on e-seal's device from different perspectives to discover their potential for improving logistics operations through the supply chain as well as at container maritime terminals as huge nodes of cargo transshipment. In the next section we list the number of the most prominent advantages of adding a small "brain" to conventional maritime containers.

2 Some Aspects of e-Seals Influence on Container Logistics

In the last years the benefits from RFID technology in supply chains have been made clear. However, there is still a lack of research about advantages from application of electronic seals on containers in logistics networks. The general benefit from RFID e-seals in container supply chains are seemed to be only for securing purposes of customs authorities. However, the container transportation process is characterized by complex interactions of numerous operating companies and organizations. While a container moves from point of origin to destination point, as many as 20 different companies have to coordinate the operations of more than 25 documents with approximately 200 data elements for only one international shipment in the global

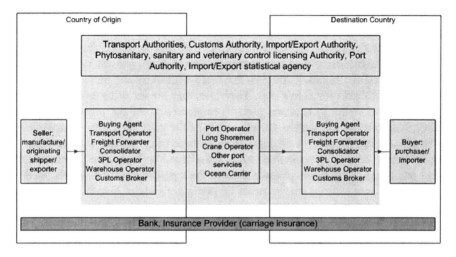

Fig. 1 Actors in container logistics networks, adapted from (Crist et al. 2005)

container network (Downey 2006). The companies might include trucking firms, terminal operators, and the shipping company, the manufacturer of the shipped goods, the purchaser, banks, and others (Fig. 1).

Currently, based on different types of container security devices, a sophisticated platform, that is capable to provide high-end security tasks and elaborated logistics management applications, can be developed (Notteboom 2004). Actually, the biggest container ports in the world, namely Singapore, Kaohsiung in Taiwan, Rotterdam in the Netherlands, Pusan in South Korea and the ports of Los Angeles and Long Beach, have already investigated RFID e-seals in projects to test their abilities for cargo tracking (Seymour et al. 2007). The projects were focused on various advantages from RFID system implementation in port environment, like greater efficiency by shortening the time for container checking and management through the port by using active RFID e-seals (Collins 2005), the issues of congestion and security in the ports (Nguyen 2006) or improving the security of containers destined for the USA, with more stringent security requirements (Clendenin 2005). Their results show that smart RFID e-seal with its multifunctional ability can be effective for logistics purposes and applications in container supply chains/container ports.

We describe below the main areas of logistics applications of smart e-seals in maritime ports and terminals.

2.1 Competitive Advantages to Connect Ocean Terminals into a Global Info-Network

Key factors for a container terminal are the efficiency of the stacking and transportation of this large number of containers to and from the ship's side. Shipping companies ask for reliability regarding adherence to delivery dates and promised handling times (Tongzon and Heng 2005). Thus, container ports are forced to provide efficient and cost-effective services. They have to invest heavily to meet the stringent demands for faster service and higher quality. The competition between container terminals has increased due to large growth rates on major seaborne container routes. Terminals are faced with more and more containers to be handled in short time at low cost. Therefore, they are forced to enlarge handling capacities and strive to achieve gains in productivity. At the present time, one more additional factor impacts the efficiency of the maritime ports—it is security issue. To secure the cargo containers more innovative companies use the smart e-seals, instead of simple mechanical seals. How does it influence the port operator? Port and terminal operators in order to keep their positions on the market need to adapt to the new security requirements for container system. Investments in RFID reading infrastructure will bring the ports the benefits by providing new kind of service for own innovative customers, that using e-seals, and open the port gates for new companies looking for the partners that can provide higher level of security and visibility for their cargo flows. The adaptation of innovative technology like e-seals is a state when no player can benefit by being the only one to change his strategy. No player can improve his position by opting out the impact of RFID seals in container system. Vice versa, no player can get better his position by adopting smart seals alone. The only way to move from the one equilibrium to another is to organize an agreement or consortium between the players (Hadow 2005) to get some of them to adapt the new technology solution (e-seals) for secure logistics processes.

2.2 Automation of Containers Passes to the Territory of Port Terminals

E-seals can contain data about goods in container, ID number of container, shippers contact information, point of cargo destination etc. By passing port gates RFID seal transmits container information to the local port network through the reading device fixed on the gate. In this case truck driver should not complete any manual formalities to get access to the terminal. This procedure takes much less time, has higher degree of accuracy, and eliminates technical mistakes (human factor).

2.3 Security and Safety of Containers

Security aspect is the strongest side of e-seals. Container terminals are usually not open territories. Nevertheless, the security of the most ports territory is provided by the simplest methods of accident prevention. For a long time containers may stay without any supervision at a terminals that increase vulnerability of the container in total. E-seal can prevent un-authorized access or theft of the container contents and inform about such accident by alarm function. On the other hand, container door with any type of seal can be removed or container can be cut and open from the top, bottom and sides. It makes a huge loss for owners of high value cargo and for insurance companies as well. To avoid container intrusion, pilferage, and thefts of containers or goods, it is necessary to combine electronic seal with different sensors (temperature, light, etc.).

2.4 Control of Access to the Containers Contents

E-seals possess a useful function to record the time of authorized access to the container contents or unauthorized access at the moment of e-seal breaking. Furthermore, the broken e-seal cannot be fabricated or changed by another one without any damages of electronic part of the seal. Hence, uniqueness of e-seal ID number and alarm function of this device provide reliable security of goods and the real time control of access to the containers contents.

2.5 Identification of Containers and their Locations

Each ISO container has unique ID number. Nonetheless, several identification numbers on the same container are the real case for many intermodal containers. In these cases terminal operators should identify which of ID numbers is correct. Such type of information can be coded to the e-seal memory. Each time a container changes the hands or documents regarding the container are transferred, the potential for miscommunication and human mistakes exists. For instance, a trucker might bring a container to a port but do not communicate to the shipping company that he has arrived. It causes a container not to be loaded onto a ship. These kinds of inconveniences may cost money to the company waiting for the shipments. Electronic information, transmitted from e-seal to the local network of the terminal, can provide companies with authorized access with required data in real time about individual container location.

2.6 Monitoring of Containers Movements

E-seal together with RTLS (Real Time Locating System) allows not only control the container location, but also to monitor every movement of the container on the port territory. RTLS is indispensable to container operators because container yards and van pools are so extremely large and store so many containers that without the support of locating systems, workers cannot find a particular container in required time limit. This informative function of e-seals can be useful to the forwarding or agent companies to inform whether the container loaded or unloaded from the ship.

2.7 Improving the Congested Situation in the Ports

The combination of dramatic increases in freight traffic and transportation systems operating at or near capacity has only recently resulted in growing visibility of freight and its role in urban congestion and environmental problems as a symptom of greater supply chain congestion (Regan and Golob 1999). The waiting time and variability of waiting time at ports can be significant. Delays might happen within container waiting for entering the ports as well as during container processing through terminal gates. Almost 44% of operators serving ports reported that their operations were often affected by congestion at the ports (Regan and Golob 2000). By creating the GreenLane—handling expedition of C-TPAT-compliant cargo at border crossing and port—RFID e-seals can play an important role in paperless information exchange. Time savings in container processing through the container terminal will influence the improvement of situation with truck congestions at the port gates; that also will have a "green" impact on the reducing of port-related truck emissions because of accelerating of truck-turn-over time at the terminals.

Another series of potential benefits belong to the logistics applications of e-seals for all businesses (Importers, Carriers, Manufacturers, and Service Providers). Regarding C-TPAT Survey 2007 (Diop et al. 2007) more than three-quarters (76.5%) of survey participants reported that it is extremely important to "reduce the time and cost of getting cargo released by U.S. Customs and Border Protection (CBP)". Next on the list of the most important motivations for joining security programs are "to reduce the time in CBP secondary cargo inspection lines" and to "improve the predictability in moving goods and services across borders".

2.8 Benefits for Private Sector from Security Enhancement—GreenLane

Membership for U.S. firms operating in the supply chain is becoming the standard rather than the exception (Silverman and Seely 2007), especially among service providers who have found it to be a relatively low cost and effective marketing tool. C-TPAT membership has certain benefits: several time less likely to be targeted for physical inspections, resulting in considerable savings in time and money as CBP increases the number of exams overall; priority for cargo inspection; reduce the handling costs by removing to their own location all containers not selected for inspection before CBP has completed its inspection of the entire shipment (Silverman and Seely 2007). At present, the customs service physically inspects only 2–4% of containers arriving at U.S. seaports (Jackson 2003). The smart-seal program can boost the number the right inspected containers. Simultaneously, by using smart e-seals or smart boxes the shippers will get the most attractive benefit—the GreenLane advantage—to accelerate their cargo clearance and expedite processing through the port. More than half of C-TPAT Survey participants in 2007 indicated that benefits from enhanced security outweighed the costs (32.6%) or the benefits and the costs were is about the same (24.2) (Diop et al. 2007).

2.9 Smooth Border Crossing and Port Gate Processing

The enhanced security system can affect as well in another manner the container flows going through the port gates or borders of the countries: a U.S. Department of Transportation program "TransCore" shows that electronic seals could help secure containers and reduce border congestion (Anonymous 2002). At the Port of Seattle a reader at the port's gate indicates that the truck has entered/left the port. The truck is tracked at six weigh stations and processing centers along a 300-mile stretch of Interstate 5. When the truck arrives at the Blaine border crossing, the e-seal is read with a handheld reader or a roadside reader. Information on the carrier, vehicle, cargo, location, and time of detection, drivers, and security status is uploaded to a secure Web site. The shipper and carrier, as well as U.S. Customs Service agents and the U.S. Department of Agriculture agents can view the information on a secure Web site. The system requirements to be the most effective is the existence of special lanes for trucks with sealed containers; otherwise, it do not help to reduce congestion at the port terminals or smooth the border crossing (Anonymous 2002).

2.10 Improve Process Flows

The improvement of process flows in container networks means that the flow of materials and the flow of information are synchronized when the information system continually displays the current status and stream of goods. The information system is thereby not just more accurate but is also up to date. E-Seals that are unaffected by weather can improve the process flow for containers at the gates (improvement of the operations at port gates by e.g. remote readability of a container number) as well as refine on the matching of the container to the manifest; or avoid typical errors made during issuing and receiving of goods, such as incorrectly logged quantities.

2.11 Protect the Brand Name and the Reputation

Looking for rewards from security programs, the huge companies like Procter & Gamble, Boeing, Starbucks, and Kmart with high-value products emphasize that they need to secure the cargo in order to "protect the brand" (Downey 2006). Electronic seals with their track-and-trace ability not only ensure the container supply chain, but also gain supply chain efficiency from automatically tracking containers. Damage to intangible assets and the contingent losses which could arise in cases where e-seals are not used are even greater, e.g. damage to reputation (contaminated goods or non-delivery of goods). In some cases the pilferage from the container or theft may lead to the loss of sensitive information or intellectual properties, therefore the container flows have to be under protection of secure environment attached to the particular container in the form of a smart device.

2.12 Anti-Temper System for Container Flows

One of the main applications of RFID technologies in the context of shipping containers is their use as e-seals. It is usually active tags that provide efficient, instant notification of container security breaches. An identification number of e-seal is protected by electronic encryption and authentication. Electronic seals have to work in harsh environments under often severe conditions. Active tags contain batteries, have more processing and operating power, and hence appear be the most promising in container logistics. Such devices provide 100% check to ensure the e-seals are not tampered with/replaced and could detect when/where container tampering occurred (this information can be useful for insurers, the police as well as for bankers). Smart seals integrate in itself useful information other than the seal number e.g. container number, destination, and consignee. The shippers get automate monitoring and tracking of containers as well as higher processing level for their shipments through the customs via eliminating of human errors in reading or visual inspection or recording the seals on the containers and more effective customs work.

2.13 Minimizing of Container Loss, Tampering/Theft or Cargo Pilferage

The risk of theft, especially if the goods have a black market value, is very real. Worldwide, the direct cost of cargo theft is estimated at about US $50 billion per year, with an indirect costs many times higher, and US $15 billion of merchandize losses in the United States alone. Cargo theft occurs in freight-forwarding yards, warehouses and during transportation in trucks, and on ships. Cargo is particularly vulnerable in the period of being loaded or unloaded from the trucks, or through documentary fraud (Mayhew 2001). To ensure the process of container operations on its different stages the cargo owner or their service providers, either the carrier and port operators should provide the reliable protection of each container during its transportation. The e-seal is a right key to solve this problem. It combines mechanical mechanism to lock the container with specific electronic components. Therefore, it can provide tamper evidence, physical security and data management as well as indicate electronically whether a conveyance has been opened or tampered with.

2.14 Loss of Insurance Claims

Theft and pilferage of the goods from containers leads to the insurance claims. If a container seal is checked at multiple points as the container moves through the supply chain, it will help a carrier, a shipper, and an insurer to determine the weak link in the chain. But if a container seal is not checked at different stages as the container moves through the supply chain, they will not be able to pin down the location where the pilferage occurred, and law-enforcement agencies will not be able to deploy their resources effectively (Armbruster 2006). Identifying the location of the breakdown by using of smart e-seals will help determine which party is responsible for the loss and thus to settle insurance claims to them.

Thereby, the range of presented e-seal's advantages for container logistics describe the most attractive and useful functions for largest container logistics providers such as port operators, shipping companies, forwarders etc. Nevertheless, there are still some challenges in worldwide adoption of e-seals' system. The first discussion point is what kind of technology should be used as a worldwide standard of e-seals. This discussion has a substantial importance for the next issues: what kind of infrastructure need to be installed and what kind of functionality one could obtain from the device (World Shipping Council 2006). The infrastructure for e-seals does not presently exist, and need to be installed on thousands of different properties. Another actual issue for e-seals global implementation is international ISO standards for the device. It is still an open question what the product needs to do; what specific events must be captured and recorded; is capturing entry through the doors enough or must it detect entry into the container through the walls, ceiling or floor; does the device have to detect conditions other than entry intrusion? Thus, the governments and industry have to archive equilibrium from security requirements and all businesses benefits, before to set all these specifications for e-seals or container security devices.

3 Conclusion

This research is based on the literature and praxis review to disclose some business-related advantages from electronic security devices (e-seals) for the global container system. The main idea of this approach is to define the contribution of security devices for locking shipping containers to the efficiency and profitability of the global logistic system. The implementation of container security devices in logistic processes provides a wide range of advantages for commercial users of e-seals in supply chains, service providers, cargo carriers, and port operators. In spite of the various challenges that face this implementation in the worldwide container network, there are sufficient set of competitive advantages from e-seals for improvement of operating efficiency in the global container system such as control, identification and monitoring of containers; automation and improvement of container processes; providing anti-tamper containers protection with minimizing of container losses and thefts; and there are also the indirect impact of e-seals on the competition factor of service providers, protection of brand names and the reputation of companies as well as improvement of congestion and ecological situations in container terminals. Therefore, smart electronic seals do feasible the achievement both supply chain security and commercial profit in the global container network.

References

Anonymous (2002) E-seals Smooth Border Crossing. http://www.rfidjournal.com/article/view/62/1/1. Accessed 15 March 2007.

Armbruster W. (2006) Security devices deter potential for cargo theft. http://www.sealock.com/pdf/Sealock_Article-Shipping_Digest.pdf. Accessed 20 March 2008.

Clendenin M. (2005) South Korea rolls out RFID for cargo port. EETimes. http://www.eetimes.com/news/semi/showArticle.jhtml;jsessionid=N0HUJIXK21UHWQSNDLPSKHSCJUNN2JVN?articleID=60407369&_requestid=75964. Accessed 9 February 2009.

Collins J. (2005) Korean Seaport Tests RFID Tracking. http://www.rfidjournal.com/article/articleview/1438/1/12/. Accessed 18 June 2007.

Crist P., Crass M., and Miyake M. (2005) Transport Security Across Modes. European Conference of Ministers of Transport and Organisation for Economic Co-operation and Development. 127.

Daschkovska K. and Scholz-Reiter B. (2008) Electronic seals for efficient container logistics, in Dynamics in Logistics. 305–312.

Diop A., Hartman D., and Rexrode D. CSR, University of Virginia (2007) Customs-Trade Partnership Against Terrorism Cost/Benefit Survey Results 2007. 152.

Downey L. (2006) International Cargo Conundrum. RFID Journal. http://www.rfidjournal.com/article/view/2120/1/82. Accessed 28 September 2006.

Hadow R. (2005) The Math Behind RFID in Logistics. RFID Journal. http://www.rfidjournal.com/article/view/1362/1/82. Accessed 12 March 2007.

ISO (2003) ISO/PAS 17712 (2003) Freight containers—Mechanical seals.

ISO (2005) ISO/DIS 18185-1 Freight containers—Electronic seals—Part 1: Communication protocol.

Jackson W. (2003) DHS plans smart seals for containers. http://gcn.com/articles/2003/12/02/dhs-plans-smart-seals-for-containers.aspx. Accessed 26 October 2006.

Mayhew C. (2001) No. 214: The detection and prevention of cargo theft. Australian Institute of Criminalogy.

Nguyen T. (2006) Port trucks get RFID'd. Fleet Owner. http://fleetowner.com/news/topstory/port_long_beach_los_angeles_rfid_truck_pierpass_011306/. Accessed 13 July 2007.

Notteboom T. E. (2004) Container Shipping and Ports: An Overview Review of Network Economics 3 (2): 86–106.

Regan A. and Golob T. (1999) Freight operators' perceptions of congestion problems and the application of advanced technologies: results from a 1998 survey of 1200 companies operating in California. Transportation Journal (38): 57–63.

Regan A. C. and Golob T. F. (2000) Trucking industry perceptions of congestion problems and potential solutions in maritime intermodal operations in California. Transportation Research Part a-Policy and Practice 34 (8): 587–605.

Seymour L., Lambert-Porter E., and Willuweit L. (2007) RFID Adoption into the Container Supply Chain: Proposing a framework. In The 6th Annual ISOnEworld Conference. Las Vegas, NY.

Silverman R. B. and Seely R. F. (2007) Global Compliance Readiness – Law Firms:The C-TPAT And CSI Supply-Chain Security Initiatives Today. The Metropolitan Corporate Counsel. http://www.metrocorpcounsel.com/pdf/2007/October/17.pdf. Accessed 25 October 2008.

Talley W. K. (2004) Guest Editor's Forward. Review of Network Economics 3 (2): 83–85.

Tongzon J. and Heng W. (2005) Port privatization, efficiency and competitiveness: Some empirical evidence from container ports (terminals). Transportation Research Part a-Policy and Practice 39 (5): 405–424.

United Nations Conference on Trade and Development (UNCTAD) (2005) Review of Maritime Transport 2005. New York and Geneva. UNCTAD. 148.

U.S. Customs and Border Protection H. S. (2008) Container Seals on Maritime Cargo. Federal Register 73, 46029–46030 http://edocket.access.gpo.gov/2008/pdf/E8-18174.pdf. Accessed 11 January 2009.

World Shipping Council. (2006) Comments on the "International Container Standards Organization" and its Effort to Propose New Standards for Container Security Technologies. http://www.worldshipping.org/icso_discussion.pdf. Accessed 12 February 2008.

Resolution of the Berth Allocation Problem through a Heuristic Model Based on Genetic Algorithms

Vanina Macowski Durski Silva, Antônio G. Novaes
and Antônio Sérgio Coelho

1 Introduction

The new global organization based on the creation of global markets requires the establishment of efficient logistic systems that are capable of allow the efficient flow of production to foreign markets. The ports, as a means of transport, should be considered as an important link in the integration of the domestic and global markets, and their modernization is one of the main activities to be prepared with the domestic costs reducing plan, as well as increasing exports.

It is perceived, therefore, that there is a gap still to be explored regarding research and methods for one of the most important operational problems found in the port system: the berth allocation problem (BAP).

One of the first studies on this subject in Brazil was written by Silva and Coelho (2007), they researched how to determine an allocation plan for the vessels that dock at the port berths, so that each vessel is placed in a berth for a period of time that is required for the loading and unloading of the cargo.

The berthing plan is to combine the best of the best possible berth to serve each vessel in order to respect the restrictions that are imposed, resulting in a lower berthing cost. The attention to these aspects makes the berthing preparation plan at the port an expensive operation and requires the use of an optimization technique to prepare this plan. Therefore, the general purpose of this article is to propose a heuristic computational model to help the decision making of a container port complex on the best way to provide the best berth-vessel allocation with the aim of reducing operational costs.

V. M. D. Silva (✉), A. G. Novaes and A. S. Coelho
Department of Production Engineering, Federal University of Santa Catarina,
Florianópolis, 88040-970 SC, Brazil
e-mail: vaninadurski@gmail.com

H.-J. Kreowski et al. (eds.), *Dynamics in Logistics*,
DOI: 10.1007/978-3-642-11996-5_42, © Springer-Verlag Berlin Heidelberg 2011

There are many papers about the berth allocation problem: Brown et al. (1994, 1997), Lim (1998), Moon (2000), Imai et al. (2003), Kim and Moon (2003), Guan and Cheung (2004) and Mulato and Oliveira (2006) and others.

Brown et al. (1994, 1997) addressed the problem of berth allocation in marine ports. The authors identified the optimal set of vessel-to-berth attributes that maximizes the sum of benefits of vessels while in port. The planning of berths in marine ports has some important differences from the planning of berths in commercial ports. According to Imai et al. (2001), in marine ports the change of berths occurs because of the ports own services, in other words, a new vessel arrives and is assigned to a berth where another vessel is already berthed. This treatment is unlikely to occur in commercial ports. Thus, the exchange of berths and other factors less relevant to commercial ports are considered in Brown et al. (1994, 1997) making the issue inappropriate for commercial ports.

In the studies done by Lim (1998), the problem was transformed into a restricted version of the two-dimensional packing problem. Tong et al. (1999) in Dai et al. (2004) solved the problem of assigning berths by using the optimization approach used by a colony of ants, but the focus of the work was to minimize the length of the required pier.

Genetic algorithms (GA) are increasingly used to solve inherently intractable problems such NP-hard problems like machine scheduling problems. A special application is from Nishimura et al. (2001) where is proposed a model for dynamic BAP that considers the arrival of vessels after the start of the elaboration of the mooring plan. For such case, GA's are applied where two different types of representation for chromosomes are utilized with satisfactory results. Despite the fact that there are many ways to resolve the BAP, the nearest literature to this article's line of study is the series of articles of Imai et al. (2001) and Nishimura et al. (2001), which will be used as basis for this research development.

The article discusses applying GA's to the berth allocation problem and is organized as follows: in Sect. 2 the formulation of the model is shown, in Sect. 3 the procedure for solving the GA is given, in Sect. 4 the numerical experiments and the final section contains the conclusion of the study and suggestions for future research is given.

2 Development of the Model

The objective of this study is to minimize the length of stay of vessels in berths/ ports to reduce operating costs. To develop a model that comes close to the maximum of the real problems are considered extremely important factors which have great impact on the operational costs, such as the cost of the vessel staying in port (Silva and Coelho 2008).

In practice it is known that a port receives vessels of various sizes, therefore, different rates are charged for the stay. For the development of this article the charges for berthing and handling of loads shall be considered, which were judged to be the greatest impact in forming the operational cost of dock side berthing.

2.1 Restrictions

Amongst the various influential restrictions in decision making for the berthing of a vessel, are: time restriction, draft restriction, length restriction, vessels scheduling restriction.

2.2 Decision Variables (When and Where to Allocate Each Vessel)

The docking plan must be capable of allocating vessels to berths in the best possible way, i.e., determine 'what' berth should moor each vessel as well as the 'moment' of berthing, analyzing if the waiting time for each vessel, accounting costs, in order to optimize the operation of the port system.

2.3 Formulation of the Problem: Heuristic Algorithm of Allocation

Initially a simplified algorithm is proposed that is able to consider a list of vessels (each containing information about the arrival moment, draft, cargo, length and differentiated tariffs), and a list of berths (release time, interval, depth, length, productivity rates). Then, it is verified whether the restrictions are met, and subsequently allocates a vessel in each given berth.

To calculate the total allocation cost (CA), consider:

$$\text{MinCA} = \sum_{i \in N} \sum_{j \in B} \left\{ \left[(E_i + A_{ij})CP_i \right] + (C_i \text{TMov}_j) + \left[(L_i \text{TAtrac}_j)P \right] \right\} x_{ij} \quad (1)$$

Subject to:

$$\sum_{i \in N} (L_j - L_i)x_{ij} \geq 0 \quad \forall j \in B \quad (2)$$

$$\sum_{i \in N} (D_j - D_i)x_{ij} \geq 0 \quad \forall j \in B \quad (3)$$

$$\text{Mcheg}_i + \sum_{i \in N} A_{ij}x_{ij} - \text{Mlib}_j \geq 0 \quad (4)$$

where CA is the objective function of the allocation cost; D_i draft of vessel i; E_i vessel i waiting time; D_j depth of berth j; A_{ij} vessel i service time in the berth j; $TAtrac_j$ wharfage duty charged by berth j to vessel i; CP_i cost of vessel i stopped; C_i load of vessel i; $Mcheg_i$ arrival moment of vessel i; $TMov_j$ movement duty charged by berth j to vessel i; $Mlib_j$ release moment of berth j; P number of executed periods of time; L_i length of vessel i; x_{ij} binary variable type 0–1, where value 1 is considered in the event of vessel i has been serviced in berth j and value 0 is assumed in the event of the contrary; $N = \{n_1, n_2, \ldots, n_k\}$, represents the entirety of vessels; $B = \{b_1, b_2, \ldots, b_m\}$, represents the entirety of berths.

Function 1 minimizes the total cost of the allocation, which is the sum of the following plots: cost of the waiting time and cost of service time, cost of cargo handling and rate of use of the berth length, per period. Equation 2 represents the restrictions in the length of the berth j which must be greater than the length of the vessel i, the restriction (3) indicates that the draft of the vessel i must be less than the depth of the berth j, the restriction (4) indicates the time the vessel berthed and the service of the vessel i and this should be more than the release moment from berth j, i.e. a vessel cannot be berth and serviced in a given berth before this is released from servicing the previous vessel.

Because of the fact of greatly variable costs in a sequence of allocation to the other, it is necessary to evaluate the various possibilities for the sequences that may exist. To be not exhaustive, it is proposed to use a genetic algorithm to perform the analysis of possible allocation sequences.

3 Genetic Algorithm for the Problem's Solution

GA's are like the heuristic methods where the optimality of the answers cannot be determined. They work in the principle of evolving a population of trial solutions over many iterations to adapt them to the fitness landscape expressed in the objective function.

As shown in the previous section, the BAP is a non-linear programming problem that is difficult to solve. To facilitate the solution of the process, a heuristic based on GA's is proposed by Silva and Coelho (2008).

3.1 Conditions of the Proposed Heuristics

In addition to the factors that were carried out, this paper presents a difference to the Nishimura et al. (2001) model. The authors work with the division of the problem in n sub-problems (SUB's) of berth allocation in terms of the time factor, as in Fig. 1.

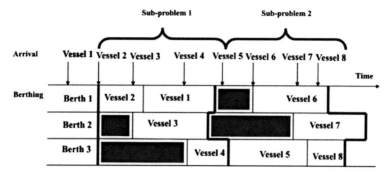

Fig. 1 Berthing schedule (Nishimura et al. 2001)

After the discharge of all the berths, the first sub-problem is solved by GA. Inheriting the solution of the first SUB (i.e. when all the berths are free again), the next SUB is solved. The process is repeated until all SUB's have been resolved, in the meantime the final solution can be affected by intermediate solutions. In the proposed model this does not occur, since it is considered as a single issue, containing all the vessels for berthing.

Another difference with the work of Nishimura et al. (2001) is the way to characterize a chromosome. The cited authors used a sequence of random berthing of vessels and then analyzed the fitness of this allocation. In the proposed study, the genes are analyzed one by one, choosing the best individual mooring to another vessel and then select another vessel for the best berthing position. So, a sequence will be formed for berthing, in other words a chromosome.

3.2 Genetic Algorithm

The GA as a heuristic method does not guarantee the optimality of the responses but search for the adequate solution of the problem. This algorithm can be understood better by following the steps below:

```
Step 1. Generate population of individuals.
Step 2. Calculate value of objective function.
Step 3. Select ancestors.
Step 4. Make the crossover between ancestors.
Step 5. Perform the mutation.
Step 6. Calculate value of objective function.
Step 7. If the value of the new objective function is better
        than the previous value of the objective function,
        go to Step 8. Otherwise, go to Step 9.
Step 8. Replace the individuals of the population by the des-
        cendants.
Step 9. If the objective function is satisfactory, END. Other-
        wise, go to Step 3.
```

Fig. 2 Presentation of the
calculus of the objective
function

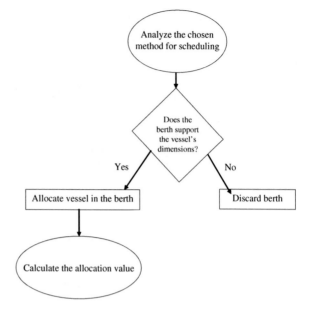

To calculate the objective function it is used the Fig. 2.

3.2.1 Representation and Genetic Operators

In the proposed method, a sequence of vessels, which can carry with themselves
the values of the variables (depth, length, cargo) is considered as a chromosome:

$$\text{Cromossome} = \{N3_1, N2_2, N1_3, \ldots, Nn_k\}$$

where Nn_k represents the kth vessel from the list, and each vessel has a position in
this list (for example: vessel $N3$ occupies position 1 in the list).

To sort the individuals of the population during the search process a measurement is used for the performance of fitness measured by (1). Being a problem
of minimizing the costs when the lower the value of the objective function, the
better the adaptation of the individual, and like this, the method always chooses the
best performing individuals causing them to reproduce.

Several rules exist in the literature to make the crossover between individuals
and therefore, for the problem of this research it was decided to differentiate and
adopt as a rule, the average of the positions occupied by individuals in the chromosomes (Silva and Coelho 2008).

Consider two chromosomes:

$$C1 - \{n3_1, n2_2, n4_3, n1_4\}$$
$$C2 - \{n2_1, n4_2, n1_3, n3_4\}$$

This is the average of the positions occupied by each individual in the two chromosomes:

$n1$ occupies the positions 4 and 3, so the average is 3.5;
$n2$ occupies the positions 2 and 1, so the average is 1.5;
$n3$ occupies the positions 1 and 4, so the average is 2.5;
$n4$ occupies the positions 3 and 2, so the average is 2.5.

Then, list them in ascending order, and when there is tie, randomly picks one of them to fill the position:

$$1,5 \leq 2,5 \leq 2,5 \leq 3,5$$
$$\downarrow \quad \downarrow \quad \downarrow \quad \downarrow$$
$$n2_1 - n3_2 - n4_3 - n1_4$$

As there was tie between the average of the individuals $n3$ and $n4$, it is chosen to insert in the list, in the second position, the vessel $n3$ and following the vessel $n4$. In this paper was not compared this method with others encountered in the literature.

After doing the crossover, the mutation takes space.

4 Numeric Validation

4.1 Parameters for the Search of the Solution

To implement the proposed algorithm to solve the PAB, a software was developed in Delphi® version 7, which was tested and the variables analyzed. To use the software it is needed to insert the files that contain information on the vessels and the berths (length, draft, duties, arrival moment, etc.) and fill other informations like the occupation period as well as the mutation rate to be considered. Three different ways can be used as stop criterion: maximum number of iterations achieved; maximum time of computing achieved and, value of *epsilon*[1] achieved. It was considered three options of resolution methods to be chosen: method 1 (based on the earliest release date of the vessels), method 2 (based on lowest cost of vessel allocation) and method 3 (based on the relationship between method 2 and the length of time the vessels remains in the port).

---

[1] *Epsilon* is the adopted parameter to evaluate if the deviation obtained between the fitness of the worst and the best chromosome is acceptable. The deviation is found by the following: Deviation = (Worst cromo fitness − best cromo fitness)/best cromo fitnessIn this manner, if Deviation > Epsilon, the algorithm continues.

Method 1 starts with a value for the Earliest Departure Date of the vessels as being infinite. Next, determine the value of Departure Date that each vessel would have, if the berth and the berthing time varied. The Departure Date is found by the maximum value between the berth release time plus the considered occupation time and the time of arrival of the vessel plus its service time. If the value found for the Departure Date is less than the value for the Earliest Departure Date, this should be replaced by the first one and the procedure be repeated. Subsequently, calculate the value of the objective function using Eq. 1.

Method 2 initially accepts a value for the minimum cost of allocation as being infinite. Then, check the best cost of allocation in each of the berths for all vessels, using Eq. 1. If the value of the Cost is less than the value of the Minimum Cost, it should be replaced by the first one. Method 3 is similar to the previous method, but differentiates itself from the moment that the value of the cost is found and multiplied by the total time that the vessel stays in the port (from arrival to departure). What is realized is that this method potentiates the time by trying to find the balance between the length of stay of a vessel in the port and cost allocation. Because this study has a heuristic nature this method was agreed to analyze the behaviour.

4.2 Simulated and Real Experiments

The proposed method was applied and analysed for some simulated cases. In every case it was considered as comparing criterion, a population of 3,000 individuals and a mutation rate of 2%, with an occupation interval of 2 h, an occupation period of 6 h and a maximum number of iterations of 500,000. The tests were carried out for the fixed processing times of 3, 5 and 10 min, also varying the 3 different methods, this totalled 540 examined cases.

In order to evaluate the proposed system in terms of quality of results, the method was applied to the case of Itajaí Port (Brazil), which currently does not have a tool for the optimization of allocation of the transaction and, uses Excel® software. The following information was considered for this test: each day from 20 November to 31 December 2007 was analyzed where 63 vessels were scheduled to berth at the port. Currently the port operates 24 h a day, 7 days a week, with 3 berths.

It is worth mentioning that Itajaí Port has a 740 m long pier and does not operate using fixed-size berths. In average there would be 3 berths × 246 m each, but in practice this length varies with the length of the vessel that is berthed. For example, if the port receives a vessel that is 260 m long, it will use the first 246 m of the first berth and another 14 m belonging to the second berth.

To test the developed model with actual data from Itajaí were discarded vessels with the greatest length, because the software works with individual berths. Whereas, the arrival of these longer vessels is not very likely, the result of the mooring plan will not be very different to the plan drawn up by the employees of the port in question.

4.3 Results Obtained

Due to the number of simulations carried out some tests were prioritized with respect to the performance obtained by the proposed tool: analysis of the number of iterations versus *epsilon* variation; analysis of the number of iterations versus the maximum processing time; analysis of the computation time difference versus the problem complexity and analysis of the percentage of differences of values of the objective function.

With regard to the number of iterations versus the *epsilon* variation, the three methods have presented a decreasing curve, in other words, as the epsilon value increases, the number of iterations is considerably reduced. In relation to the analysis of the number of iterations versus the maximum processing time required, it was found that during 75.50% of tests, there was divergence in the number of iterations between methods 1 and 2, i.e. method 1 required more iterations than method 2 compared to the same period of time. This behaviour is also repeated when compared to methods 1 and 3.

In this analysis it can also be seen the extent to which there is variation in the number of vessels for the same amount of berths (e.g. 15 berths, servicing 10, 70 and 100 vessels), increases the difference in the number of iterations between methods 1 and 2 and methods 1 and 3. It is notorious that the difference between methods 1 and 2, and 2 and 3 is around 30% and that method 1 presents twice the iterations as compared to method 3. The three methods behave in most part of the cases as for the number of iterations carried out by fixed time of analysis. As the complexity of the problem increases, the difference of behaviour between the methods also increases.

With the time convergence analysis of the results, it is possible do see that by increasing the complexity of the problem, all three methods tend to generate the optimal solution using the same computational time. See Fig. 3.

The values obtained in the objective function (cost allocation) which in most cases was the difference between methods 2 and 1 take on the behaviour of a growing curve, i.e., as the number of vessels increases, it increases the value the percentage difference between the methods, i.e. the value of the difference from method 2 to method 1 grows considerably as the number of vessels increases. Figure 4 shows this reasoning.

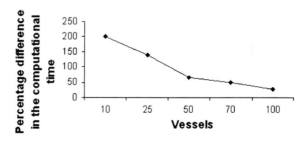

Fig. 3 Difference in the convergence time of between methods 1 and 2

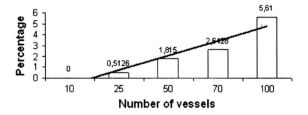

Fig. 4 Difference in the best solution found with the methods 1 and 2

In this case, when the problem has 100 vessels, the method 2 provides results in the objective function that are 5.61% greater than method 1. As this research attempts to obtain a reduction in costs, the greater the value of the solution, the poorer result is.

It should be mentioned that tests carried out in the case of Itajaí Port were satisfactory. It was possible to allocate all planned vessels and in some cases there was disagreement as to the length of stay of vessels in port (in real cases), probably due to exogenous variables (such as a delay in arrival), which were not considered in the proposed software.

Thus, an analysis was made of the difference in allocation obtained from Itajaí Port and proposed by the PAB, as shown in Fig. 5.

The graph contains the 30 cases which were analyzed and the difference obtained between the real berthing time at the port and the time proposed by the BAP. It is possible to see that for most of the cases that were examined, a gain of up to 5 h was obtained before the allocation proposed by the BAP, in other words, the BAP provided the berthing details by up to 5 h in most of dockings.

Figure 6 shows an easier way to interpret the allocations made. In approximately 70% of cases was showed a difference in allocation, between real data and

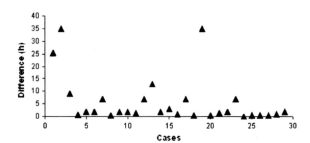

Fig. 5 Analysis of the difference in allocation at the Port of Itajaí and the PAB

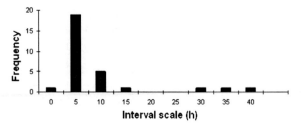

Fig. 6 Frequency difference of the allocation in the Port of Itajaí and the BAP

the proposed method, of up to 5 h; which shows that the developed model is valid and can provide satisfactory results.

This does not guarantee the efficiency of the proposed method compared to the system in Itajaí Port. For the cases tested in the BAP, it was considered that the berths were made available at the completion of the berthing plan, which in practice, in Itajaí, there were probably other vessels occupying the berths and delaying the berthing of vessels used in this analysis. However, the importance of the proposed software cannot be discarded, as the suggested dockings were similar to those that occurred in practice, achieving the proposed goal of reducing costs as well as the ease of preparing of the berthing plan.

5 Conclusions and Recommendations

A heuristic tool to support the decision-making on the best allocation of vessel-berths, using GA's, was presented in this paper. Although this study has been conducted on other references, it was not possible to compare with the proposed model, since the analyzed data were not the same and it was not possible to obtain the other author's data. This method has shown to be effective and with a low degree of implementation difficulty, and may be validated for use in container terminals.

As the behaviour of simulations varies with the value of *epsilon*, it can be confirmed that the results were not very stable but, in most cases, the number of iterations was reduced as the value of *epsilon* increased.

For the number of iterations and maximum processing time, it was concluded that method 1 is more suited to carry out the processing, since it performs more iterations than methods 2 and 3. In relation to the results obtained for the value of the objective function of the three methods was expected for method 2 the best result for the allocation cost, because this was its purpose: allocate minimizing costs, but it was seen that method 2 presented a greater allocation cost than method 1 that increased the complexity of the problem.

The decision of the best method depends on the aimed result: lesser mooring time of the vessels, or lesser allocation costs. About the analysis of convergence time of the results it is concluded that by increasing the complexity of the problem, the methods tend to have similar behaviour reducing the difference between the computational time and results for the optimal solution.

The proposed tool can have wide applications in the container port system management, facilitating the work of logistics operators that conduct, in most cases, manually the berthing plan. It also promotes the reduction of errors during this activity, as this is computerized, allowing for a reduction in operating costs.

To proceed with this research it is recommended to consider the existence of different cargoes moving around the port (bulk, liquid, etc.), the availability of equipment at the berths to receive a certain vessel for handling the cargo, and distance to be travelled by the cargo from the vessel to the warehouse or even the method of transport that will be used later. It is also recommended to change the proposed algorithm to allow a continuous berth and to consider more than one vessel berthing to optimize the available space, making a greater number of practical tests in order to increase the reliability of the system.

References

Brown G G, Lawphongpanich S, Thurman K P (1994) Optimizing ship berthing. *Naval Research Logistics* 41:1–15.

Brown G G, Cormican K J, Lawphongpanich S, Widdis D B (1997) Optimizing submarine berthing with a persistence incentive. *Naval Research Logistics* 44.

Guan Y, Cheung R K (2004) The berth allocation problem: models and solutions methods. *OR Spectrum* 26:75–92.

Imai A, Nishimura E, Papadimitriou S (2001) The dynamic berth allocation problem for a container port. *Transportation Research Part B* 3:401–417.

Imai A, Nishimura E, Papadimitriou S (2003) Berth allocation with service priority. *Transportation Research Part B* 37:437–457.

Imai A, Sun X, Nishimura E, Papadimitriou S (2005) Berth allocation in a container port: using a continuous location space approach. *Transportation Research Part B* 39:199–221.

Kim K H, Moon K C (2003) Berth scheduling by simulated annealing. *Transportation Research Part B* 37:541–560.

Lim A (1998) The berth planning problem. *Operations Research Letters* 22:105–110.

Moon K C (2000) A mathematical model and a heuristic algorithm for berth planning. Brain Korea 21 Logistics Team, July.

Mulato F M, Oliveira M M B de (2006) O impacto de um sistema de agendamento antecipado de docas para carga e descarga na gestão da cadeia de suprimentos. *Revista Produção Online* 6:3–96, set./dez.

Nishimura E, Imai A, Papadimitriou S (2001) Berth allocation planning in the public berth system by genetic algorithms. *European Journal of Operational Research* 131:282–292.

Silva, V M D, Coelho, A S (2007) Uma visão sobre o problema de alocação de berços. *Revista Produção Online* 7:2:85–98.

Silva, V M D, Coelho, A S (2008) Modelo heurístico para a resolução de um problema operacional portuário utilizando algoritmos genéticos. *Actas del XV Congreso Panamericano de Ingineria Transito y Transporte (PANAM)*. Cartagena das Indias.

Development of a Genetic Algorithm for the Maritime Transportation Planning of Car Carriers

Jae Un Jung, Moo Hong Kang, Hyung Rim Choi, Hyun Soo Kim,
Byung Joo Park and Chang Hyun Park

1 Introduction

Many car carriers are badly in need of efficient transportation planning in order to transport more cars in less logistics costs. Nevertheless, this work is usually being done manually based on experiences. Accordingly, the transportation planning, its evaluation, and its management are very inefficient, and so, if important changes are being made in terms of production schedule, production volume, and ship's arrival schedule, they have to suffer much difficulty. In order to make an efficient and systematic maritime transportation planning, we need to use all the information necessary for planning and also the changes in the information must be updated on a real-time basis, so that it may be reflected on the new transportation planning. However, the studies for the solution of these problems have not yet been made enough.

J. U. Jung (✉), M. H. Kang, H. R. Choi, H. S. Kim and C. H. Park
Department of Management Information Systems, Dong-A University, Bumin-dong,
Seo-gu, Busan 602-760, Republic of Korea
e-mail: share@donga.ac.kr

M. H. Kang
e-mail: mongy@dau.ac.kr

H. R. Choi
e-mail: hrchoi@dau.ac.kr

H. S. Kim
e-mail: hskim@dau.ac.kr

C. H. Park
e-mail: archehyun@naver.com

B. J. Park
Gyeongnam Development Institute, 152 Joongan-ro, Changwon City,
Gyeongnam, 641-060, Republic of Korea
e-mail: bjpark@gndi.re.kr

H.-J. Kreowski et al. (eds.), *Dynamics in Logistics*,
DOI: 10.1007/978-3-642-11996-5_43, © Springer-Verlag Berlin Heidelberg 2011

In this study we have tried to transform the traditional manual based system into a more systematic and optimized one, developing a genetic algorithm for optimal maritime transportation planning, so that newly updated data can be reflected on the revision of planning on a real-time basis, consequently reducing costs to a minimum.

2 Literature Review

To make an efficient transportation plan for car carriers, the maximum cargo volume should be allocated to the ships, and each ship should transport the cargoes in the shortest route and to a minimum cost, and then return to its departure port. In this respect, this is similar to a vehicle routing problem, which deals with the solution of vehicle routing, reducing logistic costs to a minimum, while satisfying the demands of both customers and shippers. The vehicle routing problem is a NP-hard, and so it takes a very long time to seek an optimal solution (Lenstra and Kan 1981).

Because of this, in order to solve a vehicle routing problem within a proper time, a variety of meta-heuristic methods have been suggested: various heuristic methods (Laporte et al. 2000; Pisinger and Ropke 2007), genetic algorithm (Baker and Ayechew 2003), tabu search (Gendreau et al. 1994; Ho and Haugland 2004), simulated annealing (Osman 1993), and ant colony optimization (Bell and McMullen 2004). The problems of vehicle routing and scheduling have been dealt with in a wide range of studies, but little study has been made on ship's routing and scheduling.

3 Routing for Car Carriers

3.1 Problem Definition

The purpose of vehicle routing is to find out an optimal voyage route that reduces the logistics cost to a minimum, in which many car carriers load the cars bound for their specific destination at the departure port, and leave the port, deliver the cars to their destination port, and return to the depot. At here, each car carrier is given an optimal car allocation and voyage route. When cars are allocated to each vessel, the voyage costs should be considered. All cars have its own destination, and so the destination of each car group determines the destination port of their car carriers.

3.2 Maritime Transportation Planning Process of Car Carriers

The maritime transportation planning of car carriers is divided into the following steps: tonnage planning, voyage planning, vessel allocation planning, and stowage

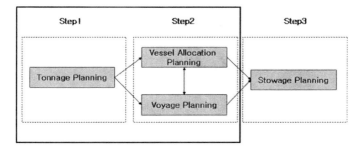

Fig. 1 Maritime transportation planning of car carriers

planning. Its process can be grouped into three steps as shown in the Fig. 1. The step 1 of tonnage planning is to obtain, maintain, and manage the necessary tonnage, so that the balance between the planned transportation volume and the tonnage (freight space) can be maintained from the mid and long term aspect. As it usually takes a long time to obtain the vessels, the tonnage planning has to be made in advance before the vessel allocation planning and the voyage planning are made.

The period of tonnage planning is usually one year, and is based on yearly production schedule. But in order to update it on an ongoing basis, the more accurate three-month production planning is being used. After this, based on the revised tonnage planning, vessel allocation planning and voyage planning are being made in order to carry the cars required to be handled in a short period of time (i.e. monthly logistics volume). The vessel allocation planning and voyage planning are to be made simultaneously in this study. Table 1 shows the necessary data for each step of the maritime transportation planning.

3.3 Integer Programming Model

The routing of car carrier is to determine the number of cars allocated to each vessel and also determine the voyage route, so that each car carrier leaves their depot, visits their destination nodes in sequence, and returns to their departure port with the minimum number of vessels and with the minimum voyage costs (i.e. maximum profits). The assumptions of car carrier routing in this study and integer programming model refer to Park et al. (2008).

Table 1 Comparison of the initial solution generation methods of vessel allocation chromosome (100 times operation performance)

Initial solution generation method	Minimum solution	Average solution
Random allocation by period	291,176	365,929
Allocation to the previous vessels by period	261,207	373,822
Allocation to the previous vessels by destination	160,761	190,921

4 Genetic Algorithm for Maritime Transportation Planning of Car Carriers

4.1 Chromosome Representation

In order to solve the maritime transportation planning problem by means of a genetic algorithm, first of all, the solution of a problem has to be represented as chromosome. The chromosome representation of the suggested genetic algorithm can be divided into two: the first chromosome is a vessel allocation chromosome to allocate the vessels according to the cargo, and the second chromosome is a voyage chromosome to determine the visiting order of each vessel. The vessel allocation chromosome is made in the form of repeating the number of vessels.

4.2 Initial Solution Generation

The genetic algorithm has usually generated an initial solution randomly, and used it. But the problem of car carriers has not only a long-term planning horizon, but also its size is very large. Because of this, this study has made a good initial solution lead to the generation and evolution of a population, so that a better final solution can be obtained in the soonest possible time. In case of the routing problem of a car carrier, the vessel allocation and the voyage route have to be determined simultaneously. Also, in order to reduce the number of times vessels have to sail for the delivery of shipper's cars, the cars with the same destination have to be properly allocated to the same vessel, while considering the car production date. Another important thing in the generation of a population is to reduce premature convergence by generation of diverse initial solutions. To this end, this study has devised diverse methods of generating initial solutions by means of allocating vessels according to the periods and destinations.

The selected initial solution generation method is based on the rule that the cargoes with the same destination have to be allocated to the same vessel, minimizing the transit ports, and consequently reducing the transportation costs, pilot charge, and stevedoring costs. As mentioned above, a good initial solution leads to a better final solution. The steps for solution generation are as follows:

Indices: i, j = node (destination), k = date, s = the number of vessels
Step 1:   k = 1
Step 2:   Select a random port of call "j" from the port of call set P = {1, 2, ..., J}. If set P = ∅, go to step 7.
Step 3:   Set Q of all the vessels capable to be loaded with cargoes produced at the date k in the port of call j: Q = {s| PS_{jk} - AL_s ≤ 0, for all s}. If Q = ∅, the production quantity is to be carried forward to the next month, and go to step 6.

Step 4: Among the set Q, preferentially select the vessel that is loaded with cargo. If there is no such vessel, you may select a vessel randomly.

Step 5: The selected vessel is to be loaded with the cargoes produced at the date k in the port of call j.

Step 6: If k < K, k = k + 1 and go to step 3. If not, the port of all j will be deleted from set P and go to step 1.

Step 7: Stop.

4.3 Genetic Operator

This study has made a comparison between 1-point crossover, 2-point crossover, and random crossover. The first two (1-point and 2-point) crossover is generally used for vessel allocation chromosome swapping, and the random crossover has been devised in consideration of the characteristics of the problem. In case of the maritime transportation planning problem, it has been confirmed that the 1-point crossover can generate a better solution than others (Tables 2, 3).

The main purpose of mutation genetic operator is to put the lost allele into a population in order to generate a new solution. To this end, many researches on a variety of mutation operations—inversion, insertion, displacement, shift, reciprocal swap, and scramble sub-list—have been made (Bagchi et al. 1991; Davis 1991; Gen and Cheng 1997). In this study a vessel allocation chromosome has picked a specific gene, and has displaced it with an anterior gene or a posterior gene, or a gene on the other side. Also it has used the method to allocate a random figure.

Table 2 Comparison of vessel allocation chromosome crossover methods (100 times operation performance)

Crossover method	Minimum solution	Average solution
Random crossover	171,072	196,731
1-point crossover	163,736	193,647
2-point crossover	167,035	193,277

Table 3 Comparison of VA chromosome mutation operation methods (100 times operation performance)

Mutation operation method	Minimum solution	Average solution
Comparison before and after 1-point mutation	164,946	191,279
1-point random mutation	177,757	204,655
Mutation before 1-point	166,064	197,683
Mutation after 1-point	165,329	193,230

4.4 Revision Process of Solutions

After carrying out genetic operations, an impracticable solution that exceeds
the vessel capacity can be generated. In this case, there should be a process
that can change it into a practicable solution. This revision process is as follows. If
and when the cargoes exceeding the vessel's capacity need to be allocated to the
other vessel, they have to be allocated to the vessel that incurs costs to a minimum. SQ means total number of cars loaded on the vessel s, and POsj indicates
total number of cars bound for destination j among the total cars loaded on the
vessel s.

Step 1:   Set Q of the vessels that exceed their space capacity and set E of the
          vessels that has still their space capacity are to be generated.
Step 2:   Select randomly a vessel from set Q, and find out PO that enables
          $SQ - PO < CP$. If there is no such PO, the least PO is to be found out.
Step 3:   From among set E, all the vessels that can be loaded with PO are to be
          found out. If there is no such vessel, it has to be carried forward to the
          next month. After loading PO on the available vessels, the profit
          increase of each vessel has to be calculated, and then the vessel that
          brings the largest profit will be selected for PO allocation.
Step 4:   If $O = \varnothing$, it stops, but if not, it moves to step 1.

Just like an integer programming model in Sect. 3.3, the evaluation function of
genetic algorithm is the sum of transportation costs, stevedoring costs, and costs
carried forward, and it uses both a seed-selection method (Park et al. 2003) and
elitism. Household formation is to be selected from current households or is to be
formed newly through genetic operator.

5 Performance Evaluation

In order to determine the crossover rates and mutation rates to be used in the
presented genetic algorithm, many tests have been conducted. The problem size in
the tests is: the number of destination port is 7, the number of vessel is 7, and the
period is 31 days. The crossover rates in the test are 0.7, 0.75, 0.8, and 0.85; the
mutation rates are 0.05, 0.1, and 0.15; the population size is 100; the number of
household is 100; the size of elitism 50; the size of seed selection 50. Repeated
tests have been performed 50 times. According to the results of tests, the crossover
rate has been given 0.85, and the mutation rate 0.15. Table 4 shows the comparison between the test results of genetic algorithm and the optimal solution from
integer programming method. In the genetic algorithm the size of a population is
500, the number of household 200, the size of elitism 50, and the size of seed 50.
ILOG CPLEX 10.0 has been used for the optimal solution from mathematical

Table 4 Comparison between an optimal solution and the results of algorithm

Ports	Vessel	Period	IP		GA		
			Optimal solution	Time (s)	Minimal solution	Time (s)	Optimality (%)
3	3	10	49,062	0.66	49,062	1.98	100
		20	54,995	0.88	54,995	2.43	100
		31	71,832	0.55	71,832	2.87	100
4	4	10	63,939	0.96	63,939	3.56	100
		20	91,381	1.64	91,381	4.11	100
		31	106,969	3.1	106,969	4.9	100
5	5	10	98,752	10.17	99,119	8.42	99.63
		20	99,057	102	99,423	10.21	99.63
		31	101,872	1,310	102,238	12.33	99.64
7	7	10	133,793	18,000	136,624	13.58	97.88
		20	131,768	18,000	134,914	15.81	97.61
		31	146,445	18,000	147,780	21.34	99.10
9	9	10	–	–	173,448	32.1	–
		20	–	–	186,575	41.34	–
		31	–	–	239,836	59.78	–
20	20	31	–	–	980,192	269	–

model, and the genetic algorithm has been developed based on Java 2 standard edition 1.5.

6 Conclusion

This paper has tried to develop a genetic algorithm for effective maritime transportation planning of car carriers. The genetic algorithm presented in this study has automated the existing manual planning method, consequently providing users with the best alternative in terms of cost. In addition, in order to respond to the sudden changes in the key information after the completion of planning, this genetic algorithm has enabled users to revise the plan manually. Because of this, the very difficult problems that cannot be solved on a manual basis have now been able to be solved by the newly presented algorithm. Furthermore, we have found out in the performance evaluation tests that the genetic algorithm proposed in this study has shown its excellence compared with the optimal solution of integer programming. Also its speedy performance has made it practicable to solve the real problems.

Acknowledgment This work was supported by the Grant of the Korean Ministry of Education, Science and Technology (The Regional Core Research Program/Institute of Logistics Information Technology).

References

Bagchi, S., Uckun, S., Miyabe, Y., and Kawamura, K. (1991) Exploring Problem-Specific Recombination Operators for Job Shop Scheduling, Proc. Fourth Int'l Conf. on Genetic Algorithms, Morgan Kaufmann, San Mateo, pp.10–17.

Baker, B.M. and Ayechew, M.A. (2003) A genetic algorithm for the vehicle routing problem. Computers and Operations Research 30, 787–800.

Bell, J.E. and McMullen, P.R. (2004) Ant colony optimization techniques for the vehicle routing problem. Advanced Engineering Informatics 18, 41–48.

Davis, L. (1991) Oder-Based Genetic Algorithms and the Graph Coloring Problem, Handbook of Genetic Algorithms, Davis, L. (ed), van Nostrand Reinhold, New York, pp. 72–90.

Gen, M. and Cheng, R. (1997) Genetic Algorithms and Engineering Design, New York, John Wiley & Sons.

Gendreau, M., Hertz, A., and Laporte, G. (1994) A tabu search heuristic for the vehicle routing problem. Management Science 40, 1276–1290.

Ho, S.C. and Haugland, D. (2004) A tabu search heuristic for the vehicle routing problem with time windows and split deliveries. Computers & Operations Research 31, 1947–1964.

Laporte, G., Gendreau, M., Potvin, J.Y., and Semet, F. (2000) Classical and modern heuristics for the vehicle routing problem. International Transactions in Operational Research 7, 285–300.

Lenstra, J. and Kan, R. (1981) Complexity of vehicle routing and scheduling problems. Networks 11, 221–227.

Osman, L.H. (1993) Metastrategy simulated annealing and tabu search algorithms for the vehicle routing problem. Annals of Operation Research 41, 421–451.

Park, B. J., Choi, H.R., and Kim, H.S. (2003) A hybrid GA for job shop scheduling problems. Computers and Industrial Engineering 45, 597–613.

Park, B.J., Choi, H.R., Kim, H.S., and Jung, J.U. (2008) Maritime Transportation Planning Support System for a Car Shipping Company. Journal of Navigation and Port Research International Edition 32, 295–304.

Pisinger, D. and Ropke, S. (2007) A general heuristic for vehicle routing problems. Computers and Operations Research 34, 2403–2435.

A Model of Wireless Sensor Networks Using Opportunistic Routing in Logistic Harbor Scenarios

Vo Que Son, Bernd-Ludwig Wenning, Andreas Timm-Giel and
Carmelita Görg

1 Introduction

With their rapid development, Wireless Sensor Networks (WSNs) have gone
beyond the scope of monitoring the environment (Timm-Giel et al. 2006; Evers
et al. 2005; Hans-Christian Müller—Wireless Networks in Logistic Systems
2005). A WSN is a wireless network consisting of spatially distributed autonomous devices using sensors to cooperatively monitor physical or environmental
conditions, such as temperature, sound, vibration, pressure, at different locations.
These sensor nodes can form a self-organizing network which fits well in mobile
environments. Having some advantages (e.g., low power consumption and
multi-hop routing), WSNs allow telemetry, control and management applications
which can be widely used in logistics, especially in autonomous logistic systems.
Multi-hop routing protocols for sensor networks can in most cases be classified
into collection and dissemination protocols (Levis et al. 2008). In this paper we
only discuss the collection protocols.

Directed Diffusion, the earliest WSN routing protocol, sets up a collection tree
based on data specific node requests (Intanagonwiwat et al. 2000). Early experiments led many deployments to move towards a simpler and less general approach
(Levis et al. 2008). Second generation protocols such as MintRoute (Woo et al.

V. Q. Son (✉), B.-L. Wenning, A. Timm-Giel and C. Görg
Communication Networks (ComNets), University of Bremen, Bremen, Germany
e-mail: son@comnets.uni-bremen.de

B.-L. Wenning
e-mail: wenn@comnets.uni-bremen.de

A. Timm-Giel
e-mail: atg@comnets.uni-bremen.de

C. Görg
e-mail: cg@comnets.uni-bremen.de

H.-J. Kreowski et al. (eds.), *Dynamics in Logistics*,
DOI: 10.1007/978-3-642-11996-5_44, © Springer-Verlag Berlin Heidelberg 2011

2003) use periodic broadcasts to estimate the transmissions per delivery on a link. MultiHopLQI is a third generation protocol which adds physical layer signal quality to the metrics. CTP (TEP123, www.tinyos.net/tinyos-2.x/doc/html/tep123. html) is a current tree-based routing protocol using information from multiple layers (Fonseca et al. 2007). ODEUR (Wenning et al. 2008) is another promising routing protocol based on detecting the movement of the sensor nodes relative to the data sink. Its disadvantage is that by design, it cannot forward the beacon over more than 2 hops; therefore its scalability in WSNs is limited. A comparison between RSSI (Received Signal Strength Indicator) and LQI (Link Quality Indicator) based on the PRR is discussed in (Kannan Srinivasan and Philip Levis 2006). In that publication, RSSI is discussed as a better indicator of delivery probability than LQI.

The inspiration for our new routing protocol, ODEUR$^+$, is motivated by the idea that applying WSNs with a combination of other technologies (RFID, GPS, etc.) to logistics can enhance an intelligent transportation system. Every item is uniquely identified, from the warehouse to the containers and at the destination, with the help of WSNs. The objective here is to define a suitable model of WSNs in logistics to provide a means for collecting the information of goods in containers during the transportation. In order to satisfy the requirements of dynamics in logistics, the goals of our routing protocol are:

- Use the RSSI, which is available in many hardware platforms, for routing. RSSI is a good indicator representing the receiving signal quality.
- Make movement estimations of the neighboring sensor nodes and use this information for routing.
- Use as little memory as possible because of the hardware resource limitations in sensor nodes.
- Support multiple sinks so that sensor nodes can automatically connect to available sinks when they move through many WSNs.
- Use a simple mechanism to synchronize the network time.
- Support a backup route in case there are problems related to loops and duplicate packets with the working route.
- Only keep the good neighbors in the neighbor table by classifying the quality of neighbors.
- Use retransmissions and acknowledgements to improve the reliability.

The contribution of this paper is not only a model of the routing protocol for sensor networks but also an application which can be used in logistic scenarios. To validate the model, a simulation environment for logistics is built and the performance of the proposed protocol is shown to match the design goals.

This paper is structured in 6 sections: Sect. 1 is the introduction to the scope of this paper, Sect. 2 describes the operation of the protocol, and Sect. 3 shows the model of this protocol as well as the functionalities of its components. Formats of beacon and data packets are proposed in Sect. 4. Simulation results are shown in Sect. 5 and finally conclusions are given at the end of this paper.

2 ODEUR⁺: An Enhanced Routing Protocol Based on ODEUR

ODEUR⁺ is a destination-based routing protocol utilizing the advantages of ODEUR to consider the movement of nodes by using a metric called Mobility Gradient (MG) and RSSI for the routing. The MG metric represents the relative movement between nodes. In order to transmit data packets from sensor nodes to the sink (destination), there are two processes: beacon broadcasting to exchange the routing information in the network and sending of data packets when the routing information is available.

2.1 Broadcasting Beacons

Our protocol ODEUR⁺ uses a periodical beacon to signal the routing information from a sink to the network to build a collection tree. Here, the original ODEUR has a disadvantage which is the 2-hop forwarding problem: beacons only are forwarded over a maximum of 2 hops. As for further hops, the mobility in relation to a sink cannot be determined reliably. So in the original ODEUR, the scalability of the network is limited. In order to overcome this problem, every node in the WSN will perform the multi-hop forwarding mechanism in ODEUR⁺. After receiving the beacon, each node extracts the RSSI and MG [the MG is calculated by subtracting two continuous RSSI values as in (Wenning et al. 2008)] of the beacon sender to update its neighbor table and then sends a new beacon with new information to its neighbor. Therefore, different from ODEUR, each sensor node knows the relative movement of all the neighbor nodes (and 2-hop neighbor nodes), but not necessarily that of a sink. In ODEUR⁺, the sink movement is not taken into account because it is not necessary in a large network where a node should keep the information of the neighbor nodes instead those far away. Unlike ODEUR, ODEUR⁺ uses a multi-hop forwarding mechanism, with the power of the sink beacons being the same as the power of data packets.

2.2 Sending Data

After building the neighbor table, each sensor node will choose the best neighbor node (BNN) to transmit data packets.

In order to describe the applied routing function, a simple topology is shown in Fig. 1 with a neighbor table of node 4 after updates. The following procedure will do the BNN selection based on the information provided by the neighbor table:

Normally, if the neighbor node or the source node is in the communication range of the sink, the source node will choose the sink or the neighbor node to communicate directly.

Fig. 1 Opportunistic routing: source RSSI (srcRSSI) and MG (srcMG) are the RSSI and MG values of the beacon sender, which are calculated from received beacons. Neighbor node RSSI (nnRSSI) and MG (nnMG) are available in received beacons through routing information exchange

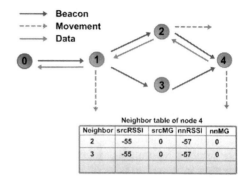

Neighbor table of node 4

Neighbor	srcRSSI	srcMG	nnRSSI	nnMG
2	-55	0	-57	0
3	-55	0	-57	0

In a network with multi-hop transmission, the source node (in the example of Fig. 1 node 4) tries to choose the neighbor which has the strongest RSSI to forward packets because this is its closest node. At the beginning, the source node will check the 1-hop neighbor nodes. Assuming that RSSI value of several 1-hop neighbor nodes (node 2 and 3 in the example) is the same, and they move, the node moving towards the source node is selected. In the example, node 2 moves towards node 4, the MG value (srcMG) will be +1 to indicate the direction of movement is "closer". Node 4 will choose node 2 as BNN. Otherwise, the MG is −1 to indicate the direction is "further", node 4 will choose node 3 as BNN because it is better. The same selection is performed if node 4 moves and node 2, and 3 do not move.

If there are no changes in 1-hop neighbor nodes (srcMG and srcRSSI do not change), the source node will check the nnRSSI and nnMG to choose the BNN. For example, if node 1 moves towards node 3, the nnMG in the table of node 4 will be +1 for node 3 and −1 for node 2 so node 4 will choose node 3 as BNN although in this case, the RSSI of node 3 and 2 is the same (assuming that node 2, 3, and 4 do not move).

3 Model

Based on the previous description, a model of the routing system (shown in Fig. 2) is proposed with many advantages. It has three components: the neighbor management, the routing, and forwarding components with separate functions so that the modification of each component does not affect the others. The neighbor management component is responsible for receiving beacons and validating these beacons to avoid loops or duplications. After that, the information in a valid beacon is used to update the neighbor table with a management policy: the insert or update rate is normally equal to the beacon rate, but the evict rate is defined by a timer using the frequency algorithm (Demaine et al. 2002) to count the number of received beacons from each neighbor and use it as a quality indicator. This can help to classify the neighbor nodes: only the high-quality nodes, from which the current node receives many update beacons, are kept in the neighbor table.

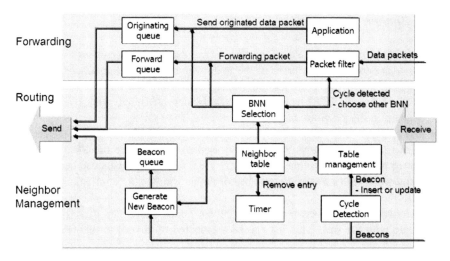

Fig. 2 Model of routing layer

The low-quality nodes, which send less update beacons, are removed step by step to save the memory for the better nodes.

In the routing component, a routing function is implemented to choose the BNN based on the information provided by the neighbor table. In addition, in order to have a higher reliability, a backup BNN is also assigned if possible. This component can easily be changed to implement different routing algorithms without affecting other components.

The forwarding component performs two tasks: sending the data packets from the application layer and forwarding the packets coming from the other nodes. It has a packet filter component to validate the forwarded packets do not create loops or are duplicated. If an incoming packet is valid, it is forwarded to a chosen BNN in the routing component, otherwise the current node can decide whether it uses the backup BNN or discards this packet.

Every node in a WSN can also be a forwarding node; therefore the forwarded data traffic volume (from other nodes) is usually higher than the originated data traffic volume. Thus, separated buffers can divide the traffic to avoid the dominance of the forwarding traffic over the originated traffic. Moreover, with the separation of buffers, it is easier to apply a user scheduling mechanism.

4 Packet Formats

Two types of packets are used in this protocol: beacons for routing and data packets for collecting information from sensor networks. The following sections show the proposed formats of these and their meanings.

4.1 Beacon Format

- *Sink ID (2 bytes):* address of the sink where the beacon originated
- *Sender (2 bytes):* the address of the node which forwards the beacon
- *Sequence (2 bytes):* a unique number to identify the beacon
- *RSSI (2 bytes):* RSSI value which the sender measured after receiving the beacon on the previous hop
- *MG (1 byte):* MG value which the sender determined after receiving the beacon
- *Network time (4 bytes):* this value is used to synchronize the local timers at sensor nodes. This assumes the delay of PHY and MAC layers can be neglected
- *Neighbor-Num (1 byte):* the number of nodes which can communicate directly with the sink
- *Neighbor-List (10 bytes):* addresses of the nodes which can directly communicate with the sink. This list can be segmented to fit well in this field

4.2 Data Packet Format

- *Mote ID (2 bytes)*: the address of sensor node which originates this packet
- *Sink ID (2 bytes)*: address of destination sink
- *Time-to-live (TTL) (1 byte)*: is used to eliminate loop problems in the network
- *Sequence (2 bytes)*: is used to recognize and avoid duplicated packets
- *Data (n bytes)*: any transmitted information

To avoid the problem of loops or duplicate packets, a cross-layer technique is used in this model. Although the fields of the data packet are on the application layer, the routing layer can use this information by cross reading the header information for its use. And these two types of packet fit well in the active message layer (message_t - TEP111, www.tinyos.net/tinyos-2.x/doc/html/tep111.html).

5 Simulation

In the simulation, an area with the size 300 × 300 m is used to represent a logistical scenario. The positions of nodes in these areas can be fixed or random. Every node can be static or mobile with a given speed between 2 and 5 m/s. The path loss of the signal in the simulation uses a free-space model with the specifications of the CC2420 radio chip. Every node will transmit the monitored temperature to the sink after a specific period. All scenarios are simulated in TOSSIM (Levis et al. 2003). Although in TOSSIM, the signal power is a constant (Rusak and Levis 2008), a matrix of gain between nodes is developed and can be varied in each step of the simulation. With that improvement, any propagation model can be applied in the simulation.

5.1 Scenario 1: Network with Mobility

In this scenario, a single WSN is used for simulation, which is representing a logistic warehouse where there are many containers needing to be transported by ships. Let us assume that each container is represented by a sensor node and the container can be moved to the warehouse to one location for storage. It can also be moved out for transportation. In order to simulate a more general mobility scenario, the movements of the containers are random paths. All the parameters are measured at the sink during the simulation time of 6 h (Fig. 3).

In order to find the optimum beacon period, some cases are simulated (static and mobility) in an area 300 × 300 m with the data rate of 1 packet/2 s (each packet has 29 bytes). It can be seen in Fig. 4 that the optimum value is in the range of 2–8 s to keep the packet reception rate at nearly 90%. If the period is too short, the physical bandwidth is mostly used for transmitting beacons instead of data packets. However, with long beacon periods, the routing will not adapt to the changes of the network.

The maximum data rate is shown in Fig. 5, in which the data period is varied from 0.5 to 60 s, and every node is static with a beacon period of 4 s. In order to achieve the PRR of nearly 90%, the data period should be greater than 2 s with a density of 20 nodes in the area 300 × 300 m. With a density of 30 nodes in this area, the maximum PRR is approximate 85% at the data period of 5 s.

The loss of packets depends on the propagation and on the buffers in each node. In order to measure the buffer loss, the number of incoming packets, outgoing packets, and the duplicate packets are monitored at each sensor node. From these, the buffer loss is calculated. Our simulations also show that the buffer loss of every node is rather low (under 1.5%) because of the separate buffers (beacon, forwarding, and originated packets). This means that the loss is mostly because of the coverage of the signal.

Moreover, in our simulations, the memory is also taken into account. The size of the routing table is 10 entries (1 entry has only 13 bytes). Hence, this low usage

Fig. 3 Random layout of nodes in the area 300 × 300 m. Every node can move in a random path. When reaching the boundary, it will change the direction following the reflecting law

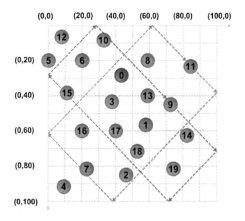

Fig. 4 The optimum beacon period is between 2 and 8 s. Any value out of this range gives a poor PRR

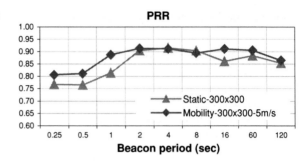

Fig. 5 The optimum data period is 2 s at the density of 20 nodes in an area 300 × 300 m

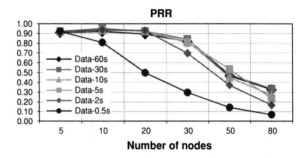

of memory is significant in many hardware platforms which have limited resources.

5.2 Scenario 2: Multiple WSNs in Logistic Harbor Scenario

In this section, a logistic scenario in a harbor is used for simulation (Fig. 6). The containers will be moved from the storage house to the crane area, and after that they will be loaded into the ship. Three WSNs can be used in this scenario, so that every container can connect to transmit the data packets which contain the environment condition information monitored by the sensors inside the container.

A mapping scenario shown in Fig. 7 is created to simulate the communication progress of the moving container. There are 4 areas: 1, 2, and 3 where the moving

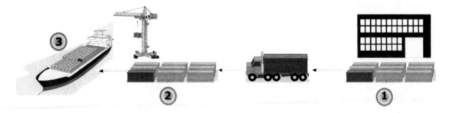

Fig. 6 Transportation of containers in a harbor

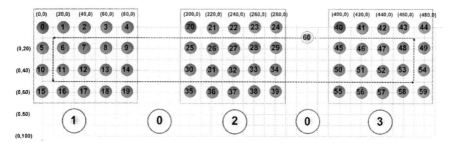

Fig. 7 Mapping scenario—multi WSNs

sensor node (container) can connect to send packets and area 4 is the range in which sensor nodes cannot connect because of a very weak signal. The sink addresses in area 1, 2, and 3 are 0, 20, and 40. The positions of sensor nodes in these WSNs can be random or deterministic. The mobile node (address 60) has the speed of 2 m/s and moves through three sensor networks.

The communication of the mobile node is recorded at each sink to which it connects by counting the number of packets that the appropriate sink receives from sensor node 60. The result (in Fig. 8) shows that when the mobile node connects to the sink (0, 20, or 40), the number of packets received by that sink increases, otherwise (when the mobile node is in other areas) this value is kept constant.

5.3 Confidence Interval

To validate the reliability of the simulation results, three cases are simulated 10 times with 10 different seeds to measure the mean of PRR shown in Table 1 to find the confidence interval. The variance is very small and similar in the cases of static networks. With a random network, depending on the random initialization, the variance is bigger than in the other cases.

Fig. 8 When connected with a sink, the mobile node will transmit packets to this sink, therefore the number of packets received by the appropriate sink increases

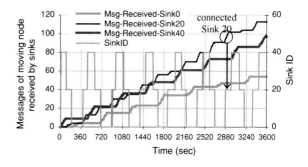

Table 1 Confidence Interval

Case	Mean	Variance	CI 90%	CI 95%	CI 99%
Static-100 × 100-21 nodes	0.92	0.00002302	0.92 ± 0.00278	0.92 ± 0.00343	0.92 ± 0.00493
Static-100 × 100-21 nodes-Mobility-Sink-2 m/s	0.91	0.0000191037	0.91 ± 0.00253	0.92 ± 0.00312	0.92 ± 0.00449
Random-300 × 300-21 nodes-Mobility-2 m/s	0.84	0.000361535	0.84 ± 0.01102	0.84 ± 0.0136	0.84 ± 0.01954

6 Conclusion

In this paper, a flexible model of the routing protocol ODEUR$^+$ is presented as well as its applications in the logistic harbor scenario. It is believed that this model is also well-suited for many applications where mobility is considered.

The simulation results show that the performance of this protocol is rather good, the PRR is nearly 90% in most cases examined. A set of parameters for configuration is also investigated to achieve the highest PRR. A small scale sensor network (12 nodes) was also deployed physically in our department and reported a significant amount of data with over 95% packet delivery.

Acknowledgments This research is partially funded by the German Research Foundation (DFG) within the Collaborative Research Centre 637 "Autonomous Cooperating Logistic Processes: A Paradigm Shift and its Limitations" (SFB 637) at the University of Bremen, Germany.

References

E. D. Demaine, A. Lopez-Ortiz, J.I.Munro (2002): Frequency estimation of internet packet streams with limited space. In Proceedings of the 10th Annual European Symposium on Algorithms ESA 2002, pages 348–360, September 2002.

L. Evers, M. J. J. Bijl, M. Marin-Perianu, R. Marin-Perianu, P. J. M. Havinga (2005): Wireless Sensor Networks and Beyond: A Case Study on Transport and Logistics, page 1, IWWAN 2005, International Workshop on Wireless Ad-hoc Networks, London.

R. Fonseca, O. Gnawali, K. Jamieson, P. Levis (2007): Four Bit Wireless Link Estimation, page 1. In Proceedings of the Sixth Workshop on Hot Topics in Networks, 2007.

Hans-Christian Müller—Wireless Networks in Logistic Systems (2005): Workshop on RFID and Wireless Networks in Logistic systems—November 16th, Duisburg, 2005.

C. Intanagonwiwat, R. Govindan, and D. Estrin (2000): Directed Diffusion: A Scalable and Robust Communication Paradigm for Sensor Networks. Sixth Annual International Conference on Mobile Computing and Network-ing (MobiCom), 2000.

Kannan Srinivasan, Philip Levis (2006) RSSI is Under Appreciated. In Proceedings of the Third Workshop on Embedded Networked Sensors (EmNets 2006).

P. Levis, N. Lee, M. Welsh, D. Culler (2003): TOSSIM: Accurate and Scalable Simulation of Entire TinyOS Applications. In Proceedings of the First ACM Conference on Embedded Networked Sensor Systems (SenSys 2003).

P. Levis, E. Brewer, D. Culler, D. Gay, A. Woo et al. (2008): The Emergence of a Networking Primitive in Wireless Sensor Networks. In Communications of the ACM, Volume 51, Issue 7, page-99-106, July 2008.

T. Rusak, P. Levis (2008): Investigating a Physically-Based Signal Power Model for Robust Wireless Link Simulation. In Proceedings of the 11th ACM International Conference on Modeling, Analysis and Simulation of Wireless and Mobile Systems (MSWiM), 2008.

A. Timm-Giel, K. Kuladinithi, M. Becker, C. Görg (2006): Wireless Sensor Networks in Wearable and Logistic Application—CRUISE Workshop, page 2, June, 2006, Greece.

B.-L. Wenning, A. Lukosius, A. Timm-Giel, C. Görg, S. Tomic (2008): Opportunistic Distance-Aware Routing in Multi-Sink Mobile Wireless Sensor Networks. ICT-MobileSummit 2008, Stockholm.

A. Woo, T. Tong, D. Culler (2003): Taming the Underlying Challenges of Multihop Routing in Sensor Networks. First ACM Conference on Embedded Networked Sensor Systems, 2003.

Logistics Service Providers in Brazil: A Comparison Between Different Developed Regions

Mônica Maria Mendes Luna, Carlos Ernani Fries and
Dmontier Pinheiro Aragão Júnoir

1 Introduction

Several studies regarding the logistics industry highlight this sector's growth. Such rapid evolution is mainly due from the outsourcing trend observed in the last decade. There is a consensus about the reasons that have led companies to out-source: the permanent search for cost reduction and the need to improve service levels. More recently, the need for large investments in technology assets as well as a well-trained labor force in the use of information and communication technologies—for the operation of more complex logistic management systems—has been one of the main reasons for contracting Logistic Service Providers (LSP) with more specific management capabilities. However, this phenomenon does not occur homogeneously across countries or regions; consequently, the LSP Industry does not show itself in the same stage of development worldwide. Recent research suggests differences between what can be called a global trend and the observed reality in some regions. The emphasis of this research is to highlight the differences in the evolving Logistics Industry by adopting operational strategies where local idiosyncrasies are prevalent.

This study examines LSP that operate in Brazil, focusing on differences between those located in the least developed region of Brazil—the Northeast—and those companies operating primarily in the Southeast, the most developed region of the country. Although some studies in Brazil show that one of the main trends for logistics and supply chain management is the improving quality of logistics

M. M. M. Luna (✉) and C. E. Fries
Department of Production Engineering and Systems, Federal University of Santa
Catarina, Florianopolis, Brazil
e-mail: monica@deps.ufsc.br

D. P. A. Júnoir
Department of Transport Engineering, Federal University of Ceará, Fortaleza Ceara,
Brazil

H.-J. Kreowski et al. (eds.), *Dynamics in Logistics*,
DOI: 10.1007/978-3-642-11996-5_45, © Springer-Verlag Berlin Heidelberg 2011

services (ILOS 2008), this research states that these trends are non-homogenous, i.e., there are significant differences in logistic service standards, including the use of up-to-date information and communication technologies by LSP of different stage-developed regions in the same country.

This paper is structured as follows: the first section of this paper deals with research on LSP in several countries, highlighting industry growth and trends identified in literature. The second section includes a more specific review related to LSP operations in regions in different stages of development. The objectives and methodological procedures are included in the third section. The results of this study are presented in the fourth section of this paper. The last section encompasses the conclusions and final considerations of this study.

2 The Growth and the Main Trends of the LSP Industry

The outsourcing of logistics services has grown significantly in recent years. Studies by Lieb and Bentz (2004) showed in a survey conducted among the 500 largest American manufacturing companies that during 2004 the percentage of companies using such services reached a record high, and that users are placing a steadily increasing percentage of their logistics operating budgets with 3PL service providers. According to the Armstrong & Associates report (2009), "Trends in 3PL/Customer Relationships—2009", the Global Fortune 500 3PL market is estimated at $187.4 billion for 2007 with growth to $199.7 billion in 2008. Studies by Langley and CapGemini (2006) in the 11th Annual Third-Party Logistics Study report project increased logistics expenditures directed to outsourcing for the years 2009–2011. This trend is also replicated in emerging countries. Approximately 89% of the Chinese respondents are optimistic about the industry's future development (Hong et al. 2007). Large American manufacturers continue to expand their use of 3PL services in the global arena, with 80% indicating they use 3PL services outside the United States. Not surprisingly, the most rapid growth reported in 3PL services in international markets has come in China and Eastern Europe (Lieb and Bentz 2004). The market for 3PL services in Malaysia has promising potential for further development; more than 80% of the shippers would at least moderately expand their companies' use of contract logistics firms (Sohail and Sohal 2003). The "Logistics Overview in Brazil 2008" report (ILOS 2008) estimates, based on data from *Revista Tecnologística*, the average Gross Revenue of Brazilian 3PL's was $65.8 million in 2005 and $81.9 million in 2006, representing more than 24% annual growth.

This growth process of the LSP industry began in the U.S. as a result of deregulation in the road freight transportation sector in the 1980s, which led to increased competition—as stated by Bowersox and Closs (1996). As a result, companies redefined their operational strategies and have developed more and more innovative value-added transportation services. In Europe, the economical and political deregulation promoted by the formation of the European Community aimed to open individual state markets and eliminate conflicting legislation. Such action caused significant reduction in entry barriers, thus providing greater competition in their transportation sector (Tixier et al. 1983). It also resulted in noticeable changes in the transportation sector, particularly road freight transport (Colin 1999).

As in the U.S. and Europe, among most LSP operating in Brazil, approximately two-fifths originated from the road transportation sector. Nearly one-fifth of logistics operators are of foreign origin (Luna and Novaes 2004). Due to the entrance of major international logistics operators as well as large industrial and commercial organizations, transportation companies changed in order to operate as logistic providers by offering other services beyond their simple road freight transport (Detoni 2003).

These changes constitute a response to increasing demand for specialized services to the extent that shippers/contractors are seeking to focus on their core competencies. Several organizations are seeking to manage their logistics operations strategically, but realize they lack the core competencies and are increasingly seeking to outsource their logistics activities (Hum 2000 apud Sohail and Sohal 2003). They have adopted outsourcing as a way to leverage their own logistic competence through competencies of third parties, i.e., replacing the internal organization for another from the market which represents affords competitive advantages (Verdin and Williamson 1994). This continuous trend of supplying new services by LSPs has led to continuous revenue growth in the sector. Outsourcing providers are becoming more capable at creating innovative solutions, using technology and software which includes the internet and specialized applications (Tompkins et al. 2005).

According to Leahy et al. (1995) and Hong et al. (2007), most typical American LSP provide a wide range of value-added services. Studies conducted by Lieb and Kendrick (2003) and Stefasson (2006) as well as Lieb and Bentz (2004) confirm the observations that in recent years LSP have been extending their service portfolios with more complex activities than ever before and serving more customers. In Taiwan, Lu (2000) and Hong et al. (2007) established that the most important strategic service element for ocean freight forwarders is value-added service, followed by promotion equipment and facilities, as well as speed and reliability.

These studies illustrate several trends in the industry and present the LSP as companies which: (1) offer a wide range of different services; (2) have logistics know-how and, (3) use technologies to allow better management of physical variables in an increasingly global market. Undoubtedly, this scenario can be considered as the last evolutionary stage of development for the LSP industry. However, some studies show that there are significant differences related to the characteristics of LSP and the dynamics of the logistics industry in different regions.

3 Regional Differences and Logistics Services Supply

Such generalization of strategies and trends in the LSP industry as a result of globalization does not occur by chance. Audretsch (2003, p. 11) states "that domestic economies are globalizing is a cliché makes it no less true". In fact, the shift in economic activity from a local or national sphere to an international or global orientation ranks among the strongest changes shaping the current economic landscape (Enright 2003). Further, he states that the numerous examples of regional clustering provide evidence that although competition and economic activity globalize, competitive advantage can be localized.

According to Enright (1993, 2000) and Enright (2003) and Scott (1998) and Enright (2003), as long as globalizing forces move at a faster pace than forces which influence the geographic sources of competitive advantage, economies will become more distinct in some ways, rather than less distinct. This phenomenon is no different in the logistics industry. Globalization influences its evolution but does not create a common standard of competition across and within countries. More recent studies attempt to identify these differences with reasons found in the developing regions of the world. Examples are studies in Mexico (Arroyo et al. 2006), China (Hong et al. 2007; Hong and Chin 2004a; Hong et al. 2004), Malaysia (Sohail and Sohal 2003) and Ghana (Sohail et al. 2004) among others (Langley and CapGemini 2006).

In the global context, Langley Jr. and CapGemini (2006) identify differences in the types of services contracted by shippers in terms of regions. They argue that transportation outsourcing is more significant in Western Europe and Asia–Pacific than in North America. Conversely, customs clearance and brokerage services are less outsourced in Western Europe, likely due to simplified requirements within the European Union. Furthermore, regarding the use of Information Technology (IT), the capabilities go hand-in-hand with the use of traditional LSP services. More sophisticated LSP services are less in demand and differ between regions.

Studies from Arroyo et al. (2006) show that Mexican firms seem to concentrate less on outsourcing tactical, integrated functions such as warehousing, inventory management, and information systems than firms in Europe and the USA. This is consistent with the finding that the latter outsource more since they buy comprehensive, integrated packages of functions. For the Mexican firms, the emphasis on routine, operational functions fits the picture of LSP as "low profile". Arroyo et al. (2006) claim that "uniform, global LSP strategies" should be carefully considered: what may work in Europe may not work in Mexico or the USA and emphasize that LSP do not appear as common practice in Mexico. The widespread practice of LSP in China follows a pattern closer than that observed in Mexico. Chinese logistics providers depend mainly on transportation and warehousing business but lack logistics information management and value-added services (Hong et al. 2007). This research has also revealed differences between LSP in the entire country and those in the Tianjin area, a relatively developed region.

Fig. 1 Geographical location of the Brazilian regions

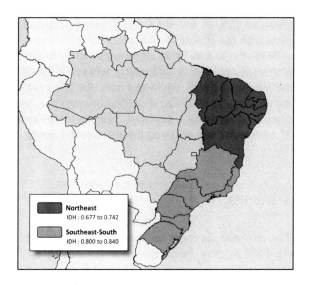

Northeast
IDH : 0.677 to 0.742

Southeast-South
IDH : 0.800 to 0.840

According to Hong et al. (2007), Tianjin logistics firms have large service geographic areas.

In Brazil, most studies highlight the growth of the logistics industry and its
development, but lack research on the differences within this national industry of
continental dimensions with significant regional differences. One of the most used
indicators to characterize different developmental levels within a country is the
Human Development Index (HDI). The HDI is measured on the basis of social
indicators. Through studies conducted by the United Nations, it appears that within
the Brazilian territory there are almost "five countries", namely five distinct
realities according to the HDI. The map of Fig. 1 shows the location of two main
regions, showing the largest (Southeast–South axis) and the lowest HDI (Northeast). The states located along the Southeast–South axis present a pattern of
development similar to southern Europe, while the states located in the Northeast
region have the equivalent level of development of the world's poorest countries
(CEPAL, PNUD and OIT 2008).

4 Methodology

The domestic industry of LSP is characterized by secondary data from the Revista
Tecnologística, published in June (Tecnologística 2008). The database provides
information on 249 companies and gives an overview of the logistics industry in
Brazil. Nine companies from the database are located and have activities outside
the predominant Southeast–South axis, showing that the database primarily reflects
the reality of the region's most developed country. The Southeast–South axis

produces 75% of national Gross Domestic Product (GDP) and is strongly correlated with the HDI of the region.

With the aim to better understand the structure of the Industry of LSP in the
Northeast and identify differences between the logistic services in that region and
the Southeast–South, additional data was collected through a survey. Considering
the difficulty in obtaining a satisfactory response rate and the need to increase the
sample of surveyed companies, personal visits were scheduled. Information on
27 companies was obtained through questionnaires.

5 Results

The collected data was subject to various statistical analyses. The differences of
the LSP industry in the two markets can be emphasized in the size of firms
(measured by their respective Gross Revenues and Number of Employees), Time
on Market, and Expected Revenue Growth. Table 1 presents the basic statistics for
these variables for both markets.

The average Gross Revenue and Number of Employees of Northeast are,
respectively, equivalent to 12 and 40% of those found for the Southeast and South,
and shows a strong difference due to firm size. Furthermore, the growth of the LSP
market—which refers to the rate of Growth Revenue of the last 2 years—shows a
significant increasing Industry in both regions. However, the growth rate in the
Northeast region is less than 70% of that found for the most developed regions.
Incorporating the relationship between regional GDP into the analysis, let's deduce
that there is a restriction in the development of the LSP industry and that this is
related to the degree of development of local economy.

Regarding the variable Time on Market (or age) of the LSP companies, we
can see that this variable assumes higher values for the Northeast region, whose
average age is 17 years. On the other hand, the Southeast and South have a
significant number of new LSP entries. It is worthwhile to mention that many
foreign LSP came to Brazil during the 1990s and settled mainly in the
Southeast and South region, enticed by the presence of other large, foreign,
industrial and commercial firms. Indeed, the entry of new international operators in this region continues, with this high growth rate justifying the interest of
these new entrants, which has therefore led to the drop in the regional average
LSP company age.

Figures 2 and 3 are the scatter plots of the transformed into normal distributed
variables for both markets and their respective linear regressions. The transformation was necessary because the original measures did not fit normal distributions. All statistical tests were performed at a 5% level of significance.

In the Southeast and South, we identify a strong positive correlation between
Time on Market, Number of Employees, and Gross Revenue, showing that firms
tend to grow with age. In contrast in the Northeast, it can be observed that there is

Table 1 Basic statistics for variables of interest for both regions

Variables	Northeast				Southeast–South			
	Mean	Minimum	Maximum	Std deviation	Mean	Minimum	Maximum	Std deviation
Time on market	17	2.5	50	13	12	1.0	79	12
Number of employees	265	9.0	2,000	486	668	5.0	11,000	1,457
Gross revenues	17,353,149	101.0	185,000,000	39,546,991	136,861,634	7,800.0	3,200,000,000	424,413,374
Growth of revenue	30	0.0	101	30	44	−6.5	1,087	114

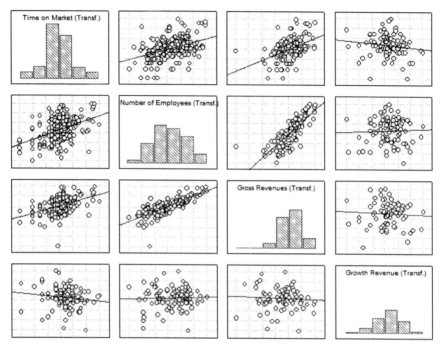

Fig. 2 Scatter plot matrix of select variables for Southeast–South region

only a strong correlation between Number of Employees and Gross Revenue. Perhaps the difficulty of increasing the market share in the Northeast is due to the lack of supply of more value-added logistics services or new technologies geared to logistics management.

When comparing Northeast' companies with those of the more developed region, it can be seen that there is a large gap using these technologies in their daily activities. Of the companies in Southeast–South, 94% operate their own warehouse or those of their contractors, 84% use warehouse management systems, and 83% offer tracking on internet.

For the Northeast, of the companies surveyed, more than 50% have warehouse capability and only 2 companies offer services beyond simple storage. Considering the asset factor, 21 companies maintain their own fleet. However, a low degree of complexity, customization of services, and a low use of technology are observed. Thirty percent of these LSP utilize bar codes and WMS and 25% use EDI and provide online tracking. The consequence is the low range of management services among these Northeast companies. It is important to note the little concern regarding the use of management systems, lack of training programs, or desire to obtain some type of certification (quality, security, or social responsibility, among others).

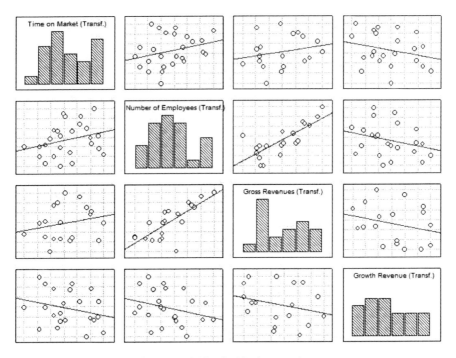

Fig. 3 Scatter plot matrix of select variables for Northeast region

6 Conclusions

This study shows that the logistics services in Brazil are heterogeneous. In the Northeast, only a few companies offer differentiated, high value logistics solutions. The vast LSP majority in that region still competes on the basis of ownership of assets, vehicles and warehouse. The widespread use of information and communication technologies such as WMS, bar code or tracking on internet—and widely employed in the Southeast–South axis—is incipient. This observed pattern is partly justified by the socioeconomic characteristics of the region which leads to a lower demand for higher value-added services or management services. The absence of large shippers and negligible competition with competitors who have know-how in advanced logistics appears to contribute to longevity of these "low profile" companies.

According to Oliveira (2007), the strong socioeconomic inequality among Brazilian regions has caused differences in the growth of activities of companies. However, the rates of GDP growth in recent years have been higher in the Northeast as compared to rates in South and Southeast, which had the lowest rates in the country during 1996–2004. The value of GDP per capita in the Northeast region increased from 47% of the national GDP per capita in 1996 to 51% in 2004 while in the Southeast region decreased from 136% in 1996 to 129% in 2005

(RIPSA 2008). The Northeast region's economy should still grow above the GDP in Brazil in 2009, according to study done by Datamétrica Consulting. For 2008, it is projected that the Northeast's GDP grows 5.5%, up from 5.18% to the projected GDP (RECID 2009). This can push the logistics market to more rapid evolution of demand, which in the region active companies may not be prepared to respond.

The results outline an analysis of two distinct regions of Brazil with regard to the degree of development and the dynamics of their logistics industries. Understanding the regional differences for logistics services demand and supply is crucial for defining strategies among companies which operate or intend to operate in these markets. Companies located in the Northeast should be aware of the region's shortcomings in addressing new demands of the logistics services market, since regional economic growth should attract better prepared potential competitors, especially those who operate in more developed areas as identified in this study. The low HDI may indicate a strong restriction in acquiring new skills towards facing logistical external competitors. It is recommended that personal development be encouraged by local authorities; especially those related to formal and continuing education of the population.

The inertia of local participants combined with higher economic growth of the Northeast region could lead to creating a hierarchy within the sector. This results in the entry of external firms which count upon higher logistics know-how, technology management, and a broader range of services, as evidenced with the entry of foreign logistic providers in Brazil in the 1990s.

Acknowledgments This research was supported by the Conselho Nacional de Desenvolvimento Científico e Tecnológico (CNPq), Ministério de Ciência e Tecnologia, Brazil.

References

Armstrong & Associates (2009) 3PL Customers Report Identifies Service Trends, 3PL Market Segment Sizes and Growth Rates. http://www.3plogistics.com/PR_3PL_Customers-2009.htm. Accessed February 2009.

Arroyo, P., Gaytan, J., Boer, L. (2006) A Survey of third party logistics in Mexico and a comparison with reports on Europe and USA. *International Journal of Operations & Production Management*, 26(2), 639–667.

Audretsch, D.B. (2003) Globalization, Innovation and the Strategic Management of Places. In J. Bröcker, D. Dohse, R. Soltwedel (Ed.), *Innovation Clusters and Interregional Competition* (pp. 11–27) Heidelberg: Springer.

Bowersox, D.J., Closs, D.J. (1996) *Logistical Management*. New York: MacGraw-Hill.

CEPAL, PNUD, OIT (2008) *Emprego, desenvolvimento humano e trabalho decente: a experiência brasileira recente*. Brasília: CEPAL/PNUD/OIT.

Colin, J. (1999) Les Mutations du Marché et leurs implications sur les processus logistiques. *Logistiques Magazine*, 138: 60–64.

Detoni, M.M.M.L. (2003). A Evolução da Indústria de Prestação de Serviços Logísticos no Brasil: Uma Análise de Mercado, *Tese de Doutoradodo Programa de Pós-Graduação em Engenharia de Produção*. Florianópolis, Brasil: UFSC.

Enright, M.J (2003) Regional Clusters: What we know and what we should know. In J. Bröcker, D. Dohse, R. Soltwedel (Ed.), *Innovation Clusters and Interregional Competition* (pp. 99–129) Heidelberg: Springer.

Hong, J. Chin, A.T.H. (2004a) Firm-specific characteristics and logistics outsourcing by Chinese manufactures. *Asia Pacific Journal of Marketing and Logistics*, 16(3), 23–36.

Hong, J., Chin, A.T.H., Liu, B. (2004b) Logistics outsourcing by manufactures in China: a survey of the industry. Transportation Journal, 43(1), 17–25.

Hong, J., Chin, A.T.H., Liu, B. (2007) Logistics Service Providers in China: Current status and future prospects. *Asia Pacific Journal of Marketing and Logistics*, 19(2), 168–181.

ILOS 2008 (2008) The Logistics Overview in Brazil 2008. Instituto de Logística e Supply Chain. http://www.ilos.com.br/site/index.php?option=com_docman&task=cat_view&gid=10&Itemid=44. Accessed July 2008.

Langley and CapGemini (2006). 2006 Third-Party logistics: Results and Findings of the 11th Annual Study. http://origin.at.capgemini.com/m/at/tl/Third-Party_Logistics_2006.pdf. Accessed July 2008.

Lieb, R.C., Bentz, B.A. (2004) The use of third party logistics services by large American manufacturers, the 2004. http://web.cba.neu.edu/~rlieb. Accessed July 2008.

Lieb, R.C., Randall, H.L. (1999) Use of third-party logistics services by large US Manufacturers in 1997 and comparisons with previous years, *Transport Reviews*, 2: 103–115.

Luna, M.M.M., Novaes, A.G. (2004) Estrutura de Classificação dos prestadores de Serviços Logísticos. In CNT (Ed.), *Transporte em Transformação VIII: Trabalhos vencedores do Prêmio CNT Produção Acadêmica*, 177–194. Brasília: LGE.

Oliveira, E.F. (2007) Um Estudo Exploratório sobre a Homogeneidade do Crescimento das Empresas Brasileiras no Período 1995–2003. RAC. 11(2), 135–150.

RECID (2009) PIB do Nordeste pode crescer acima do brasileiro em 2008 e 2009. Rede de Educação Cidadã. http://www.recid.org.br/index.php?option=com_content&task=view&id=1017&Itemid=2. Accessed February 2009.

RIPSA (2008) Produto Interno Bruto (PIB) per capita. Rede Interagencial de Informações para a Saúde. http://www.ripsa.org.br/fichasIDB/record.php?node=B.3&lang=pt. Accessed February 2009.

Sohail, M.S., Sohal, A.S. (2003) The use of third party logistics services: a Malaysian perspective. *Technovation*, 23, 401–408.

Sohail, M.S., Austin, N.K., Rushdi, M. (2004) The use of third party logistics services: evidence from a sub-Sahara African nation. International Journal of Logistics: research and Applications, 7(1), 45–57.

Stefasson, G. (2006) Collaborative logistics management and the role of third-party service providers. *International Journal of Physical Distribution & Logistics Management*, 36 (2), 76–92.

Tecnologística (2008) Mercado Brasileiro de Operadores Logísticos, Revista Tecnologística, Junho, 76–126.

Tixier, D., Mathe, H., Colin, J. (1983). *La Logistique au Service de l'entreprise*. Paris: Dunod.

Verdin, P., Williamson, P. (1994) Successful Strategy: Stargazing or Self-examination? *European Management Journal*, 12 (1), 10–19.

Adapting Dynamic Logistics Processes and Networks: Advantages Through Regional Logistics Clusters

Ralf Elbert, Hans-Dietrich Haasis, Robert Schönberger and Thomas Landwehr

1 Introduction

Concerning the topic of "adapting dynamic logistics processes and networks", this paper follows the research question: what are the advantages for companies to rapidly and flexibly adapt logistics processes and networks to continuously changing conditions through an involvement in regional logistics clusters? Obviously, it is often easier for a single company, as a "lone standing fighter", to meet the challenges of responsiveness and flexibility on the short-run. But co-operation in networks can offer in a long run an additional potential of responsiveness and flexibility. As a special form of inter-organisational co-operation clusters were discussed intensively in recent years.[1] Almost 20 years ago it was Michael E. Porter who transferred the term "clusters" into the managerial economics and connected it with competitive advantages and so set the focal point for the current enthusiasm on clusters.[2] Nevertheless clusters are still discussed intensively,

---

[1] Often industrial districts (see Markusen (1996) and Marshall (1920)), industry cluster (see Cortright (2005), p. 8), Innovative Milieus (see Franz (1999), p. 112), hot spots (see Pouder/St. John (1996), p. 1194), sticky places, regional innovation networks and hub-and-spoke-districts (see Markusen (1996), p. 296) are used synonymously.
[2] See Porter (1990).

R. Elbert
Chair of Logistics Services and Transportation, University of Technology Berlin, Berlin, Germany

H.-D. Haasis and T. Landwehr
Institute for Shipping Economics and Logistics Bremen (ISL), Bremen, Germany

R. Schönberger (✉)
Chair of Clusters & Value Chain, University of Technology Darmstadt, Darmstadt, Germany
e-mail: schoenberger@tud-cluster.de

H.-J. Kreowski et al. (eds.), *Dynamics in Logistics*,
DOI: 10.1007/978-3-642-11996-5_46, © Springer-Verlag Berlin Heidelberg 2011

because a wide variety of clusters arisen in the last years in all economic sectors and areas. But if clusters are a favourable economic structure, how can they practically help the logistics industry? This paper should give an idea about what logistics clusters in reality are and what they can do to realize advantages.

Clusters as a specific inter-organisational form were recognized by the logisticians immediately. They quickly realized—because of their general knowledge about inter-organisational co-operation e.g. in networks and supply chains—the opportunities of this kind of co-operation. So it is no surprise that until today more than 40 logistics clusters have been established in Germany.[3] To find out how logistics clusters can increase the responsiveness and flexibility of the involved companies, a model is being developed in this paper on the basis of the ideas of Michael E. Porter (clusters) and Hans-Christian Pfohl (logistics system). The model is being revised by means of the "Logistics Cluster Bremen" and by dint of a longitudinal study over the last 15 years. A textual analysis with three milestones in the years 1993, 2003 and 2007 will show the possibilities for advantages through regional logistics clusters.

The paper is structured in a theoretical part (Sect. 2) giving an overview over the relevant bases concerning clusters. A research framework in Sect. 3 is describing the procedure for the upcoming case study in Sect. 4. The paper finishes with an outlook in Sect. 5.

2 Responsiveness and Flexibility: Logistics Clusters and Their Advantages

Triggered by the holistic view of global supply chains, economic competition is addressing regional networks within logistics clusters.[4] In 1990, the term "cluster" was firstly introduced into business administration and management literature by Michael E. Porter, as "geographic concentrations of interconnected companies, specialized suppliers, service providers, firms in related industries, and associated institutions [...] in particular fields that compete but also co-operate".[5]

Porter's often cited scientific concept of clusters can be traced back to Marshall's agglomeration theory—describing the impact of firm proximity on productivity and economic growth—by Becattini's and Piore/Sabel's theory of industrial districts.[6] It is the initial point for a multitude of other similar concepts like the idea of sticky places by Markusen or the theory of hot spots by Pauder/St. John.[7] According to Porter, clusters can effect the competition in three ways: they

---

[3] See Elbert/Schönberger/Tschischke (2009), pp. 61–67.

[4] See Haasis/Elbert (2008), p. 22.

[5] Porter (1998), pp. 197–198.

[6] See Marshall (1920); Becattini (1990) and Piore/Sabel (1985).

[7] See Markusen (1996) and Pauder/St. John (1996).

boost the productivity, they increase the innovative ability and they stimulate the appearance of new business formations.[8] These competitive advantages of clusters are due to a free flow of information, the discovery of value-adding exchanges or transactions and strong motivation for collaborative strategies.[9]

Manifold studies reveal and document that clusters are a source for success of corporations.[10] So far, cluster development emerges as a legitimate instrument of economic policy—the economic potential of clusters receives attention from policy-makers at all political levels throughout Europe: The European Commission, national, regional as well as city governments are promoting regional clusters. On the European policy level see e.g. "The European Cluster Memorandum" from December 2007 that is supported by national and regional agencies for innovation and economic development and addressed to policy makers at the national and European levels. In Germany, the Federal Ministry of Education and Research spends 600 million Euros for the development of 15 clusters all over the country in the next years. Furthermore, there are cluster development programs on the federal state level e.g. in Bavaria ("Bayerische Clusterpolitik"), Hesse ("1. Cluster-wettbewerb des Landes Hessen") and North Rhine-Westphalia ("RegioClus-ter.NRW") where cluster development is supported with financial means as well as training and consulting of cluster managers.

In this way clusters become a myth promising positive effects on employment, productivity and innovation. The positive effects of clusters were formulated by Michael E. Porter in his work "The competitive advantages of nations".[11] For the analytic description of the competitive edge of clusters Porter transferred his model of the competitive advantages of nations—the so called Porter Diamond—into the field of regional clusters. His diamond model represented in the illustration below covers the following four determinants: The factor conditions describe the position of a region concerning the availability of production factors. This covers all relevant factors for an industry such as work force, raw materials and services. The demand conditions describe the kind of regional demand for products or services of an industry. The related and supporting industries mark the presence of international competitive related industries or suppliers. The firm strategy, struc-ture and rivalry determine finally the conditions of a cluster, how companies are organized and led, as they co-operate and how the regional competition looks like. As further two determinants the chance and the government were later added, which can shape the regional competition sustainable and give at the same time important impulses for the cluster development.

The development of clusters is either founded by particularly favourable con-ditions in one of the four original determinants of the diamond model or it can be

---

[8] See Porter (2003), pp. 562–571 and Porter (1998), pp. 213–225.

[9] See Porter (1998), pp. 213–214.

[10] See e.g. Sölvell/Lindqvist/Ketels (2003); Rosenfeld (1997); Porter (1998); Roelandt/den Hertog (1999) and Porter (2003).

[11] Porter (1990).

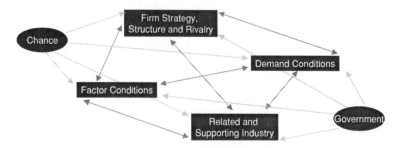

Fig. 1 The Porter diamond model (source: Porter (1990), p. 72)

released by acts of business, which cannot be traced to special local conditions. The further development of clusters explains Porter with a reciprocal process, which runs between the determinants of the diamond model (Fig. 1). He understands the diamond as a strengthening model, in which a positive development in one determinant leads to an improvement of the other competitive conditions. He assumes that a certain number of participants in the cluster—a critical mass—must be achieved, thus the interactions in the cluster initiate an inherent dynamism and automatically lead to a settlement of specialized subcontracting firms, special training and research institutions, as well as to a further development of the infrastructure.[12]

The mostly macro-oriented analysis of Porter and others is very important in order to set up the topic "cluster" in the mind of scientists, politicians and company leaders. But these mostly economically and economically-geographically driven studies do not explain how the success impact of clusters can be generated. Even if regional agglomerations and also clusters in various industries have been a research subject for decades, it was until now no focus on how clusters can adapt the dynamic of logistic processes and networks and could support companies in a rapid and flexible reaction.

Bringing clusters and logistics together, it is definite to say that logistics clusters are clusters, where logistics services are the central attribute of the connection between the participants.[13] Following the argumentation of Porter, logistics clusters can be described as concentrations of logistics companies, specialized suppliers for logistics companies, logistics service providers, firms in logistics-intensive industries as well as public and private institutions related to logistics, which can compete and co-operate with one another. The term logistics company covers both, the logistics industry (engineering and plants for the intra-logistics and material handling, software producer for logistic applications and system integrators, etc.) and the logistics service providers (players like contract logistics, shipping companies, warehouse operators, IT service providers, consulting firms, etc.). Referring to the related devices, research institutions with an emphasis on

---

[12] See Porter (1990), p. 216.

[13] See Elbert/Schönberger/Müller (2008), p. 315.

logistics as well as logistics associations and logistics-relevant institutes for standardization could be named as potential cluster participants. Depending upon the development of a logistics cluster, vertical interdependences by supplier-customer relations on the one hand and horizontal interdependences by companies on the same stage in the value chain on the other hand can occur.[14]

Logistics clusters at large, likewise every other logistics company or service provider, are acting in logistics systems.[15] Pfohl differentiates the logistics system by two types of variables which affect the management's room for manoeuvre: influenceable organisation variables and un-influenceable context variables.[16] The organisation variables can be classified into the four variables of the logistics management—task, staff, technology and organization—which has to be taken into account while designing logistics processes. These logistics organisation variables, which could be swayed by the logistics management, are very closely related. Changes in the logistics organization in order to increase the efficiency of the logistics system must be accompanied by changes in the technology and by the staff. Interdependences also exist with logistics tasks, why also these can be named as organization variables. But the possibilities to increase the efficiency of the logistics system are limited by a number of unswayable context variables. The context variables which could not be influenced by the logistics management are differentiated by the un-influenceable company-external variables[17] and the un-influenceable company-internal variables.[18] Both kind of variables have to be accepted as the given, external and internal, restrictions of the logistics system—so far.[19] Companies within the logistics cluster increase their adaptability and the chance to turn the given restrictions into influenceable variables.

3 Research Framework

Following up the exemplified organisation and context variables, it is theoretical traceable how logistics clusters can help to rapidly and flexibly adapt logistic processes. To get an idea how in the logistics clusters practice the dynamic can be assimilated, an exploration over time is necessary. An eased description model illustrates, on a longitudinal perspective with three cuts in the years 1993, 2003 and 2007, the general cluster effect.

---

[14] See Elbert/Schönberger/Müller (2008), p. 316.

[15] See Pfohl (2004a), pp. 14–20.

[16] See Pfohl (2004b), p. 27.

[17] E.g. production-managerial, logistics-technological and -institutional as well as demand-oriented conditions for the logistics process.

[18] E.g. existing organization, existing manufacturing plants, business size and the company's policy.

[19] See Elbert/Schönberger/Müller (2008), pp. 315–316.

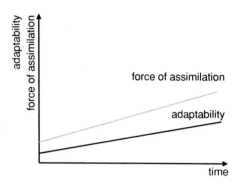

Fig. 2 Force of assimilation and adaptability with reference to time

It is obvious that the environment and its own dynamics cause a certain force for adjustments by every company. This force of assimilation—to set an example, this could be the required reactions to changes in the legislation—increases by and necessitates consecutive reactions by every player within the environment, because of a constantly changing environment. Every company is able to take the dynamic changes of the environment into account and to encounter it with assimilation, up to a certain degree. Since it is to be assumed that companies only could follow the changes of the environment and cannot anticipate it without special efforts, the relationship between adaptability of a company and time as well as the environmental-induced force of assimilation and time could to be outlined as the following (Fig. 2).

Over the time, the gap between the force of assimilation and the company's adaptability is getting bigger; the dynamic is getting more powerful and could be described from two sides. On the one hand the requirements for a company to assimilate could be of exogenous nature, in the sense of an only to the environmental-induced requirements following reaction. But on the other hand, from the opposite perspective, the requirements can be also endogenous, if in addition the company itself takes part in the formation of environmental-induced requirements and so thereby is necessitating adjustments of other participants in the environment. In both cases bundling of forces and joint acting is a success-promising approach and corresponds to Porter's cluster principles. To think further, a company acting in a cluster, which can realize the cluster advantages and generates an information lead and an edge in flexibility, is to be assumed to reach a better position in comparison to a single firm. And in the course of time, anticipation and co-designing of the environmental-induced assimilation can be expected. Then comparing the company's adaptability, on the one hand acting as a single firm and on the other hand as a cluster participant, a cluster effect is rising, starting from the same level of adaptability. Putting these two curves of adaptability together with the environmental-induced force of assimilation, Fig. 3 can be deduced.

The explanation therefore is that logistics clusters now allow the logistics management to transfer the un-influenceable context variables into organization variables and to thus create an additional scope of action, because:

Fig. 3 Over the time the cluster effect leads to a cluster arbitrage

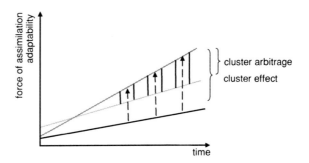

- The production-oriented and the demand-oriented basic parameters can be affected by the cluster.
- Close co-operation between politics, economy and science promotes the development of the region—and is useful to every single player within the cluster.
- Through an exchange of knowledge and experience with other cluster participants and thereby an enhanced transfer of best practice solutions in different areas of logistics, structures and processes impulses changes in the whole organization.
- By the joint construction of real estate and manufacturing plants the companies in the cluster can access common resources and also can react more flexible on requirement changes.

Through trainings in line with the market and making staff available in all stages of qualification according to the companies needs, completed by an early recruiting of junior staff, a more favourable development—independent of the businesses size—can succeed to the enterprises.

The scope of action which was limited to the own company and the size of the own business, so far, can be extended by the cluster dynamic.

By the connection of supplementing products and services, logistics can offer itself as a cross-company system supplier and thus exert influence on the company's policy.[20]

If so a better and faster ability of transformation can be achieved in the cluster, it leads not only to a cluster effect; it even generates a delta, an extra advantage, to the force of assimilation. This advantage enables a totally new scope of activity. This additional potential for every single company within the cluster can be defined as the cluster arbitrage.

The Porter diamond model gives a good explanation that clusters offer competitive advantages in a macro-economic reflection. On the micro-economic level it does not answer the question, how the realization of the cluster arbitrage in the day-to-day business and by single firms can be realized. Numerous authors give answers and pursue the discussion.[21] They focus on interactions, relations and therefore on the social infrastructure—e.g. shared meaning systems, norms of

---

[20] See Elbert/Schönberger/Müller (2008), pp. 315–316.

[21] See amongst others e.g. Giddens (1984), Rosenfeld (1997).

reciprocity and sufficient levels of trust that are embedded in professional, trade and civic associations and in informal socialisations patterns[22]—in order to enable the exchange and flow of information and knowledge in clusters.[23] Those micro-economical approaches of interaction lead to a foundation for the case study within this paper which will concentrate on a textual analysis to detect the increase of adaptability of logistic processes and networks and the cluster arbitrage.

4 Case Study: Logistics Cluster Bremen

The logistics cluster in the region of Bremen grew out of its own long history. The driver of the first historical development in Bremen as a logistics centre was the city's traditional function as a seaport. From the early beginnings in Hanseatic shipping during the Middle Ages, Bremen's role as a seaport developed continuously throughout the successive eras. On May 6, 1966 the first shipping containers to arrive in Germany were unloaded in Bremen.

The nowadays called logistics competence center describes the situation like: "The state of Bremen is today the second largest logistics site in Germany, integrated into the North-West metropolitan region, itself characterized by its logistics competency. Next to these flagships stand the growing container turnover in Bremerhaven, Europe's leading car terminal, the freight transhipment centre with its intermodal network in Bremen, important logistics and distribution centres, as well as excellent business and educational institutions. Due to the combination of maritime location and high-quality hinterland connections, Bremen and Bremerhaven are well positioned within the European freight area. This advantage will be secured into the future through the Jade-Weser-Port for post-Panamax container ships. The logistics industry of Bremen and Bremerhaven encompasses almost 1,300 businesses, with a combined revenue of 4 billion euro, employing 24,000".[24]

Cluster management seeks to optimize, under market conditions, the historically-rooted integration of services in the region in order to improve the performance of logistics processes and minimize costs. Moreover, the use of further resources should be minimized, while high-quality jobs are created. This triad of economic, ecological and social sustainability goals shaping the logistics industry

---

[22] See Rosenfeld (1997), p. 8–10. Rosenfeld argues that the "flow of information, technological advances, innovations, skills, people, and capital into, out of, and within the cluster, from point to point" are for achieving the cluster's economies equally important as are scale or critical mass. Following this, he defines clusters as "a geographically bounded concentration of interdependent business with active channels for business transactions, dialogue, and communications, [...] that collectively shares common opportunities and threats".

[23] See Sydow et al. (2007), p. 5.

[24] See Kompetenzzentrum Logistik Bremen (2009): Logistikstandort Bremen, Bremerhaven und die Nordwest-Region: http://www.klb-bremen.de.

can only be reached with the involvement of various decision makers from different competency areas. Cluster management is usefully applied in this merging of interests.

The logistics function of the co-operation between various actors in the strategic development of the Bremen region as an interface between the micro-logistics of single businesses and super-regional logistics systems at the macro-logistics level is reflected in the description as cluster.[25] The following examples from the Bremen industry demonstrate the cluster effect:

- The co-operation between the BLG and Tchibo to successfully operate Europe's largest high-bay warehouse for non-food articles in Bremen;
- The decision to expand the Kaiserschleuse lock, allowing for future growth in automobile transhipment in Bremerhaven, and securing the long-term competitiveness of that harbour and the jobs it creates;
- The creation of a brand new regional cluster for wind energy logistics and the assembly of wind energy turbines in Bremerhaven at Luneort;
- The development of the Jade-Weser-Port in Wilhelmshaven with a Bremen operator.

All these examples show the interdependencies of the decisions between partners and actors of different institutions on the effective design of the logistics region.[26] The economic effect of the interdependencies of decisions as described is referred to as the cluster effect, which sets the ratio between the pressure to adapt to changes in the business environment and the ability to adapt. In theory, the ability to adapt to changes in the business environment is heightened by clusters, in that the actors are able to mutually adjust their reaction to the others' reactions through communication and co-operation. This cluster effect is also described as cluster arbitrage.

The following case demonstrates the cluster development in Bremen in three stages for the years 1992, 2003 and 2007. In the framework of developing a cluster management strategy for the state of Bremen, a conceptual study "Logistics Centre Bremen" was undertaken in 1992/93 by the "German Foreign Trade and Transport Academy" (Deutschen Außenhandels- und Verkehrs-Akademie DAV), under the direction of the "Bremen Commission for Economic Research" (Bremer Ausschusses für Wirtschaftsforschung BAW). This study set the goals for a first approach towards a co-operation of various logistics actors within that network.[27] Although the cluster effects of this study are hardly measurable in terms of implementation in industry, it formed a basis for the development of the logistics cluster in the state of Bremen.

This thought process has been pursued in the years 2003/2004 with the study "Bremen—The Logistics Company", which was conducted by the Institute for

---

[25] Cluster also could be described as meso-logistics; see amongst others Haasis (2008); Haasis (2007), pp. 98–107; Haasis/Fischer (2007).

[26] See Haasis (2008); Haasis/Fischer (2007).

[27] See DAV (1992), p. 4.

Shipping Economics and Logistics Bremen (ISL) under the direction of the Senator for Education and Science and the Senator for Economics and Ports. Within a framework of this study a competency map was drawn up for the state of Bremen. Even though the concept of the cluster management strategy was specified in greater detail in the study, it became clear that a single initiator group (in this case the selected ministerial administration), was not able to implement a successful cluster management strategy. Moreover, it was especially evident that a "caretaker" for the knowledge management aspect was missing.

As a reaction to this weakness, the cluster management strategy in Bremen was revived as the non-profit Kieserling Foundation was founded by Karsten Kieserling in 2004. "The Kieserling Foundation has, according to the objectives of its charter, the goal to promote science and research as well as education and training with the focus on transport and logistics. To this end the means, which the foundation has provided, under the protection of the requirements of the rights of the public interest are to be efficiently and sustainably used".[28] Since 2006 the Kieserling Foundation has organized the "Logistics Day Bremen", a yearly meeting of logistic professionals and educators in Bremen, which it supports by it own publication series on logistics themes. What is more, the Logistics Competence Centre Bremen (KLB) has been created and installed as of 2007 as the operative institution ("caretaker") for implementing a logistics cluster management strategy in Bremen.

In the form of an advisory board for communication, co-operation and innovation, as well as the political environment, experts from various areas develop strategies within the KLB, to ensure the future sustainability of the logistics region Bremen. Besides infrastructure development, the improvement of qualified employment opportunities will especially play a decisive role in the region. These qualified opportunities are tied, however, to the offerings for education and training available within the region, that meet the needs of the employment market, and focus on building for the future. The transfer of innovative products out of research and development projects and into businesses stands as a further important goal, as much as a clear recognition of the logistics cluster Bremen and a unified, structured marketing of the region with an international focus.

The common goals of the expert advisory board of the KLB were published in a "Masterplan" for logistics in the state of Bremen in April 2008. These common goals were derived from an inventory analysis of the competencies of Bremen as a logistics region. For this, 68 logistics companies in the Bremen region were interviewed. First through the involvement of an authority with close ties to industry as a "caretaker" (the KLB), as well as through the beginnings of a co-operation between different actors from the transport and logistics industry, government, science, and research, as well a teaching (co-operation within the advisory boards of the KLB), a positive cluster effect could finally be recognized in 2007/2008.

---

[28] See Scope and objectives of the Kieserling Stiftung, in: http://www.kieserling-stiftung.de.

To illustrate, how un-influenceable logistics variables turned into influenceable logistics variables, as described in the research framework, the following table gives some examples of the developments in Bremen in form of case examples and a short discussions (Table 1).

The described examples show how a participation in a regional logistics cluster can lead to additional advantages for each cluster company and the whole logistics cluster region. These cluster effects can generate a cluster arbitrage as discussed in Sect. 3. Bringing the three longitude cuts of the case study and the idea of the cluster arbitrage together, Fig. 4 can be drawn to show the development of the cluster effect and the cluster arbitrage over years in the Logistics Cluster Bremen.

Adopting technologies for communication, co-operation and knowledge transfer in cluster developing processes allows changing un-influenceable logistics variables into influenceable logistics variables. For example the application of RFID-technology allows the impact of time-structure of commodity allocation (in particular regarding to small parts-logistics), because the information processes per unit charge along the supply chain can be optimized (acceleration of the availability of cargo related data), so the un-influenceable logistics variable "time-structure of commodity allocation" can be changed into an influenceable variable. A further example for this theory are the co-ordination processes in workgroups (for example regarding to investments for infrastructure in the region), which can be supported by technologies like "round tables" or "learning arenas". As a result according this approach the "traffic infrastructure" as an un-influenceable logistics variable can be changed into an influenceable variable. Further implementations of technologies for communication, co-operation and knowledge transfer for changing un-influenceable logistics variables into influenceable logistics variables are by all means conceivable, but further research is required. The application of technologies for communication, co-operation and knowledge transfer in pressure groups can raise the transparency regarding the manifold thinking processes of the different actors and decision-makers in the cluster. The implementation of such technologies for the co-ordination processes between the actors allows the achievement of the cluster arbitrage. Sustainable learning effects protect the development of a capable logistics region.

Bremen is a naturally grown cluster and recently a real cluster management was implemented. The Porter Diamond factors—which are necessary to call Bremen a Logistics Cluster—could be found in the region. There are logistics companies, suppliers, service providers and, of course, important universities and research departments. All working together which leads to the strengthening process described by Porter. The growth of the logistics sector in Bremen was enormous during the last years—mostly all involved companies are in a better situation today then years ago. This development would not have been comparable without the described factors and interaction. The logistics clusters in the Bremen region gained a real cluster arbitrage.

Table 1 Cases and discussion of how uninfluenceable logistics variables turned into influenceable logistics variables in the Logistics Cluster Bremen

Un-influenceable logistics variables for logistics processes			1993	2003	2007
			Logistics Centre Bremen	Bremen—the logistics company	Logistics competence centre Bremen
Company-external variables	Production-managerial conditions	Cases	Turnover in manufacturing industry in Bremen: 13.5 billion EUR.[a]	Turnover in manufacturing industry in Bremen: 20.5 billion EUR.[a]	Turnover in manufacturing industry in Bremen: 17.6 billion EUR.[a]
		Discussion	Value added services are increasing up to an important competitive factor. The cluster is recognizing and forcing this development.		
	Logistics-technological and -institutional conditions	Cases	Regional and physical formation of a cluster of transport-, handling-and storage-services in the freight village of Bremen.	Companywide co-operation in supply chain management, vertical co-operation between retail market and logistics service provider. Example: launch of the BLG-high rack warehouse for non-food-articles (BLG—Tchibo).	Regional co-operation between production and logistics, e.g. Stute-Airbus.
					Especially in the automobile sector suppliers are strongly bound to production sites (just-in-time, just-in-sequence).
				Foundation of the German Maritime Competence Network and participation of the ISL since 2004.	International co-operation, exchange of information and mutual learning (Know how-transparency and -transfer) between the players in the Northwestern Logistics Cluster.
		Discussion	The player's competences appear as main productive factors in cooperation of companies; in the cluster each company focuses on its core competence and in the regional network the turnover increases through co-operation. E.g. complete production tasks are planned and managed through cluster partners.		

(continued)

Table 1 (continued)

Un-influenceable logistics variables for logistics processes		1993	2003	2007
		Logistics Centre Bremen	Bremen—the logistics company	Logistics competence centre Bremen
Demand-oriented conditions	Cases	Container handling: 1.36 million TEU.	Container handling: 3.19 million TEU.	Container handling: 4.89 million TEU.
	Discussion	To meet the international developments and especially the market development in the BRIC-countries (Brazil, Russia, India and China) the cluster is working together with local partners and the state on an extension of the infrastructure. In 2008 the expansion of the A1 motorway to 6 lanes is started and the A281 motorway between the freight village and the airport is being opened.		
Company-internal variables	Existing organization	Cases	Participation of processes in companies because of participation in regional networks of freight villages.		Participation of logistics service providers in the implementation of knowledge groups.
		Discussion	Vertical and horizontal co-operation and networking in the logistics cluster leads to a stronger positions of each cluster participant.		
	Company's policy	Cases	Optimization of processes in companies because of participation in regional networks of freight villages.	Co-operative- and collaborative management because of a beginning of local networking of logistics players in interest groupings.	A jointly cluster management leads to a participation Participation of companies in the Logistics Competence Centre Bremen and the Bremen Logistics Day.
		Discussion	An increase of importance of communication and co-operation between companies in the region is being recognized in local companies through the regional cluster activities.		

[a] Source: Statistical State Office Bremen

Fig. 4 Development of the cluster effect in Bremen

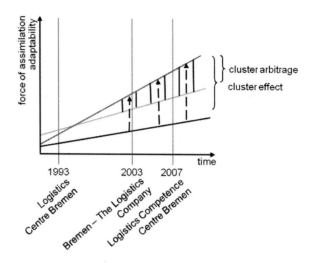

5 Further Research and Outlook

During the last six years about 40 logistics initiatives and competence centres have been established (in Bremen operative since 2007). The consequent request for the studies on cluster development since 1993 till today shows the impact of cluster development. Moreover, an implied knowledge of positive effects in conjunction with co-operation and transfer of knowledge between the logistics actors within the region exists in different institutions in Bremen like logistics companies, political and qualifying institutions, administration, and logistics associations.

But the qualitative character of these three studies (Sect. 4) and the lack of measurability affect the argumentation regarding the significance of positive cluster effects. Unfortunately, the appropriate standardized performance indicators for cluster development were not deducible from the proceeding of these three case studies. Therefore further research on standardized key performance indicators for cluster development is needed.

For example the only measurable performance indicators today are the number of members in logistics initiatives and competence centres, the total of aid money per year, the number of round-table meetings, as well as the number of initiated projects per year. But the KLB exists only since 2006 (operative since 2007). Till today the quantitative performance indicators in the time-series data are not available. Therefore we only give the first theory on these effects (cluster effect and cluster arbitrage) in this paper.

But as it is obvious from the above discussion the cluster arbitrage is based on two mechanisms. First, clusters can adopt changes in the logistics processes and networks rapidly and flexibly and second, clusters can take part in the formation of their own environment. Both allow companies to produce a superior output with similar or lower costs thereby improving their competitive position. Whereas agglomerations lead to shared commonalities across companies they still act independently in the market place. It is only through co-operation that companies engage in joint regional value creation systems, in which an independent transformation process takes place. The company's input, consisting of joint configuration of value activities as well as their combined resources and capabilities, are transformed by the reinforcing effects of the diamond model and lead to upgraded products and innovations.

Thus, it gives an idea and first examples how a cluster generates competitive advantages through processes of value creation and creating a cluster arbitrage by turning the un-influenceable logistics variables into influenceable variables. For the further research the Bremen Logistics Cluster example has to be analysed at the three described points of time; on the one hand regarding the Porter diamond model and on the other hand regarding the logistics system respectively the logistics variables. Through a comparison of both perspectives it can be determined how and when the un-influenceable variables become influenceable variables. This should even allow evaluating the adaptability and rating the cluster arbitrage of the cluster companies.

References

Becattini, G. (1990): The Marshallian industrial districts as a socio-economic notion. In: Pyke, F./Becattini, G./Sengenberger, W. (Eds.): Industrial districts and interfirm co-operation in Italy. Genf 1990, pp. 37–52.

Cortright, J. (2005): Making Sense of Clusters: Regional Competitiveness and Economic Development. Available on: http://www.brook.edu/metro/pubs/20060313_Clusters.pdf. Last visit June 18th 2009.

Deutsche Aussenhandels- und Verkehrs-Akademie (1992): Konzeptstudie Logistik Zentrum Bremen, Bremen 1992.

Elbert, R./Schönberger, R./Müller, F. (2008): Regionale Gestaltungsfelder für robuste und sichere globale Logistiksysteme. Strategien zur Vermeidung, Reduzierung und Beherrschung von Risiken durch Logistik-Cluster. In: Pfohl, H.-Chr./Wimmer, T. (Eds.): Wissenschaft und Praxis im Dialog. Robuste und sichere Logistiksysteme. Hamburg 2008, pp. 294–322.

Elbert, R./Schönberger, R./Tschischke, T. (2009): Wettbewerbsvorteile durch Logistik-Cluster. In: Wolf-Kluthausen, H. (Hrsg.): Jahrbuch Logistik 2009. Korschenbroich 2009, pp. 61–67.

Franz, P. (1999): Innovative Milieus: Extrempunkte der Interpenetration von Wirtschafts- und Wissenschaftssystem. In: Jahrbuch für Regionalwissenschaft 19(1999) 2, pp. 107–130.

Giddens, A. (1984): The Constitution of Society. Outline of the Theory of Structuration. Cambridge 1984.

Haasis, H.-D. (2007): Mesologistik: Leistungsoptimierung in der maritimen Logistikregion Nord-West-Deutschland. In: Kieserling Stiftung (Ed.): Quo vadis Netzwerk – Evolution der Logistik. Bremen, 2007, pp. 98–107.

Haasis, H.-D. (2008): Produktions- und Logistikmanagement. Wiesbaden 2008.

Haasis, H.-D./Fischer, H. (Eds.) (2007): Kooperationsmanagement. Eschborn 2007.

Haasis, H.-D.: Elbert, R. (2008): Bringing regional networks back-into global supply chains: Strategies for logistics service providers as integrators of logistics clusters, in: Kersten, W./Blecker, T./Flämming, H. (Eds.): Global Logistics Management, Berlin 2008, pp. 21–31.

ISL – Institut für Seeverkehrswirtschaft und Logistik (Ed.) (2004): Bremen – die Logistik Company. Ergebnisse einer Konkretisierung. Bremen 2004.

Markusen, A. (1996): Sticky places in slippery space: A typology of industrial districts. In: Economic Geography 72(1996)3, pp. 293–313.

Marshall, A. (1920): Principles of Economics. London, Basingstoke 1920.

Pouder, R./St. John, C.H. (1996): Hot Spots and Blind Spots: Geographical Clusters of Firms and Innovation. In: Academy of Management Review, 21(1996)4, pp. 1192–1225.

Pfohl, H.-Chr. (2004a): Logistiksysteme. Betriebswirtschaftliche Grundlagen. 7., korrigierte und aktualisierte Auflage. Berlin.

Pfohl, H.-Chr. (2004b): Logistikmanagement. Konzeption und Funktionen. 2., vollständig überarbeitete und erweiterte Auflage. Berlin.

Piore, M.J./Sabel, C.F. (1985): Das Ende der Massenproduktion. Berlin 1985.

Porter, M.E. (1990): The Competitive Advantage of Nations. London, Basingstoke 1990.

Porter, M.E. (1998): On Competition. Boston 1998.

Porter, M.E. (2003): The Economic Performance of Regions. In: Regional Studies, 37(2003)6/7, pp. 549-578.

Roelandt, T.J.A./den Hertog, P. (1999): Cluster Analysis and Cluster-Based Policy Making in OECD Countries: An Introduction to the Theme, Ch 1. In: OECD (Eds.): Boosting Innovation: The Cluster Approach. Paris 1999, pp. 9–23.

Rosenfeld, S. A. (1997): Bringing Business Clusters into the Mainstream of Economic Development. In: European Planning Studies, 5(1997)1, pp. 1–15.

Sölvell, Ö./Lindqvist, G./Ketels, C. (2003): The Cluster Initiative Greenbook. Stockholm 2003.

Sydow, J./Lerch, F./Huxham, C./Hibbert, P. (2007): Developing Photonics Clusters. Commonalities, Contrasts and Contradictions. Advanced Institute of Management Research. London 2007.

Quotation Behaviour of Profit Centres for Offers on Dynamic Logistic Services

Model and Govern Corporate Offer Calculation through an Approach Derived from Poker Game Strategies

Bernd Pokrandt, Marcus Seifert and Stefan Wiesner

1 Introduction

Today, technological progress leads to changes in market scenarios, which affect the purchase, production and sales structures of consigning companies. Suppliers, manufacturing sites and customers are located internationally. At the same time, simplification of processes, enhancement of flexibility and the continuous requirement for cost reduction results in more and more tasks being outsourced. Freight forwarders have to react to those changes when organising the transportation of goods for their customers (Zimmermann 2004). In general, the traditional brokerage of single shipments is more and more replaced by the provision of complex logistic services with direct control of logistic means and facilities (Fig. 1).

The left arrow shows the extension of the logistical tasks from shipping to the earlier and later phases of the product life-cycle. Previously, freight forwarders were only required to broker the shipment of goods from the production site to the distribution area. Logistics in procurement, production, distribution or disposal were handled by the consignor itself. With those tasks becoming more and more complex due to international cooperation, manufacturing companies are looking to outsource them to reduce costs and enhance flexibility.

As can be seen from the right graphic, freight forwarders are expected to offer a range of different (value added) logistics services in those areas to the customer, such as the management of warehouses or customs formalities. They also have to take direct control or ownership of the required transportation means and facilities

B. Pokrandt
Besselstraße 88, 28203, Bremen, Germany

M. Seifert and S. Wiesner (✉)
Bremer Institut für Produktion und Logistik GmbH – BIBA, Hochschulring 20, 28359, Bremen, Germany
e-mail: wie@biba.uni-bremen.de

H.-J. Kreowski et al. (eds.), *Dynamics in Logistics*,
DOI: 10.1007/978-3-642-11996-5_47, © Springer-Verlag Berlin Heidelberg 2011

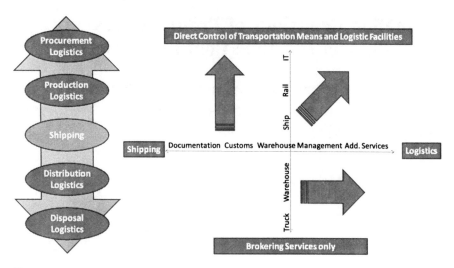

Fig. 1 Development from "shipping" to "logistics services"

to have better access to their capacities. But they are also expecting appropriate yields from the additional services.

In the past, however, international air and sea transportation tended to result in higher returns than other logistics services. As a result, the freight forwarders are in the dilemma that, while their customers ask for the provision of additional logistic services, they want to limit their portfolio for economic reasons. The market development however, especially the migration of production sites, had positive effects on the demand for global transports. Consequently for the global air and sea freight sector, value added logistics services were less relevant than in domestic markets (Stahl 1995).

In order to provide complete international air and sea transport services, freight forwarders usually try to establish a network of branch offices covering the im- and exporting countries. As the origin and destination of a transport can be any country, a new branch office (Country n) added to the network results in (n − 1) new transport routes to be considered as further markets for the logistics provider (Fig. 2). Uncovered areas have to be served with independent forwarding agents, mostly associated with a lack of customer control and profitability.

At the same time customer proximity, detailed knowledge about the markets and low response times have to be achieved. Freight forwarders traditionally react to those challenges by setting up self dependent profit centres for specific geographic areas. Thus, a given or potential customer is always assigned to a regional company contact which is aware of the needs of the regional market and which is fast and flexible enough to respond to the customer's requests. The network itself is created through cooperation of the individual profit centres. Interestingly, the work flows between the profit centres are quiet similar to that proceeded between independent logistic providers (Hess 2002).

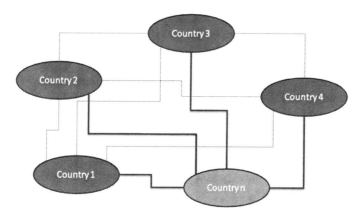

Fig. 2 Adding countries to the network leads to rapid increase in complexity

From	To	Quantity 1	Quantity 2	Quantity 3
Site A	Site X	Price	Price	Price
Site B	Site Y	Price	Price	Price
Site C	Site Z	Price	Price	Price
...	...	...	...	...

Purchase Profit Center 1   Purchase Profit Center 2

Site A   Site X

Sales Profit Center 1   Sales Profit Center 2

Fig. 3 Request for quotation and task distribution among profit centres

Large customers with high global freight volumes tend to centralise their transport management to increase demand concentration. Typically, requests for quotation are issued periodically to ensure adequate market prices. Often the prices are fixed over the contracting period. However, according to the International Commercial Terms (Incoterms), processing of export and import procedures is still handled independently.

In the forwarder's network the exporting profit centre may organise the pre-carriage as well as the global shipment. The importing profit centre controls the on-carriage and the delivery (Fig. 3). While both profit centres involved try to maximise their profit from the services provided locally, the yield of the global transportation is typically split (equally) among the profit centres.

In typical freight forwarding organisations, the local purchase conditions and consequently the profitability per shipment is kept confidential even amongst the branch offices within one group of companies. This behaviour can be described as using information asymmetry in order to compete with profit centres often within the same group of companies. This information asymmetry causes frustrations when combining local offers of individual profit centres to fulfil global transport demands of mentioned large customers (Welge and Andreas 2008).

A global forwarding group should seek to maximise its profit expectation in respect to a global customer's request for quotation. However, the individual profit centres within the forwarding group try to maximise their domestic yield. The combination of such local quotes to a price for a global logistics service may result into an unattractive offer for the customer, thus generating no benefit for the company at all (Botta 1997).

2 Problem and Objectives

The central problem given attention in this paper is the information asymmetry that influences the behaviour of the profit centres when quoting for collaboratively offered logistic services. The profit centres are independent for quotation purposes, but tied together for order handling. Every profit centre decides individually about its profit margin. An unattractive partial performance may therefore be answered with a defensive quote to either avoid receiving the order, or in the other case getting a high compensation for it. Resulting from information asymmetry concerning regional costs and market prices, excessive profit margins in partial quotes are not easily identified. Bids for international transports are lost, because of unattractive pricing resulting from the quotes of the involved profit centres. The customer rejects the offer on the basis that other freight forwarders can fulfil the task with less cost (Gudehus 2007). The market price for transport services is the average price which is paid for the offered service (Gutermuth 2004).

The main objective of this paper is to create a model of the decision processes in profit centre quotations. The effects of the existing information asymmetry and its influence on the quotation behaviour of the profit centres will be investigated through a simplified role game. In a second step, game theory will be introduced as a possible means to govern the profit centre actions in the interest of the whole company through rewarding and penalisation. It is tested, if principles of Poker games are suitable to reduce the effects of the information asymmetry.

3 Research Approach

The offered price for the overall transport is based on the collected quotes for the partial performances. The main-carriage is organised by the exporting profit centre, but the profit will be shared with the importing profit centre. This approach assumes that the profit centres involved in the export and import for a requested transport route have agreed on a fixed profit distribution, the Profit Share. While the costs for each profit centre are also set by the regional market conditions, its individual profit can be influenced by changing the quotation for the partial performance, the (potential) earnings. The following equation for the overall gain of a shipping order is derived from this simplified approach:

$$G = \sum_i^n (e_i - c_i) - PS + \sum_j^m (e_j - c_j) + PS$$

G:=   Gain of a shipping order
i:= 1,…,n   Partial service provided by profit centre i
j:= 1,…,m   Partial service provided by profit center j
e_i, e_j:=   Earnings for partial services i or j
c_i, c_j:=   Costs for partial service i or j
PS:=   Profit Share

As can be easily derived from the formula, the overall gain from a shipping order consists of the individual yields of the involved profit centres. Those yields are made up of the respective earnings less the costs. A Profit Share (PS) is introduced as a correction factor, transferring parts of the yield from one profit center to another. The arrangement and the level of this Profit Share have a direct influence on the quotation behaviour of the profit centres, thus influencing the chances of success for the bidding. The approach will be tested in a simplified role game.

4 Application of Poker Game Principles

The information asymmetry during the cooperation among profit centres can be characterised as of the type "Hidden Action". One profit centre cannot keep track of the actions of another profit center after the start of the cooperation. The ex-post evaluation of its actions is impossible because of limited knowledge on environmental factors, such as the regional market. Because the information asymmetry cannot be compensated, regulations have to ensure collaborative behaviour. In this case, the moral hazard of following individual interests can be handled by the introduction of incentive schemes and control systems (Ossadnik 2003).

Because of high surveillance costs, control systems are not feasible to guide the actions of the profit centres. Instead, incentive schemes involve the profit centres in the outcome of their actions with rewarding and penalisation. The theoretical background for situations, where the choices of others determine an individual's success in making choices, is game theory (Nalebuff 2007). A game which is quite similar to the situation of cooperating profit centres is Poker:

> [Poker]…is a game of imperfect information, where multiple competing agents must deal with probabilistic knowledge, risk assessment, and possible deception, not unlike decisions made in the real world. (Billings et al. 2002)

The mentioned reference illustrates, why the principles of Poker gaming might be applicable to the problem of profit centre quotations. In Poker as well as during the offer calculation for international transports, the actors have to deal with imperfect (asymmetrically distributed) information. The competing agents (profit centres) rely on probabilistic knowledge for risk assessment and have to expect possible deceptions (excessive profit margins) from the other agents. This makes the underlying decision process similar for both cases.

Poker is the name for a family of card games. In most of them, the Players have to form a *hand* out of playing cards and bet according to their perceived chance of winning the game. At the beginning of most variations of the Poker game, some kind of forced bet (*ante*, *blind*) is collected. On this basis, the first round of betting commences. The Players have either to match the previous bet, raise the bet or fold and lose their chance of winning. If all but one player fold, he collects the pot (of bets). Otherwise the *hands* are shown and the highest *hand* wins the pot (Fig. 4).

The forced bets create an initial cost for the players. It assures that all involved parties behave careful when betting, in spite of information asymmetry. The forced bet is definitely lost when folding. If the profit centres are required to place a bet with each quotation, which is only returned in case of a successful offer and otherwise collected in a pot, motivation to act in the common interest is strengthened.

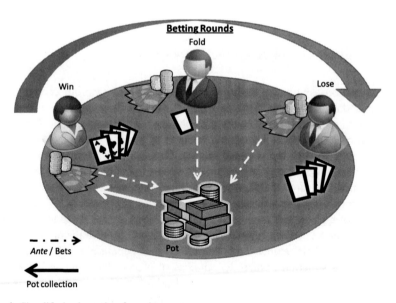

Fig. 4 Simplified schematic of a poker game

The opportunity to not only get the initial stakes back, but to participate proportionally on the payback of the "pot" of bets on unsuccessful offers, represents an incentive for the profit centres to quote adequate market prices. Additional yield is no more realised by excessive pricing, but by placing higher bets. A profit center, which is confident that its quotation will lead to a successful offer, will raise its bet to participate disproportionately high in profit sharing. Even profit centres which do not believe in the success of an offer will quote as low as economically justifiable to maximise the chance of return for the forced bet.

For a second round, the role game has been modified to incorporate the principle of forced bets, or *antes*, from Poker gaming to test this hypothesis. The players will now have to place a forced bet with every quotation they are asked for. The results of this modification are analysed for the desired effects. Applicability in a real company environment is discussed.

5 Simplified Role Game

The basis for the role game is a strongly simplified shipping network with three locations of dispatch. A customer requests quotes for three different routes with 100 transports each (Fig. 5). Three players (A, B, C) represent the involved profit centres and give three quotes, respectively for the overall offer: one for the on-carriage when importing and two for the precarriage and the main-carriage when exporting (A → B, B → C, C → A). The acceptance of offers is decided randomly by a computer program.

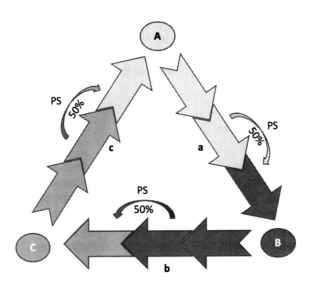

Fig. 5 A simplified Shipping Network

Fig. 6 Example for an Offer

Shipping Route	Pre-Carriage	Main-Carriage	On-Carriage
a (A -> B)	350	2.300	6
b (B -> C)	210	4.300	4
c (C -> A)	450	3.200	2
Total			

The profit for the main run is shared equally (50%) among the involved centres. Purchase prices for the partial shipments are disclosed known to the responsible player only. The acceptance of the combined offer is randomly determined on the basis of the aggregated bid price (Fig. 6). The profit of each player is the margin of his scope of responsibility plus/minus his Profit Share for 100 transports. The only measure of success for the players is the yield of their respective profit center. The experiment is repeated 30 times.

Results of the simplified role game were altogether as expected. The players tried different strategies to maximise their individual profit. While some tried to win the maximum number of bids by quoting even below costs when the losses are shared, others increased their profit margins steadily. The information asymmetry supports this "egoistic" behaviour. An optimal solution for the whole company is not achieved. Bids, which are altogether attractive for the whole company, are lost.

6 Modified Role Game

For the second run of the role game, the rules have been modified according to Poker Game principles. Quotation for partial performances is still done by the local profit centres. To ensure competitive pricing, a certain bet (*ante*) has to be supplied with every quotation. The stake can be voluntarily increased. In the case of an unsuccessful offer, the *antes* are retained. If the offer is successful, the collected *antes* are distributed according to the current bets. Thus the players have the opportunity to generate an additional surplus from the pot. This is an incentive for the profit centres to quote a competitive price, independent from the attractiveness of their partial performance, to win back their bet.

Fig. 7 Profit Expectation—
Cost of Information
Asymmetry

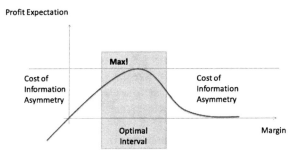

The results of the modified role game show that maximisation of the profit expectation for the whole company (Fig. 7) is still not achieved. There are still costs of the information asymmetry left. The sample however was quite small. Further investigations with a greater number of players and variations of the forced bet principle might yield more conclusive results.

7 Conclusions

Within the scope of this paper, quotation behaviour of the profit centres of a freight forwarder has been modelled by a role game approach. The influence of information asymmetry and profit sharing on the generation of competitive offers has been investigated. Additionally, forced bets derived from Poker game principles have been tested as a means to govern the self-interest of the profit centres towards successful bidding.

The specific difficulties during offer calculation for international logistic services have been described. Customers with high freight volumes call for tenders for a multitude of transport routes and logistic services. Such an offer can only be provided, if all involved profit centres calculate quotations for their partial performances. Because not every partial performance may be as attractive as the whole potential order, it is probable that some profit centres quote non-competitive prices, endangering the success of the offer. This is possible because of asymmetric information distribution about transport costs, overheads and profit margins. The loss of an offer is negative for the whole company.

The decision processes for the scenario described above have been modelled in a simplified role game. Three players represent three different profit centres, quoting for three different transport routes. Offers generated from the quotation are accepted or rejected at random on the basis of the overall profit margin. This role game has shown that players acting in their own self-interest do not automatically contribute to the overall benefit of the company. A high number of offers have been lost due to excessive margins of the individual profit centres. Thus, the need for new methods of profit sharing and motivation for behaviour in the interest of the company has been identified.

Poker game principles are a possible source for such methods, because players of this card game also deal with imperfect information, competing agents and possible deception. Forced bets are an incentive for the player to participate in the game, while raising the bet creates the chance of a higher individual benefit. This approach has been integrated into a modified role game, where the players have to place a forced bet with each quotation with an optional raise. Results have been variable, with no clear trend towards the maximum profit expectation for the whole company.

Altogether, it has been shown that information asymmetry and profit sharing prevent the success of individual offers for international logistic services. Incentives for behaviour in the interest of the whole company could be a possible solution. However, the introduction of forced bets yielded no clear result. Further tests of this and other approaches are still required to propose a useful strategy in this respect.

References

Billings, D., Davidson, A., Schaeffer, J., and Szafron, D. 2002: "The challenge of poker", Artificial Intelligence, Volume 134, Issues 1-2, January 2002, pp. 201-240

Botta, V. 1997: "Vom Cost-Center zum Profit-Center", in Roth A & Behme W: Organisation und Steuerung dezentraler Unternehmenseinheiten, pp. 221–238, Gabler, Wiesbaden

Gudehus, T. 2007: "Dynamische Märkte", p. 274, Springer, Berlin

Gutermuth, J. 2004: "Güterverkehr, Spedition und Logistik", p. 372, Bildungsverlag EINS, Troisdorf

Hess, T. 2002: "Netzwerkcontrolling—Instrumente und ihre Werkzeugunterstützung", Deutscher Universitätsverlag, Frankfurt/Main

Nalebuff, B. 2007: "Coopetition—kooperativ konkurrieren", in Boersch C. & Elschen R.: Das Summa Summarum des Managements, pp. 217-230, Gabler, Wiesbaden

Ossadnik, W. 2003: "Controlling", Oldenbourg, München

Stahl, D. 1995: "Internationale Speditionsnetzwerke", Vandenhoeck & Riprecht, Göttingen

Welge, M. and Andreas, A.-L. 2008: "Strategisches Management—Grundlagen, Prozesse, Implementierung", Gabler, Wiesbaden

Zimmermann, B. 2004: „Kontraktlogistik als Zukunftsmarkt der Logistikdienstleistungswirtschaft. Mittelstandskongruenz und Entwicklung eines mittelstandsgerechten Vertriebsmodells", Privatdruck, Nürnberg

Long Haul Trucks Dynamic Assignment with Penalties

Antonio Martins Lima Filho and Nicolau D. Fares Gualda

1 Introduction

The problem considered in this work falls on the concept of operational planning. According to Gualda (1995), the operational planning presupposes the existence of a physical system implanted and looks for the operational optimization of this system. Powell (2003) considers the operational planning of transportation systems associated to the management of operations, in real time, and analyzes, for each type of transport service, the resources to be managed and the decisions, which are the form of management of these resources; for the author, the decisions, in turn, should take into account the procedures to which they are related, that is, the dynamics and the restrictions of the system.

The operational planning, for a long distance road transportation system, is performed by the management staff acting in a highly dynamic environment, where the time factor plays an important role and details on the types of vehicles, edifications and activities should be taken into account. The planning may be defined as a bridge between knowledge (data, theory) and the action, or decision making (Gualda 1995).

In the case in study, the system is composed of full-truck-loads (a single load on the vehicle, which is carried from a single origin to a single destination). The demand for transportation occurs, in a random way, in several points in space, throughout time. A basic feature of this type of transport, defined by the expression Long Haul, is the fact that the vehicle does not return immediately to the operational base of origin, after fulfilled the delivery of its load. Thus, for each vehicle that becomes available, for each point in space-time, a decision of operational character between three possible alternatives must be taken: to attend a load at that

A. M. L. Filho and N. D. F. Gualda (✉)
Departamento de Engenharia de Transportes, Escola Politécnica da Universidade de São Paulo, São Paulo, Brazil
e-mail: ngualda@usp.br

H.-J. Kreowski et al. (eds.), *Dynamics in Logistics*,
DOI: 10.1007/978-3-642-11996-5_48, © Springer-Verlag Berlin Heidelberg 2011

point, to wait for a future load at the location, or to displace the vehicle, unloaded, to another point.

Section 2 presents a literature review. In Sect. 3, the methodology proposed for the problem solution is shown. In Sect. 4 a typical problem is solved. Section 5 presents conclusions of the research.

2 Resources Allocation in Load Transportation

In the case of allocating vehicles to loads in long distance road transportation it is necessary to distinguish between regular services, where the tactical planning is the most important, and the non-scheduled services, where the operational planning is prominent. For regular services, the works by Crainic and Roy (1992) and Powell and Sheffi (1989) may be cited. The non-scheduled services, in turn, are the scope of this work. This subject has somehow been treated by Haghani (1989), Frantzeskakis and Powell (1990), Powell et al. (1995a), Hane et al. (1995), Powell and Carvalho (1998), Godfrey and Powell (2002a, b), and Topaloglu and Powell (2006), among others.

Operations planning models in transportation may be static or dynamic. Static models are proper for tactical planning, where the time factor may be left in the second plan. In the case of long distance road transportation operational planning, dynamic allocation is a real need imposed by the characteristic of the problem: the resource (transportation equipment), after serving a task, is available at a different location from the original, in which new tasks waiting to be served may or may not exist. Resources allocation considered here is "dynamic" in three different senses (Powell et al. 1995b): the problem is dynamic, since information (position of the resources, availability of loads) changes with time; the model used to represent the problem is also dynamic, since it incorporates explicitly the interaction of activities over time; and, finally, the method of solution (Dynamic Programming) solves the model repeatedly as new information is received.

A dynamic model may capture the temporal stage of physical activities, but not the temporal stage of information; in this case the model is called deterministic (Powell 2003).

Powell (1986) proposes representing the problem of resources allocation in long distance transportation in the form of a space-time network. This form of modeling produces a network of acyclic structure, facilitating the operation of optimization algorithms. The difficulty with this type of network approach is the great extent of the problems involved, in terms of quantity of variables. Furthermore, any details of the operation, as time windows, for example, involve new generations of a great quantity of variables. Therefore, models based on pure Linear Programming approaches, as presented by Haghani (1989) and Hane et al. (1995), are limited to solve real dynamic problems.

A way to bypass the above difficulties is presented by Powell et al. (1995a), based on a discrete event dynamic system (DEDS), where demands are queued at terminals while waiting for available capacity. The authors call it a logistics queuing network (LQN). When a vehicle arrives at its destination, it becomes empty and available for its next movement; the flow of demands into a location may not equal the flow out of it, leading to the necessity to reposition empty vehicles from one location to another.

The proposal, then, is to replace the optimization approach based on Linear Programming with a solution by Recursive Dynamic Programming, allowing decomposing the general solution of the network problem in a series of subproblems solutions at each location level. The general approach consists of a series of steps forward and backward, across time: at each iteration, the forward step assigns vehicles to loads and performs, in a limited way, re-positions of vehicles; the backward step calculates gradients, and an adjustment phase modifies the potential values and the limits of empty movements. The procedure continues until it ensures a desired level of convergence (Crainic 2003).

Powell et al. (1995a) and Powell and Carvalho (1998) introduce modifications on that model and apply it to deterministic problems, utilizing a linear estimation of the value function. The development of CAVE (Concave Adaptive Value Estimation) algorithm by Godfrey and Powell (2001) allows to performing an accurate estimate of concave functions values based on the results of a succession of experiments, without depending on the specific function of probabilities associated with the events. The algorithm is based on gradients generation to the amount of resources available, taking into account the result obtained in view of the demand actually occurred. It permits to treat probabilistic demands. Thereafter, the algorithm executes a smoothed change in the form of the curve for use in following iteration.

Godfrey and Powell (2002a, b) propose a solution focused basically in two reductions of complexity: the problem is separated by location/period (concept of logistical queuing network, as described above) and then it is solved using an estimate of the value associated with the future state (the algorithm CAVE, described above).

Topaloglu and Powell (2006) extend the model for the treatment of heterogeneous fleets, utilizing a hybrid linear/concave estimation for the value function. As the previous works, demands not attended are considered lost (or attended by unlimited available chartered vehicles).

3 Methodology

This work applies to the Operational Planning of a Long Haul Load Transportation System, aiming to maximize the economic result of the System in function of the allocation of transportation equipments to loads. The adopted methodology is based in the Stochastic, Approximate and Adaptive Dynamic Programming Model described by Godfrey and Powell (2002a, b), and introduces a network modeling proposal which allows to manage demands not attended, with the payment of a penalty, or the utilization of third parties transportation firms to avoid the payment of such penalties. The stochastic dynamic program model adopted is presented in Sect. 3.1; Sect. 3.2 presents the adaptive strategy of solution; and Sect. 3.3 presents the network modeling approach.

3.1 Stochastic Dynamic Programming Model

Let:

T = number of planning periods in the planning horizon
$\mathcal{T} = \{0, 1, \ldots, T - 1\}$ = periods at which decisions are taken
$\mathcal{J}$ = set of physical locations in the network, indexed by i and j
L_{it}^{+} = set of loads to be serviced at time t from location i
R_{it}^{+} = total number of resources in location i available to be acted on time t

Decision variables, for each $t \in \mathcal{T}$, $i, j \in \mathcal{J}$, $l \in L_{it}^{+}$, may be described as:

$$x_{lt} = \begin{cases} 1 & \text{if are source at } i \text{ is assigned to a task } l \in L_{it}^{+} \text{ at time } t \\ 0 & \text{otherwise} \end{cases}$$

y_{ijt} = quantity of resources repositioned from i to j beginning at time t

Initially, for a particular period of time t, the following reward function may be defined:

$$g_t(x_t y_t) = \sum_{i \in J} \sum_{l \in L_{it}^{+}} r_{lt} x_{lt} - \sum_{i \in J} \sum_{j \in J} c_{ij} y_{ijt} \tag{1}$$

where r_{lt}, "reward" related to task $l \in L_{it}^{+}$ beginning in time $t \in \mathcal{T}$; c_{ij}, "cost" of repositioning one resource from location i to location j.

Considering the entire planning horizon, the objective function, in terms of stochastic dynamic programming, is given by:

$$\max_{x_0, y_0 \in H_0} g_0(x_0 y_0) + E \left\{ \sum_{t \in T \backslash 0} \max_{x_t, y_t \in H_t} g_t(x_t y_t) \right\} \tag{2}$$

where H_t is the σ-algebra (subset of events of the sample space Ω for which probability measure P is defined)

There are two types of constraints, relative to decisions and to the system dynamics: (1) constraints applied to decisions taken at a point in time t (flow conservation) and (2) constraints which control the system dynamics across time (resources and loads quantity).

3.2 Adaptive Dynamic Programming Strategy of Solution

In practical applications for this class of problems, the most natural solution strategy is to use a rolling horizon procedure, solving the problem at time t, using what is known at time t and a forecast of future events over some time horizon (Godfrey and Powell 2002a, b). For this, it is necessary to define a function of value for period t, based on an *estimate* of value for period $t + 1$:

$$\tilde{V}_t(R_t) = \max_{x_t, y_t \in H_t(R_t)} g_t(x_t, y_t) + \hat{V}_{t+1}(R_{t+1}) \tag{3}$$

Godfrey and Powell (2002a) use an estimate of Value Function decomposed by location $\hat{V}_t(R_t) = \sum \hat{V}_{it}(R_t)$, where functions $\hat{V}_{it}(R_{it})$ are estimated utilizing the CAVE algorithm (Godfrey and Powell 2001).

The problem is solved performing a "forward step" in time, establishing a set of decisions $(x_t)_t \in T$, for a particular sample ω, and using a particular set of value functions approximations.

At the end of the "forward step", the dual variables obtained for each subproblem are used to update the function approximations using the CAVE algorithm (Godfrey and Powell 2001). The solution is found after successive approximations of the decision variables ("forward steps"), followed by the respective updates of the value functions. That is, known the value function estimates and a random result $\omega \in \Omega$ for future demands, the following sequence of network subproblems is solved, starting with $t = 0$ and continuing until $t = T - 1$:

$$\max_{x_t, y_t \in H_t} g_t(x_t y_t) + \sum_{\tau=1}^{\tau_{max}} \hat{V}_{t+1, t+\tau}\left(R_{t+1, t+\tau}\right) \tag{4}$$

Subproblem (4) in time t may be viewed as a two-stage network, as illustrated in Fig. 1. For the 1st stage, arrows in continuous line indicate load movements, which have "cost" (reward) $= r_{lt}$ and upper bound $L_{ijt}^+(\omega)$; arrows in dotted line represent empty vehicles repositioning, with cost $= -c_{ij}$ and upper bound $= +\infty$. The 2nd stage decisions (times $t' = t + \tau$, $\tau > 0$) incorporate the Value Function approximation $\hat{V}_{t+1, t'}\left(R_{t+1, t'}\right)$.

3.3 Network Modeling

Modeling methodology adopted in this research involves applying the dynamic programming tool previously described to a relatively complex real situation. The assumption adopted by Godfrey and Powell (2002a, b) that not serviced loads are simply lost is here enriched by incorporating to the model the utilization of third party fleets and also penalties for not attending requested loads.

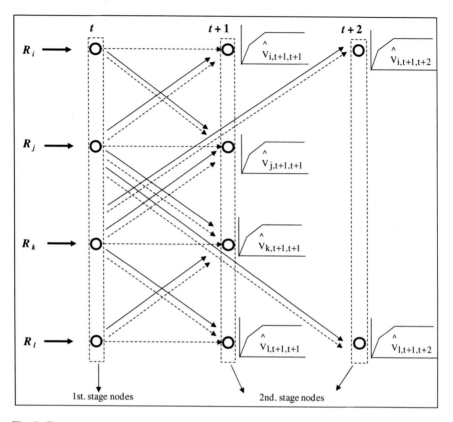

Fig. 1 Two-stage network for time t subproblem

The modeling showed in this work is built taking into account that models are representations of something real, with characteristics and complexities that depend of the process to be represented and of the representation's purpose (Gualda 1995). According to Pidd (1999) models may have two main uses: (1) explore consequences of actions before take them and (2) serve of base for decision support computational systems. In this line, the model developed here allows to achieve system optimization through available software utilization, such as Excel and the Jensen Network (Jensen 2008).

Once the algorithm utilized by Jensen (2008) applies to problems modeled as circulation networks of minimal cost (or maximum profit), the 2-stage network shown in Fig. 1 must be translated to a circulation network. For this purpose the network shall be designed to incorporate a super sink, at the end of the second stage, from which the resources return to first stage nodes, in controlled quantities, in accordance with its availability.

From this generic scheme it is possible, then, to consider actual conditions, that is, to incorporate new forms of resources (third carriers, for example), as well as conditions of compulsory tasks (through the lower and upper limits of the arcs) and penalties (in the form of cost). Particularly, the penalties were modeled, in this work, like resources attending the tasks at a cost equivalent to the penalty value plus the corresponding reward, once the model considers the task "accomplished" and so its reward is computed. Item 4, below, details the adopted model and experimental results.

4 Model Application

In order to ensure the applicability of the model and the effectiveness of its results, an experiment was conducted using a problem dimensioned in such a way that it was possible its overall optimization, through a commercial software, in addition to the analysis of the behavior of the system on a visual way. This problem consists in three locations and planning horizon of six periods. A total of up to 56 vehicles were considered on the performed experiments.

Aiming to simplify the model, but without prejudicing the evaluation of its effectiveness, the transit times were considered as a single unit and the availability of loads as deterministic ones. It is possible to say that the extension of the experiment for problems with multiperiod travel times and probabilistic demand will be guaranteed, since, for both, it is sufficient to alter the characteristics of the network and the way to allocate quantities of demand for the next iteration.

The problem may be represented as a circulation network of maximum profit, consisting of 343 nodes (including the sink) and 534 arcs. The optimal solution was obtained through Jensen Network Solver (Jensen 2008), for the global network, using a linear-integer model. Ten configurations of value for the same basic structure of the network were simulated, varying the upper limit of some arc (quantity of loads), the value of recompense for attendance some of the loads (the cost of arc), the distribution of vehicles among the three locations in the initial period.

The problem was decomposed into six subproblems, one for each period. Each subproblem was configured as a circulation network with 63 nodes and 139 arcs and the solution was obtained with the use of the same Network Solver, in the same structure of linear-integer model. Iterations were performed until perceiving no improvement in the solution for the first stage. Then, at least ten iterations were conducted to obtain the overall result (first plus second stages), looking for an overall solution improvement of at least 1% as compared to the previous solution. The quantity of iterations varied, in accordance with the adopted configuration, from a minimum of 27 to a maximum of 93.

The adopted smoothing factor varied, among the various experiments, from 0.5 to 0.05. So the value function has been updated, from one iteration to the next, using 50 to 95% of the value accumulated on the previous iteration, and, respectively, 50 to 5% of the value for the new iteration. This factor must be calibrated for each particular case, taking into account that higher values will lead to a quicker solution, but of lower quality.

Table 1 Experiments summary

Experiment	Smooth factor	Iterations	Own fleet vehicles	Third party vehicles	Available loads	Own fleet loads	Third parties loads	Penalties
1	0.5	27	26	36	159	133	21	5
2	0.1	51	26	36	159	133	21	5
3	0.1	53	26	36	159	133	21	5
4	0.05	64	26	36	159	133	21	5
5	0.05	57	27	36	159	137	18	4
6	0.05	93	25	36	159	129	23	7
7	0.05	67	25	37	159	129	23	7
8	0.05	74	55	90	203	202	0	1
9	0.05	84	35	108	203	162	34	7
10	0.05	60	35	108	183	165	18	0

Table 1 shows a summary of the experiments. The cargo values, in all experiments, were identical to the ones supplied by the process of global optimization of the network, showing the effectiveness of the subproblem model.

Although the tested problem is relatively small in size, the computational processing time was in the order of 1 s, which allows to consider its applicability to cases of substantially higher dimensions. It is possible to suggest that the solution in stages, typical of Dynamic Programming, does not increase the size of the problem to be solved at each stage if the time horizon to consider is extended. There is, quite simply, a greater quantity of subproblems to compute. Thus, if the solved problem had a horizon of 30 periods, the solution of the global optimization problem would spend a computational time greater that the solution for six periods, although the time of solution for each subproblem would remain the same.

5 Conclusions

An Adaptive Dynamic Programming Method was presented to solve the problem of resource allocation in long distance road transportation. A network modeling to manage demands not attended was introduced, considering the payment of penalties and/or the utilization of third parties transportation firms. The method consists of solving subproblems corresponding to each period of time, using an estimate of value for the optimization of the network throughout the planning horizon.

Tests show that the processing time for subproblems does not depend on the magnitude of the planning horizon, while the processing time for the solution of the overall problem grows as larger problems are considered. The results also indicate that the method may be applied to real world problems.

References

Crainic, T.G. (2003) *Long-haul freight transportation.* In: Handbook of transportation science, Randolph W. Hall (ed.). 2nd. edition. Boston: Kluwer Academic Publishers.

Crainic, T., Roy J. (1992) Design of regular intercity driver routes for the LTL motor carrier industry. *Transportation Science*, Baltimore, v. 26, pp. 280–295.

Frantzeskakis, L.F., Powell, W.B. (1990) A successive linear approximation procedure for stochastic, dynamic vehicle allocation problems. *Transportation Science*, Baltimore, v. 24 (1), pp. 40–57.

Godfrey, G.A., Powell, W.B. (2001) An adaptive, distribution free algorithm for the Newsvendor Problem with censored demands, with applications to inventory and distribution problems. *Management Science*, v. 47 (8).

Godfrey, G.A., Powell, W.B. (2002a) An adaptive dynamic programming algorithm for dynamic fleet management, I: single period travel times. *Transportation Science*, Baltimore, v. 36 (1), pp. 21–39.

Godfrey, G.A., Powell, W.B. (2002b) An adaptive dynamic programming algorithm for dynamic fleet management, II: multiperiod travel times. *Transportation Science*, Baltimore, v. 36 (1) pp. 40–54.

Gualda, Nicolau Dionísio Fares (1995) *Terminais de transportes: contribuição ao planejamento e ao dimensionamento operacional.* Tese (Livre-Docência) – Departamento de Engenharia de Transportes, Escola Politécnica da Universidade de São Paulo, São Paulo.

Haghani, A. (1989) Formulation and solution of a combined train routing and makeup, and empty car distribution model. *Transportation Research*, v. 23B (6), pp. 433–452.

Hane, C., Barnhart, C., Johnson, E., Marsten, R., Nemhauser, G., Sigismondi G. (1995) The fleet assignment problem: solving a large-scale integer program. *Mathematical Programming*, Amsterdan, v. 70, pp. 211–232.

Jensen, P.A. (2008) *Operations Research Models and Methods.* Internet: http://www.me.utexas. edu/~jensen/ORMM/. Acesso em 13 fev. 2008.

Pidd, M. (1999) Just modeling through: a rough guide to modeling. *Interfaces*, v. 29 (2), pp. 118–132.

Powell, W.B. (1986) A stochastic model of the dynamic vehicle allocation problem. *Transportation Science*, v. 20 (2), pp. 117–129.

Powell, W.B. (2003) *Dynamic models of transportation operations.* In: Handbooks in Operations Research and management science, 11: Supply Chain Management: design, coordination and operation. A.G. de Kok e Stephen C. Graves (eds.), Amsterdam; Boston: Elsevier.

Powell, W.B., Carvalho, T.A. (1998) Dynamic control of logistics queueing networks for large-scale fleet management. *Transportation Science*, v. 32 (2), pp. 90–109.

Powell, W.B., Sheffi, Y. (1989) Design and implementation of an interactive optimization system for network design in the motor carrier industry. *Operations Research*, v. 37 (1), pp. 12–29.

Powell, W.B., Carvalho, T.A., Godfrey, G.A., Simão. H.P. (1995a) Dynamic fleet management as a logistics queueing network. *Annals of Operations Research*, v. 61, pp. 165–168.

Powell, W.B., Jaillet, P., Odoni, A. (1995b) *Stochastic and dynamic networks and routing.* In: Handbook in Operations Research and Management Science, 8: Network Routing. M.O. Ball et al. (eds), Amsterdam; Boston: Elsevier, pp. 141–295.

Topaloglu, H., Powell, W.B. (2006) Dynamic programming approximations for stochastic time-staged integer multicommodity-flow problems. *INFORMS Journal on Computing*, v. 18 (1), pp. 31–42.

Lightning Source UK Ltd.
Milton Keynes UK
01 March 2011

168453UK00002B/2/P